T0388232

VOLUME ONE HUNDRED AND TWENTY SEVEN

ADVANCES IN PROTEIN CHEMISTRY AND STRUCTURAL BIOLOGY

Proteomics and Systems Biology

VOLUME ONE HUNDRED AND TWENTY SEVEN

ADVANCES IN PROTEIN CHEMISTRY AND STRUCTURAL BIOLOGY

Proteomics and Systems Biology

Edited by

ROSSEN DONEV
MicroPharm Ltd,
United Kingdom

TATYANA KARABENCHEVA-CHRISTOVA
Michigan Technological University,
United States

Academic Press is an imprint of Elsevier
50 Hampshire Street, 5th Floor, Cambridge, MA 02139, United States
525 B Street, Suite 1650, San Diego, CA 92101, United States
The Boulevard, Langford Lane, Kidlington, Oxford OX5 1GB, United Kingdom
125 London Wall, London, EC2Y 5AS, United Kingdom

First edition 2021

ISBN: 978-0-323-85319-4
ISSN: 1876-1623

For information on all Academic Press publications
visit our website at https://www.elsevier.com/books-and-journals

Publisher: Zoe Kruze
Acquisitions Editor: Ashlie Jackman
Developmental Editor: Naiza Ermin Mendoza
Production Project Manager: Abdulla Sait
Cover Designer: Miles Hitchen

Typeset by SPi Global, India

Contents

Contributors

Leila Abdelrahman
Bascom Palmer Eye Institute, University of Miami Miller School of Medicine; Department of Electrical and Computer Engineering, University of Miami, Miami, FL, United States

Suruchi Aggarwal
Translational Health Science and Technology Institute, NCR Biotech Science Cluster, Faridabad, Haryana; Department of Molecular Biology and Biotechnology, Cotton University, Guwahati, Assam, India

V. Anu Preethi
School of Computer Science and Engineering, Vellore Institute of Technology, Vellore, Tamil Nadu, India

Jennifer Arcuri
Bascom Palmer Eye Institute, University of Miami Miller School of Medicine; Molecular and Cellular Pharmacology Graduate Program, University of Miami; Miami Integrative Metabolomics Research Center, University of Miami, Miami, FL, United States

Sanjoy K. Bhattacharya
Bascom Palmer Eye Institute, University of Miami Miller School of Medicine; Molecular and Cellular Pharmacology Graduate Program; Miami Integrative Metabolomics Research Center; Vision Science and Investigative Ophthalmology Graduate Program, University of Miami, Miami, FL, United States

Nicolas Borisov
Moscow Institute of Physics and Technology, Moscow, Russia; OmicsWay Corporation, Walnut, CA, United States

Anton Buzdin
Moscow Institute of Physics and Technology; Shemyakin-Ovchinnikov Institute of Bioorganic Chemistry; World-Class Research Center "Digital biodesign and personalized healthcare", Sechenov First Moscow State Medical University, Moscow, Russia; OmicsWay Corporation, Walnut, CA, United States

Emidio Capriotti
Department of Pharmacy and Biotechnology, University of Bologna, Bologna, Italy

Tara Cornet
Bascom Palmer Eye Institute, University of Miami Miller School of Medicine; Miami Integrative Metabolomics Research Center, University of Miami, Miami, FL, United States

Pedro M. Costa
UCIBIO—Applied Molecular Biosciences Unit, Departamento de Ciências da Vida, Faculdade de Ciências e Tecnologia da Universidade Nova de Lisboa, Caparica, Portugal

Magdalena Djordjevic
Institute of Physics Belgrade, University of Belgrade, Belgrade, Serbia

Marko Djordjevic
Computational Systems Biology Group, Faculty of Biology, University of Belgrade, Belgrade, Serbia

Piero Fariselli
Department of Medical Sciences, University of Torino, Turin, Italy

Nurshat Gaifullin
Department of Pathology, Faculty of Medicine, Lomonosov Moscow State University, Moscow, Russia

Andrew Garazha
OmicsWay Corporation, Walnut, CA, United States

C. George Priya Doss
School of BioSciences and Technology, Vellore Institute of Technology, Vellore, Tamil Nadu, India

Srishti Gupta
Translational Health Science and Technology Institute, NCR Biotech Science Cluster, Faridabad, Haryana; School of Biosciences and Technology, Vellore Institute of Technology, Vellore, India

Bojana Ilic
Institute of Physics Belgrade, University of Belgrade, Belgrade, Serbia

Anna K. Junk
Bascom Palmer Eye Institute, University of Miami Miller School of Medicine; Miami Integrative Metabolomics Research Center, University of Miami; Miami Veterans Affairs Health Care System, Miami, FL, United States

Vibhaa Kumar
School of BioSciences and Technology, Vellore Institute of Technology, Vellore, Tamil Nadu, India

Richard K. Lee
Bascom Palmer Eye Institute, University of Miami Miller School of Medicine; Vision Science and Investigative Ophthalmology Graduate Program; Miami Integrative Metabolomics Research Center, University of Miami, Miami, FL, United States

Carolina Madeira
UCIBIO—Applied Molecular Biosciences Unit, Departamento de Ciências da Vida, Faculdade de Ciências e Tecnologia da Universidade Nova de Lisboa, Caparica, Portugal

N. Madhana Priya
Department of Biotechnology, Sri Ramachandra Institute of Higher Education and Research (DU), Porur, Chennai, Tamil Nadu, India

R. Magesh
Department of Biotechnology, Sri Ramachandra Institute of Higher Education and Research (DU), Porur, Chennai, Tamil Nadu, India

Sean D. Meehan
Molecular and Cellular Pharmacology Graduate Program; Miami Integrative Metabolomics Research Center, University of Miami, Miami, FL, United States

Giulia Menichetti
Center for Complex Network Research, Department of Physics, Northeastern University; Department of Medicine, Brigham and Women's Hospital, Harvard Medical School, Boston, MA, United States

Alessandra Merlotti
Department of Physics and Astronomy, University of Bologna, Bologna, Italy

Ognjen Milicevic
Department for Medical Statistics and Informatics, Faculty of Medicine, University of Belgrade, Belgrade, Serbia

Astha Mishra
Amity Institute of Biotechnology, Amity University Uttar Pradesh, Noida, India

Jada Morris
Bascom Palmer Eye Institute, University of Miami Miller School of Medicine; Vision Science and Investigative Ophthalmology Graduate Program, University of Miami; Miami Integrative Metabolomics Research Center, University of Miami, Miami, FL, United States

Sergey Moshkovskii
Pirogov Russian National Research Medical University; Federal Research and Clinical Center of Physical-Chemical Medicine, Moscow, Russia

Ciara Myer
Bascom Palmer Eye Institute, University of Miami Miller School of Medicine; Miami Integrative Metabolomics Research Center, University of Miami, Miami, FL, United States

Dhanushya Nagarajan
School of BioSciences and Technology, Vellore Institute of Technology, Vellore, Tamil Nadu, India

Priyanka Narad
Amity Institute of Biotechnology, Amity University Uttar Pradesh, Noida, India

G. Naresh
Amity Institute of Biotechnology, Amity University Uttar Pradesh, Noida, India

Diksha Parashar
Amity Institute of Biotechnology, Amity University Uttar Pradesh, Noida, India

Kevin K. Park
Bascom Palmer Eye Institute, University of Miami Miller School of Medicine; Miami Integrative Metabolomics Research Center; Miami Project to Cure Paralysis, University of Miami, Miami, FL, United States

Daniel Remondini
Department of Physics and Astronomy, University of Bologna, Bologna, Italy

Andjela Rodic
Computational Systems Biology Group, Faculty of Biology, University of Belgrade, Belgrade, Serbia

Aisha Saleem
School of Biosciences and Technology, Vellore Institute of Technology, Vellore, Tamil Nadu, India

Igor Salom
Institute of Physics Belgrade, University of Belgrade, Belgrade, Serbia

Mohammad Samarah
Carroll University, Waukesha, WI, United States

Abhishek Sengupta
Amity Institute of Biotechnology, Amity University Uttar Pradesh, Noida, India

Maksim Sorokin
Moscow Institute of Physics and Technology; I.M. Sechenov First Moscow State Medical University, Moscow, Russia; OmicsWay Corporation, Walnut, CA, United States

Maria Suntsova
Moscow Institute of Physics and Technology; World-Class Research Center "Digital biodesign and personalized healthcare", Sechenov First Moscow State Medical University, Moscow, Russia; OmicsWay Corporation, Walnut, CA, United States

Iftikhar Aslam Tayubi
Faculty of Computing and Information Technology, King Abdul-Aziz University, Rabigh, Saudi Arabia

D. Thirumal Kumar
Department of Bioinformatics, Saveetha School of Engineering, Saveetha Institute of Medical and Technical Sciences; Meenakshi Academy of Higher Education and Research, Chennai, Tamil Nadu, India

Victor Tkachev
Moscow Institute of Physics and Technology, Moscow, Russia; OmicsWay Corporation, Walnut, CA, United States

Priya Tolani
Translational Health Science and Technology Institute, NCR Biotech Science Cluster, Faridabad, Haryana, India

S. Udhaya Kumar
School of BioSciences and Technology, Vellore Institute of Technology, Vellore, Tamil Nadu, India

Amit Kumar Yadav
Translational Health Science and Technology Institute, NCR Biotech Science Cluster, Faridabad, Haryana, India

Kirti Yadav
Translational Health Science and Technology Institute, NCR Biotech Science Cluster, Faridabad, Haryana; Department of Pharmaceutical Biotechnology, Delhi Pharmaceutical Sciences and Research University, New Delhi, India

Salma Younes
Translational Research Institute, Women's Wellness and Research Center, Hamad Medical Corporation; Department of Biomedical Sciences, College of Health and Sciences, QU Health, Qatar University, Doha, Qatar

Hatem Zayed
Department of Biomedical Sciences, College of Health and Sciences, QU Health, Qatar University, Doha, Qatar

Dusan Zigic
Institute of Physics Belgrade, University of Belgrade, Belgrade, Serbia

Marianna Zolotovskaia
Moscow Institute of Physics and Technology, Moscow, Russia; OmicsWay Corporation, Walnut, CA, United States

CHAPTER ONE

Using proteomic and transcriptomic data to assess activation of intracellular molecular pathways

Anton Buzdin[a,b,c,d,*], Victor Tkachev[a,b], Marianna Zolotovskaia[a,b], Andrew Garazha[b], Sergey Moshkovskii[e,f], Nicolas Borisov[a,b], Nurshat Gaifullin[g], Maksim Sorokin[a,b,h], and Maria Suntsova[a,b,d]

[a]Moscow Institute of Physics and Technology, Moscow, Russia
[b]OmicsWay Corporation, Walnut, CA, United States
[c]Shemyakin-Ovchinnikov Institute of Bioorganic Chemistry, Moscow, Russia
[d]World-Class Research Center "Digital biodesign and personalized healthcare", Sechenov First Moscow State Medical University, Moscow, Russia
[e]Pirogov Russian National Research Medical University, Moscow, Russia
[f]Federal Research and Clinical Center of Physical-Chemical Medicine, Moscow, Russia
[g]Department of Pathology, Faculty of Medicine, Lomonosov Moscow State University, Moscow, Russia
[h]I.M. Sechenov First Moscow State Medical University, Moscow, Russia
*Corresponding author: e-mail address: buzdin@oncobox.com

Contents

Abstract

Analysis of molecular pathway activation is the recent instrument that helps to quantize activities of various intracellular signaling, structural, DNA synthesis and repair, and biochemical processes. This may have a deep impact in fundamental research, bioindustry, and medicine. Unlike gene ontology analyses and numerous qualitative methods that can establish whether a pathway is affected in principle, the quantitative approach has the advantage of exactly measuring the extent of a pathway up/downregulation.

Advances in Protein Chemistry and Structural Biology, Volume 127
ISSN 1876-1623
https://doi.org/10.1016/bs.apcsb.2021.02.005

This results in emergence of a new generation of molecular biomarkers—pathway activation levels, which reflect concentration changes of all measurable pathway components. The input data can be the high-throughput proteomic or transcriptomic profiles, and the output numbers take both positive and negative values and positively reflect overall pathway activation. Due to their nature, the pathway activation levels are more robust biomarkers compared to the individual gene products/protein levels. Here, we review the current knowledge of the quantitative gene expression interrogation methods and their applications for the molecular pathway quantization. We consider enclosed bioinformatic algorithms and their applications for solving real-world problems. Besides a plethora of applications in basic life sciences, the quantitative pathway analysis can improve molecular design and clinical investigations in pharmaceutical industry, can help finding new active biotechnological components and can significantly contribute to the progressive evolution of personalized medicine. In addition to the theoretical principles and concepts, we also propose publicly available software for the use of large-scale protein/RNA expression data to assess the human pathway activation levels.

1. Molecular pathways

Intracellular molecular pathways (IMPs) are specific networks that combine gene products involved in common molecular processes (Junaid, Akter, Afrose, Tania, & Khan, 2020; Ma & Liao, 2020; Zheng et al., 2020). Research of IMPs is hot topic in life sciences for several decades because they mediate all major events in the normal or pathological cell (Aliper, Jellen, et al., 2017; Aliper, Korzinkin, et al., 2017; Borisov et al., 2009; Kholodenko, Demin, Moehren, & Hoek, 1999; Kiyatkin et al., 2006; Marshall, 1995), and knowledge of pathway regulation is crucial for understanding all major intracellular processes including cell survival, growth, differentiation, motility, proliferation, senescence, malignization, and death (Buzdin et al., 2018). IMPs are affected during organism development and growth, aging, malignization and disease progression (Parkhitko, Filine, Mohr, Moskalev, & Perrimon, 2020). The most commonly mentioned IMPs are metabolic, signaling, DNA repair and cytoskeleton pathways, but this list can be further expanded (Buzdin et al., 2018; Zolotovskaia, Sorokin, Roumiantsev, Borisov, & Buzdin, 2019).

Metabolic pathways orchestrate chains of biochemical reactions by organizing the corresponding enzymes in biochemically meaningful networks. Signaling pathways comprise gene products dealing with specific aspects of signal recognition, modulation and transduction. The signals sensed and transduced can be very different. For example, they can be initiated by ligand binding with the cell surface or intracellular receptor molecule; by altered concentrations of specific proteins, metabolites, oxygen, pH or regulatory ions such as Ca^{2+}; by the physical factors such as certain temperature, irradiation exposure or viscosity conditions; by tight protein-protein interactions, deconstructing or aggregation of higher-order molecular complexes, or by their remodeling (Borger et al., 2020; Borisov, Sorokin, Garazha, & Buzdin, 2020; Negro et al., 2020; Tkachev, Sorokin, Garazha, Borisov, & Buzdin, 2020; Zolotovskaia, Sorokin, Garazha, Borisov, & Buzdin, 2020). The signal is initially detected and then transduced to the downstream pathway nodes (each containing one or several proteins) until it initiates one or several relevant effector molecules that can execute further molecular functions. The intermediate steps in signal transduction are very diverse and may deal with biochemical modifications, oligomerization, protein or cofactor binding, shifted molecular localization (e.g., from cell nucleus to cytoplasm or vice versa), release or sequestering of ions and metabolites (Borger et al., 2020; Borisov et al., 2020; Negro et al., 2020; Tkachev, Sorokin, Garazha, et al., 2020; Zolotovskaia, Sorokin, Garazha, et al., 2020; Zolotovskaia, Sorokin, Petrov, et al., 2020; Zolotovskaia, Tkachev, Seryakov, et al., 2020).

Consequently, the signaling pathways may have various outputs (one or multiple per each pathway). These include targeted alterations of gene expression, modified permeability of channel proteins or their complexes, biochemical modifications of macromolecules such as phosphorylation, assembly or demolition of molecular complexes, and modulating other molecular pathways. For example, one of the most frequent types of signaling pathway outputs is regulation of one of the cytoskeleton remodeling pathways, which reflects their strong connectivity in eukaryotic interactomes (Borisov et al., 2020; Disanza, Frittoli, Palamidessi, & Scita, 2009; Filteau et al., 2015). DNA repair pathways are also included in the context of complexes with other molecular pathways, e.g., repair of DNA lesions to allow progression of the cell cycle, where DNA repair, cytoskeleton and signaling pathways are tightly interconnected (Branzei & Foiani, 2008; Malumbres & Barbacid, 2009; Vermeulen, Van Bockstaele, & Berneman, 2003).

Different pathways may have tens or hundreds of nodes (Wishart, Mandal, Stanislaus, & Ramirez-Gaona, 2016). Representatives of different types of molecular pathways may have similar organization and can partly overlap in their nodes or node's components. The individual pathway nodes are most frequently built not by just products of one gene, but rather by groups of gene products. Those node-forming groups can be various molecular complex subunits or/and can be homologous protein families with common functions (Apweiler et al., 2011; Mathivanan et al., 2006). Thus, a molecular pathway may accumulate far more different gene products than the number of its nodes (Croft et al., 2014; Elkon et al., 2008; Nakaya et al., 2013; Nikitin, Egorov, Daraselia, & Mazo, 2003).

Contemporary methods of large-scale molecular and genomic analyses enabled to catalog millions of protein-protein, genetic and other molecular interactions (e.g., Universal protein resource (UniProt, https://www.uniprot.org)) and thousands of molecular pathways were deduced (Wishart et al., 2020). These databases accumulate information on pathway architecture, and nature and functional meaning of the enclosed molecular interactions, e.g., Reactome (Croft et al., 2014), Kyoto Encyclopedia of Genes and Genomes (KEGG) (Nakaya et al., 2013), QIAGEN SABiosciences (http://www.sabiosciences.com/) (Apweiler et al., 2011), Ariadne Pathway Studio (Nikitin et al., 2003), Human protein reference database (HPRD) (Mathivanan et al., 2006), Signaling Pathways Integrated Knowledge Engine (SPIKE) (Elkon et al., 2008), HumanCyc Encyclopedia of Human Genes and Metabolism (https://www.humancyc.org) and WikiPathways (Bauer-Mehren, Furlong, & Sanz, 2009). Some of the databases are specifically devoted to the pathways with specific functions, such as MetaCyc (Caspi et al., 2020) and SynSysNet (von Eichborn et al., 2012).

However, every individual pathway can be regarded as just a fragment of an overall molecular interacting network in a cell, called interactome (Vidal, Cusick, & Barabási, 2011), Fig. 1A and B. On the other hand, many pathways have two or more terminal branches with different functional impact(s) and can be further subdivided into smaller molecular networks (Fig. 1C). These "micropathways" are sub-graphs of full-size IMPs which execute smaller and more specific molecular functions (Fig. 1C). In many cases, the micropathways may show molecular processes in more detail by separately characterizing different terminal branches of the pathways leading to different, sometimes contradictory functional consequences (Sorokin et al., 2021).

It was found using various experimental and bioinformatic approaches that many IMPs are affected during growth, development, aging (Aliper et al., 2016, 2015; Aliper, Jellen, et al., 2017; Aliper, Korzinkin, et al., 2017; Buzdin, Prassolov, Zhavoronkov, & Borisov, 2017), onset and progression of acute and chronic diseases (Alexandrova et al., 2016; Makarev et al., 2016; Wirsching et al., 2017), and during carcinogenesis (Petrov et al., 2017, 2016; Shepelin et al., 2016; Shtam et al., 2019). However, it

A

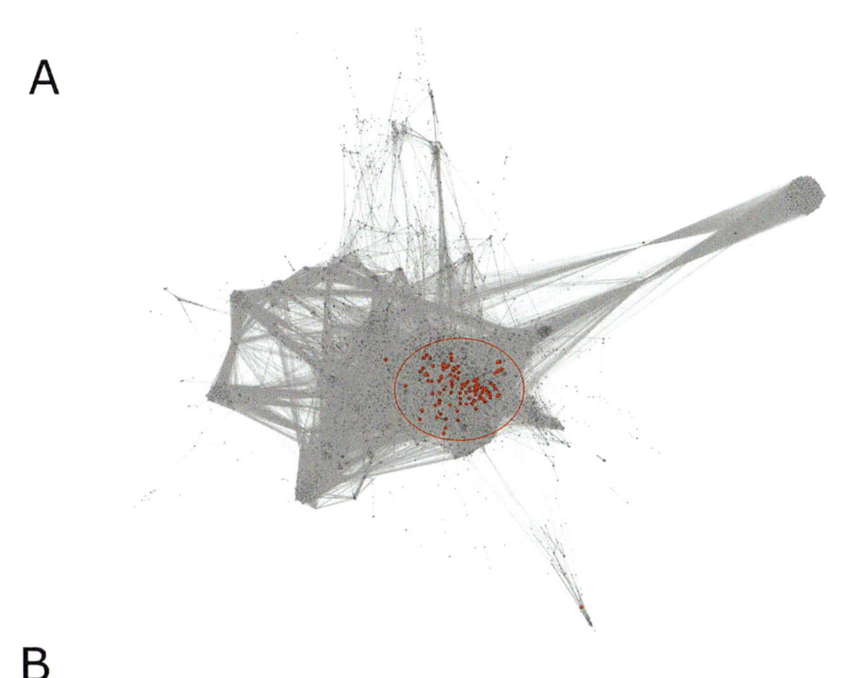

B

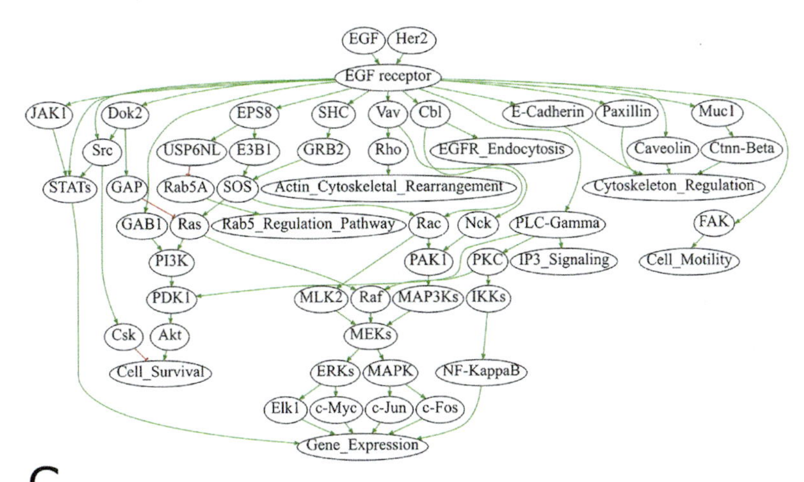

C

Fig. 1 See figure legend on next page.

is technically challenging to quantitatively measure extents of pathway activation in a high-throughput manner (Buzdin et al., 2018; Zolotovskaia, Sorokin, Garazha, et al., 2020; Zolotovskaia, Sorokin, Petrov, et al., 2020; Zolotovskaia, Tkachev, Seryakov, et al., 2020). This task cannot be solved without using various kinds of OMICS data such as gene expression profiles at the RNA and/or protein levels (Borisov et al., 2020; Tkachev, Sorokin, Garazha, et al., 2020). Structures of thousands of IMPs have been reconstructed to the date, and knowledge of their molecular architecture can help interrogating the pathway activation levels using bioinformatic algorithms and large-scale gene expression data (Buzdin et al., 2019; Schulze & Downward, 2001; Shih, Chetty, & Tsao, 2005; Sorokin, Ignatev, Poddubskaya, et al., 2020; Sorokin, Ignatev, Barbara, et al., 2020; Willier, Butt, & Grunewald, 2013).

2. Quantitative analysis of gene expression

In various experimental settings, activities of individual IMPs can be measured using two major strategies. The first strategy deals with measuring any of the pathway terminal molecular outcomes. This can be achieved by the screening of protein phosphorylation, by measuring genomic binding of transcription factors, by using specific gene expression or protein aggregation biosensors, by measuring physiological outputs and by many other approaches (Otto & Brar, 2018; Yang, Sathiyaseelan, Ho, & Gorski, 2018). It may be problematic, however, to interrogate complex IMPs with multiple lines of molecular outcomes with these approaches. In turn, the second strategy utilizes agnostic rationale of calculating pathway activities based on the concentrations of as many types of pathway member molecules as possible (Buzdin et al., 2014).

In the latter case, using high throughput transcriptomic and/or proteomic gene expression techniques is required, followed by profound bioinformatic analysis (Buzdin et al., 2018). Such large-scale gene expression

Fig. 1—Cont'd Model of total human interactome built for 302,234 protein-protein interactions for 9022 genes involved in 3044 molecular pathways cataloged in (Sorokin et al., 2021). (A) Gene products from EGFR signaling pathways are highlighted and circled. (B) EGFR signaling pathway shown separately as an interacting network. Color of arrows indicate functional type of molecular interactions: green—activation, red—inhibition. (C) Graphs of two *micropathways* included in EGFR pathway: *EGFR* (*Cytoskeletal regulation*) (left) and *EGFR* (*Rab5 regulation*) (right). Interacting network indications are as above.

screening techniques must have qualities of high throughput analysis, strong reproducibility, high performance, reasonable costs, and sufficiency of minute amounts of biomaterials (Buzdin, Skvortsova, Li, & Wang, 2021). Thus, several existing quantitative proteomic and transcriptomic approaches can be regarded as the best candidates (Li & Zhan, 2021; Navajas, Corrales, & Paradela, 2021; Wang et al., 2020). Both types of input data can be used for further pathway analytic algorithms (Borisov et al., 2020; Tkachev, Sorokin, Garazha, et al., 2020).

Gene expression can be assessed on both RNA and protein levels while both types of the analysis have their *pros* and *contras*. The protein level is obviously closer to the biological phenotype because most of terminal phenotype-related molecular functions in a living cell are executed by the proteins. The approaches of measuring expressions of single proteins include immunoassays, Western blotting, and daughter multiplexing methods (Painter, Clayton, & Herbert, 2010; Stephen, 2017; Stephen & Guest, 2018). At the same time, the high throughput protein expression analyses can be performed by using protein arrays (Duarte & Blackburn, 2017) and mass spectrometry assays (Aebersold & Mann, 2016).

- *Protein arrays*. Since early 1980s protein arrays (solid or reverse phase (Stetson, Dazard, & Barnholtz-Sloan, 2016)) are based on the principle of protein recognition by specific monoclonal antibodies or other binders like oligonucleotide aptamers, where each spot on an array is reserved to a specific protein (Chang, 1983; Mirus et al., 2014; Sreekumar et al., 2001). This analysis, however, has several technical difficulties and limitations. First, a large panel of highly specific and sensitive monoclonal antibodies is required to do proteome-wide analyses. This is problematic because conformation/modification features of individual proteins in the real tissues may have little in common with those used for *in vitro* production and validation of the analytic antibodies (Duarte & Blackburn, 2017). However, the problem can be solved in part by using of aptamers as multiple protein binders (Witt, Walter, & Stahl, 2015) like in Somascan™ platform (Duo et al., 2018; Webber et al., 2014).

  Another major problem is drastically different physico-chemical properties of different proteins extracted from the biosamples like hydrophobicity and solubility (Reymond & Schlegel, 2007; Spisak & Guttman, 2009; Zhou, Li, Wang, Huang, & Nice, 2016). In addition, different proteins can be degraded with different rates in different biosamples, and all the technical procedures like sample isolation and processing must be tightly controlled to exclude differential degradation-linked artifacts

(Reymond & Schlegel, 2007; Rosenberg & Utz, 2015; Spisak & Guttman, 2009; Zhou et al., 2016).

- *Mass-spectrometry–MALDI profiling.* Another relevant large group of methods is a mass-spectrometry-based proteomics (Mann, 2016). An approach of distinguishing biospecimens based on their protein composition using matrix-assisted laser desorption/ionization time-of-flight (MALDI-TOF) mass spectrometry is known for two decades (Karpova, Moshkovskii, Toropygin, & Archakov, 2010). In this approach, specimen is applied without separation, and tens to hundreds of proteins can be detected. The resulting profiles may be compared between the control and case samples accurately and reproducibly, so that technically they can serve for clinical applications (Fidler et al., 2018; Petricoin et al., 2002; Zhang et al., 2004). However, these profiles frequently lack analytical sensitivity because they are formed by highly abundant peptides and proteins (Moshkovskii et al., 2007). However, multiple efforts were made to translate the MALDI-TOF based assays into clinical practice as prognostic and/or predictive tests (Zhang et al., 2004).
- *Shotgun proteomics.* In the shotgun approach, the proteins extracted from a biosample are digested to completion by proteolytic enzymes with known specificities thus generating a characteristic pool of peptides that can be resolved as specific peaks during high-resolution mass spectrometry (Aebersold & Mann, 2016). These peaks may reflect protein contents and thus serve to characterize their abundances in biosamples. With this approach, up to 10,000 gene products can be characterized in 1 sample (Coscia et al., 2016). This technique has wide applicability for finding and tracing protein biomarkers (Betancourt et al., 2019; O'Neill et al., 2017; Principe et al., 2018), and for characterizing biochemical modifications such as protein phosphorylation (Kuenzi et al., 2017; Yang, Freeman, & Kyprianou, 2018).
- *Proteogenomic approach.* Combining genomic and proteomic data can significantly increase accuracy and repertoire of the peptides resolved by mass-spectrometry. Such approach of combining DNA analysis with shotgun proteomics is termed proteogenomics. This approach provides better resolution, but also makes it possible to identify and quantify mutant peptides, thus being useful for prognostic and predictive decisions (Zhang et al., 2019) and identification of clinically actionable mutations (Polyakova, Kuznetsova, & Moshkovskii, 2015).
- *Ribosomal profiling.* This approach also termed "translatomics" occupies intermediate position on the interface between transcriptomics

and proteomics (King & Gerber, 2016). Ribosomal profiling rationale is based on the deep sequencing of mRNA molecules directly attached to the ribosomes (Ingolia, Ghaemmaghami, Newman, & Weissman, 2009). In this way, it enables to quantitatively estimate profiles of mRNA translation in a biosamples (Andreev et al., 2017; Anisimova et al., 2020; Michel & Baranov, 2013). Although obviously closer to molecular phenotype than the transcriptomic approach itself (Zhao, Qin, Nikolay, Spahn, & Zhang, 2019), this method demands significant amounts of fresh biomaterials and is laborious and experimentally challenging (Bartholomäus, Del Campo, & Ignatova, 2016).

– *Transcriptomic methods*. At the RNA level, gene expression can be quantified by RNA sequencing, microarray hybridization and NanoString-based techniques (Buzdin, Sorokin, Garazha, et al., 2019; Eastel et al., 2019; Wang et al., 2020). Transcriptomics directly analyzes concentrations of RNA molecules, including mRNAs of protein coding genes (Ma, Liang, Zhou, & Qu, 2018). Standing not so close to the phenotype like proteomics and translatomics, transcriptomics, however, remains unparalleled in terms of reproducibility, simplicity and cost-effectiveness (Bossel Ben-Moshe et al., 2018). It allows for the use of virtually all types of biosamples including formalin-fixed, paraffin-embedded clinical biomaterials stored for several years at room temperature (Sorokin, Ignatev, Barbara, et al., 2020; Sorokin, Ignatev, Poddubskaya, et al., 2020; Sorokin, Kholodenko, Kalinovsky, et al., 2020; Sorokin, Poddubskaya, Baranova, et al., 2020; Suntsova et al., 2019).

Interestingly, relatively high statistically significant correlations with R range 0.59–0.89 have been observed for the comparison of ribosomal and RNA sequencing profiles of mammalian mRNAs (Barry, Ingolia, & Vance, 2017; De Klerk et al., 2015; Dunn, Foo, Belletier, Gavis, & Weissman, 2013). These findings evidence that the use of transcriptomic profiles for functional annotation of gene expression can be justified.

There are three major directions of quantitative transcriptomics: RNA sequencing (Byron, Van Keuren-Jensen, Engelthaler, Carpten, & Craig, 2016), hybridization of microarrays (Tao et al., 2017), and quantitative polymerase chain reaction (qPCR) (Denis et al., 2018). In qPCR, multiplex profiling of gene expressions can be made at a very high accuracy (Carlson et al., 2013). However, it has strong technical limitations that prevent measuring expressions of thousands of genes, and limits number of analytes to few hundreds (Nault, Fader, & Zacharewski, 2015).

Microarray hybridization utilizes basepairing of nucleic acids in solution, where labeled tester fraction of nucleic acids is hybridized with the gene-specific probes attached to a surface of an array, and hybridization signal is proportionate to concentration of the respective gene in a biosample (Watson, Mazumder, Stewart, & Balasubramanian, 1998). Several popular expression microarray platforms can be mention, owned by Agilent (Wolber, Collins, Lucas, De Witte, & Shannon, 2006), Illumina (Teumer et al., 2016), Affymetrix (Dalma-Weiszhausz, Warrington, Tanimoto, & Miyada, 2006) and CustomArray (Petrov et al., 2017) manufacturers. They utilize different reagents and library preparation protocols, but also different equipment and physical principles for detection of the hybridization signals (Borisov et al., 2017).

These differences also exist for the different active RNA sequencing platforms like Illumina (Lahens et al., 2017), BGI (Li et al., 2019), Oxford Nanopore (Kono & Arakawa, 2019), and Ion Torrent/Proton (Workman et al., 2019). This results in a dramatic lack of compatibility for the results obtained not only using different sequencing platforms, but also using the same platform, but different reagents and kits (Borisov et al., 2019, 2017; Buzdin, Sorokin, Poddubskaya, & Borisov, 2019; Buzdin et al., 2014). This is one of the main reasons why experimental gene expression results are generally compared only for the same platform (Borisov et al., 2019; Suntsova et al., 2019; Vladimirova et al., 2021).

Chronologically, expression microarrays were introduced one decade earlier than first RNA sequencing protocols (Lin, Wang, & Cheng, 2010; Schena, Shalon, Davis, & Brown, 1995). This led to strong penetration of microarrays in various research protocols, and significant portion of current public gene expression datasets was obtained using various microarray platforms (Castillo et al., 2017). However, technical comparisons of RNA sequencing and expression microarrays performances (Rai, Tycksen, Sandell, & Brophy, 2018; Sîrbu, Kerr, Crane, & Ruskin, 2012; Zhang et al., 2015) has demonstrated superior precision of RNA sequencing methods in measuring gene expression and lower false discovery rate for differentially expressed genes (Nault et al., 2015; Su et al., 2014). Nowadays, RNA sequencing is generally regarded a method of choice for high throughput screening of gene expression (Su et al., 2014).

- *Data normalization and harmonization.* The problem of gene expression data incompatibility dramatically complicates comparison of expression datasets obtained using different experimental platforms and protocols—at both RNA and protein levels (Lin et al., 2013; Maouche et al., 2008; Wen et al., 2010; Zhang et al., 2013). Consequently, this makes problematic further levels of data analysis, first of all finding differentially expressed genes and profiling molecular pathway activation (Aliper, Jellen, et al., 2017; Aliper, Korzinkin, et al., 2017; Borisov et al., 2017).

Solving this problem could dramatically enhance quality and contents of the datasets available for the reciprocal comparisons (Borisov et al., 2019). Accomplishing this task may require different levels of data conversion: technical normalization for datasets obtained using the same experimental platform, or harmonization for datasets obtained using different experimental platforms (Borisov et al., 2019). The major normalization techniques include quantile normalization (QN) (Bolstad, Irizarry, Åstrand, & Speed, 2003), DESeq (Anders & Huber, 2010), and DESeq2 (Love, Huber, & Anders, 2014) methods. There are also many methods developed for data harmonization including cross-platform normalization (XPN) (Rudy & Valafar, 2011; Shabalin, Tjelmeland, Fan, Perou, & Nobel, 2008), and the others (Borisov et al., 2019), where XPN method showed the best performance (Rudy & Valafar, 2011). XPN method acts by restructuring distributions of gene expression levels for the samples under analysis to organize them in a uniformly comparable way. XPN uses data clustering to identify blocks of similarity between the expression profiles obtained using different platforms, and then expands them by reshaping other regions of the expression profiles (Shabalin et al., 2008). However, XPN and all but one harmonization methods have an important limitation that they are capable to do harmonization for only two expression datasets of a comparable sample size, thus complicating harmonization of the real-world data (Borisov et al., 2019; Rudy & Valafar, 2011). Furthermore, the resulting new hybrid format of data will be incompatible with any other existing format of gene expression data (Borisov et al., 2019). This is a fundamental problem that prevents converting different expression datasets into a uniform data format enabling reciprocal direct comparisons.

However, this problem was partly solved in a recent cross-platform data harmonization method termed Shambhala which is independent on number of datasets and/or of experimental platforms under harmonization, and also independent on number of samples in every dataset (Borisov et al., 2019). Shambhala converts all the profiles in a universal pre-defined format, thus making them ready for further direct comparisons (Borisov et al., 2019). So far, this technique was successfully tested only for few model datasets obtained from several major microarray and RNA sequencing platforms (Borisov et al., 2019; Su et al., 2014). The initial version of this method was optimized for the analysis of human transcriptomes, but it could be further updated to create a more universal tool for the data analysis also in the other species.

- *Different levels of gene expression data analysis.* Gene expression data can be analyzed in many ways. For example, depending on the type of biomaterials under investigation, very different RNA sequencing results can be obtained. For fresh biosamples, high-integrity RNAs can be isolated, and longer, up to few kb sequencing reads can be achieved. For formalin-fixed, paraffin-embedded samples, much more degraded RNAs can be obtained, resulting in short 50 bp-long single end reads (Suntsova et al., 2019). In fact, both types of those output data are suitable for the molecular pathway analysis. To this end, one quantifies RNA sequencing reads that were unambiguously mapped on known genes and calculates relative gene expression characteristics.

Depending on the platform and object investigated, there is a minimal number of gene-mapped RNA sequencing reads that ensures an adequate depth and clustering of a library. For example, this threshold value is 2.5 million mapped reads for human RNA sequencing libraries obtained using random primers and Illumina sequencers (Suntsova et al., 2019). We stress this protocol because it was confirmed in a recent study where it resulted in comparable gene expression data at RNA and protein levels measured by RNA sequencing and immunohistochemistry, respectively (Sorokin, Ignatev, Barbara, et al., 2020; Sorokin, Ignatev, Poddubskaya, et al., 2020; Sorokin, Kholodenko, Kalinovsky, et al., 2020; Sorokin, Poddubskaya, Baranova, et al., 2020; Suntsova et al., 2019).

After mapping, gene expression profiles can be compared with the control samples to establish differentially expressed genes and to find case-to-normal ratios (fold-change values) (Buzdin, Sorokin, Poddubskaya, & Borisov, 2019; Tkachev et al., 2018). The differential gene sets then can be analyzed in several ways.

First, one can be curious in case-to-normal expression ratios for the genes of a special interest. Second, the differential sets of upregulated and downregulated genes can be analyzed in a systemic way. This type of analysis may include assessing enrichment of the Gene Ontology (GO) terms (Eden, Navon, Steinfeld, Lipson, & Yakhini, 2009; Huang, Sherman, & Lempicki, 2009b) or it may deal with the analysis of molecular pathways (Artcibasova et al., 2016; Buzdin et al., 2018).

- *Control gene expression datasets.* When finding case-to-normal ratios, it can be important to have a pool of control (normal) samples to calculate statistical significance of these measurements. From the technical point of view, it is very important to compare only comparable expression data. Thus, one can benefit from the collections of normal expression profiles obtained using the same or similar equipment and protocols, especially when no such data can be profiled experimentally by the researcher (Borisov et al., 2019). This is the reason why we briefly describe here several datasets that could be used as the source of reference human tissue gene expression profiles.

First of all, the Clinical Proteomic Tumor Analysis Consortium (CPTAC) database can be mentioned (Edwards et al., 2015) that includes not only proteomic, but also genomic and transcriptomic profiles for various human cancer and normal tissue samples. More data had been published at the transcriptomic level. The biggest published dataset that includes RNA sequencing profiles for 11,688 samples of human postmortal tissues was released by The Genotype-Tissue Expression (GTEx) consortium (Lonsdale et al., 2013). It has no publicly available data on the donor's age. The normalized expression data are publicly available, but an access to the primary sequencing data requires complicated registration steps. It was shown recently that at least part of the GTEx profiles have signs of contamination by the other tissues which could be due to problems with using the same equipment for the different types of biosamples (Nieuwenhuis et al., 2020).

Four other relevant databases with the technically uniform RNA sequencing profiles have smaller sizes but they include the donor's age information and are freely accessible:

- The Cancer Genome Atlas (TCGA) project database with 625 normal samples (Weinstein et al., 2013);
- two ENCODE project databases (Davis et al., 2018) where RNA sequencing libraries were prepared using poly(A)-specific primer (41 samples), or using random primers to sequence total RNA following ribosomal depletion (92 samples);
- Oncobox Atlas of Normal Tissue Expression (ANTE) database (Suntsova et al., 2019) represents 159 healthy human tissue samples for 20 human tissues or organs.

All these databases have their advantages and limitations. In TCGA dataset, the normal profiles correspond to the samples that were adjacent to the tumor tissue (Huang, Stern, & Zhao, 2016). Thus, they can be not physiologically normal because tumors influence neighboring tissues in many ways including altered vascularization (Zhao et al., 2013), inflammation (Casbas-Hernandez et al., 2015), and biased concentrations of growth factors and cytokines (Jones et al., 2015). The ENCODE datasets were obtained for the autopsy material of post-mortal normal human tissues using two different library preparation methods, but they only include 1–4 samples per tissue type (for both genders together) and in many cases can't support forming statistically significant reference groups. However, in the ANTE database such groups can be formed for 20 human tissues/organs. Importantly, it has solid tissue profiles for the human healthy donors killed in road accidents instead of autopsy materials for the donors who died after chronic diseases as in the other datasets. Profiling healthy donor tissues makes this database unique (Suntsova et al., 2019).

Nevertheless, it is very important to compare the library preparation and sequencing protocols for the experimental and control datasets in order to identify the best possible reference groups.

3. Quantization of pathway activities

Currently, there are several alternative approaches to characterize molecular pathway activities. Intracellular molecular pathways (IMPs) can be assessed using mRNA, protein, microRNA, transcription factor binding sites, and even histone modification patterns (Artcibasova et al., 2016; Borisov et al., 2017; Buzdin et al., 2016, Buzdin, Prassolov, et al., 2017; Buzdin, Sorokin, et al., 2017; Igolkina et al., 2019; Jovčevska et al., 2017; Knyazeva et al., 2020; Nikitin et al., 2019b, 2019, 2018; Shtam et al., 2018). Previously they could be classified in three major groups: Pathway Topology (PT)-based approaches, Functional Class Scoring (FCS), and Over-Representation Analysis (ORA) (Khatri, Sirota, & Butte, 2012). Like in Gene Ontology (GO) analysis, ORA methods calculate enrichment of a pathway by differentially expressed genes (Khatri & Drăghici, 2005). These methods don't make difference for the up/down-regulation of gene products with enhancer/inhibitory functions. They also ignore all non-differential genes. These limitations are partially tackled by FCS methods that calculate fold change-based scores for all genes and then combine these gene-scores in an overall pathway enrichment score (Tian et al., 2005). However, these methods also don't consider molecular functions of the affected genes in a pathway. In turn, PT methods in addition to concentration enrichment also analyze pathway topology and assign weight characteristics to the participating gene products according to the pathway graph connectivity characteristics (Mitrea et al., 2013). Many popular bioinformatic tools like GO analysis software (Huang, Sherman, & Lempicki, 2009a, 2009b), Metacore (Ekins, Nikolsky, Bugrim, Kirillov, & Nikolskaya, 2007) and Pathway Studio (Thomas & Bonchev, 2010) software can utilize the abovementioned principles to analyze pathways enrichment by differentially regulated genes (Dubovenko, Nikolsky, Rakhmatulin, & Nikolskaya, 2017).

However, those techniques and tools have a very important limitation. They cannot identify inhibited or enhanced status of a regulated pathway (Khatri et al., 2012) because pathways may have not only positive, but also negative regulatory nodes and may have negative feedback loops (Buzdin, Zhavoronkov, Korzinkin, Venkova, et al., 2014; Zhavoronkov, Buzdin, Garazha, Borisov, & Moskalev, 2014). Thus, the pathway nodes involve gene products with both pathway-activating and inhibitory functions (Borisov et al., 2020), where upregulation of an inhibitory node means pathway downregulation, and vice versa (Buzdin et al., 2018).

This problem was solved by the next generation of pathway analysis tools. In 2014, a new approach was published that was based on kinetic models using the "low-level" approach of mass action law to perform both quantitative and qualitative pathways enrichment analysis (Buzdin, Zhavoronkov, Korzinkin, Venkova, et al., 2014). For every sample of interest, it does a normal-case pairwise comparison to find the Pathway Activation Level (PAL), a value that was introduced to serve as the quantitative measure of pathway activation (Buzdin, Zhavoronkov, Korzinkin, Venkova, et al., 2014). Unlike previous methods, this approach could identify if a pathway is significantly activated or inhibited in comparison to the controls, but it and also could return PAL as the quantitative measure of this deregulation.

The first-generation PAL-based method OncoFinder (Buzdin, Zhavoronkov, Korzinkin, Venkova, et al., 2014) has evolved by adding analysis of pathways topologies like for iPANDA approach (Ozerov et al., 2016), and to assign for each node specific weighing coefficients that depend on its algorithmically-curated function in a pathway like in Oncobox method (Artcibasova et al., 2016; Borisov et al., 2020; Sorokin et al., 2018). Interestingly, this approach can be also expanded to analyze DNA mutation data and to assess pathways with respect to their mutation burden (Zolotovskaia et al., 2019; Zolotovskaia, Sorokin, Garazha, et al., 2020; Zolotovskaia, Sorokin, Petrov, et al., 2020; Zolotovskaia, Sorokin, Roumiantsev, et al., 2019; Zolotovskaia, Tkachev, Seryakov, et al., 2020).

Performance of PAL approach OncoFinder (Buzdin, Zhavoronkov, Korzinkin, Venkova, et al., 2014) was compared with four other pathway analysis methods (Borisov et al., 2017): Topology-Based Score (TBscore) (Ibrahim, Jassim, Cawthorne, & Langlands, 2012), Topology analysis of pathway phenotype association (TAPPA) (Gao & Wang, 2007), Signal pathway impact analysis (SPIA) (Tarca et al., 2009), and Pathway-Express (PE) (Draghici et al., 2007). There were several biosamples profiled with different experimental platforms, and the performance of pathway scoring methods was assessed. It was investigated whether these methods can improve correlations between the expression data from different platforms. Expression profiles obtained for the same biosamples using different experimental platforms were clustered at the gene- and pathway levels. It was then assessed if the pathway level could provide advantage in data-clustering according to biological nature of a biosample, but not the experimental platform. In this test, the PAL-based approach significantly outperformed the competitor methods (Borisov et al., 2017).

Importantly, the PAL approach has also demonstrated ability to strongly suppress levels of experimental noise in the input expression data (Borisov et al., 2017; Buzdin, Zhavoronkov, Korzinkin, Roumiantsev, et al., 2014). This noise can be introduced during experimental profiling of gene expression, especially for the gene products with low expression levels (Liu, François, & Capp, 2016; Sanchez, Choubey, & Kondev, 2013). For both transcriptomic and proteomic expression datasets obtained using several experimental platforms, the PAL methods returned more stable results compared to the analysis of expression of individual genes (Fig. 2) (Borisov et al., 2017). At the single gene level, RNA sequencing and microarray hybridization data for the same biosamples showed rather low correlation (<0.2) which raised to as high as ~ 0.9 at the level of activation of 90 molecular pathways (Buzdin, Zhavoronkov, Korzinkin, Roumiantsev, et al., 2014).

The phenomenon of improved robustness was studied in a mathematical model simulating error acquisition in the gene expression-based and in the PAL approaches. In agreement with the experimental findings, simulated use of the PAL approach led to significantly more stable results. This model also enabled to deduce that the noise-suppressing effect of PAL methods becomes significant when a pathway includes least 30 gene products and grows with the number of pathway members (Buzdin, Zhavoronkov, Korzinkin, Venkova, et al., 2014).

In all PAL methods, the pathway scoring is based on the specific algorithm that utilizes the following main concepts. First, in each pathway

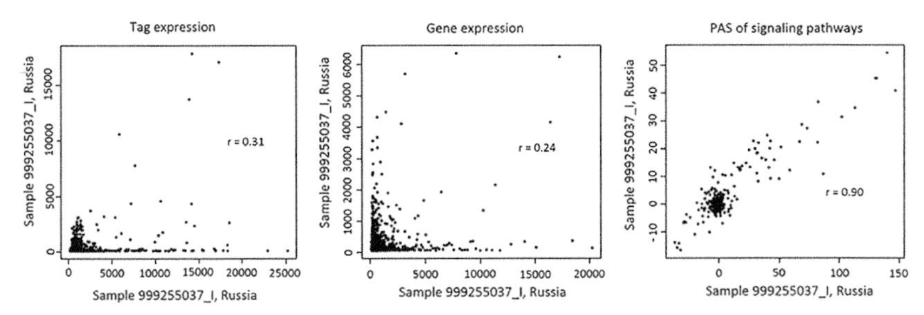

Fig. 2 Correlation between RNA expression data obtained for the same representative human renal cell carcinoma specimen using the Illumina HT12 (ordinate) and CustomArray (abscissa) transcriptomic microarrays. The panels represent (left to right) correlation between the signals sensed by individual oligonucleotide probes, correlations at the level of individual genes, and correlation at the level of molecular pathway activation levels.

the graph of molecular interactions can be presented as two parallel chains of events, one leading to its activation and another—to its suppression. Second, expressions of all gene products having "activator" roles in a pathway are presumed to be higher when the pathway is activated, and lower when the pathway is suppressed. Reciprocally, expressions of "inhibitor" gene products in a pathway are presumed to be higher when the pathway is suppressed, and lower when the pathway is upregulated (Buzdin, Zhavoronkov, Korzinkin, Venkova, et al., 2014). This principle is based on the published data that the deeply unsaturated states of each of the proteins-signal transducers in an individual molecular pathway are congruent with the low pathway activity states (Aliper, Jellen, et al., 2017; Aliper, Korzinkin, et al., 2017). The algorithm requires to functionally annotate to each node in a pathway by assigning activator or repressor roles to each participating gene product (Buzdin, Zhavoronkov, Korzinkin, Venkova, et al., 2014).

The resulting PAL values can delineate intracellular molecular regulatory landscape in a biosample tested. Positive value means increased activity of a pathway compared to the group of control samples; in contrast, the negative value indicates suppression of a pathway; zero value evidences that the pathway regulation is not affected between the case and the control samples (Buzdin, Zhavoronkov, Korzinkin, Venkova, et al., 2014). We will consider here in detail Oncobox, the recent clinically validated PAL scoring algorithm (Borisov et al., 2020).

3.1 PAL calculation algorithm

The algorithm assumes that all gene products participating in a pathway have equal possibilities to cause its activation or inhibition. For PAL calculation, the following formula is used:

$$\mathrm{PAL_p} = \sum\nolimits_{\mathrm{n}} \mathrm{NII_{n,p}} \cdot \mathrm{ARR_{n,p}} \cdot \ln\, \mathrm{CNR_n} / \sum\nolimits_{\mathrm{n}} |\mathrm{ARR_n}|,$$

where $\mathrm{PAL_p}$—pathway p activation level; $\mathrm{CNR_n}$—case-to-normal ratio; ratio of concentrations of gene product n in the sample of interest and in the controls (simple average or geometric mean of concentrations in the control group can be used); ln—natural logarithm; $\mathrm{NII_{np}}$—Boolean index of gene product n participation in pathway p, NII is equal to one for gene products included in a pathway and is equal to zero for gene products not included in a pathway; ARR_{np}—activator/repressor role, a discrete value that is determined for gene n in pathway p as follows:

$$ARR_{np} = \begin{cases} -1;\ \text{protein } n \text{ is a signal repressor in a pathway } p \\ -0.5;\ \text{protein } n \text{ is more likely a signal repressor in a pathway } p \\ 0;\ \text{the role of a protein } n \text{ in a pathway } p \text{ is either ambivalent or netral} \\ 0.5;\ \text{protein } n \text{ is more likely a signal activator in a pathway } p \\ 1;\ \text{protein } n \text{ is a signal activator in a pathway } p \end{cases}$$

$\Sigma|\mathrm{ARR_n}|$ parameter is used to normalize PAL value to the number of participating gene products with known functional roles in a pathway.

3.2 Annotation of functional roles for pathway members

Accurate annotation of the pathway member roles reflected by the ARR parameter are crucial for the correct PAL calculations. For most of published applications of PAL, five types of discrete roles for gene products were comprised. They can be formulated as follows: activator (ARR = 1), rather activator (ARR = 0.5), repressor (ARR = − 1), rather repressor (ARR = − 0.5), and finally gene product with inconsistent or unknown role (ARR = 0). In many previous investigations, ARR values were assigned by manually curating pathway graphs (Borisov et al., 2014; Lezhnina et al., 2014; Ozerov et al., 2016; Shepelin et al., 2016; Spirin et al., 2014). This is however not realistic for annotating thousands of molecular pathways. To solve this problem, we recently developed an algorithm that can

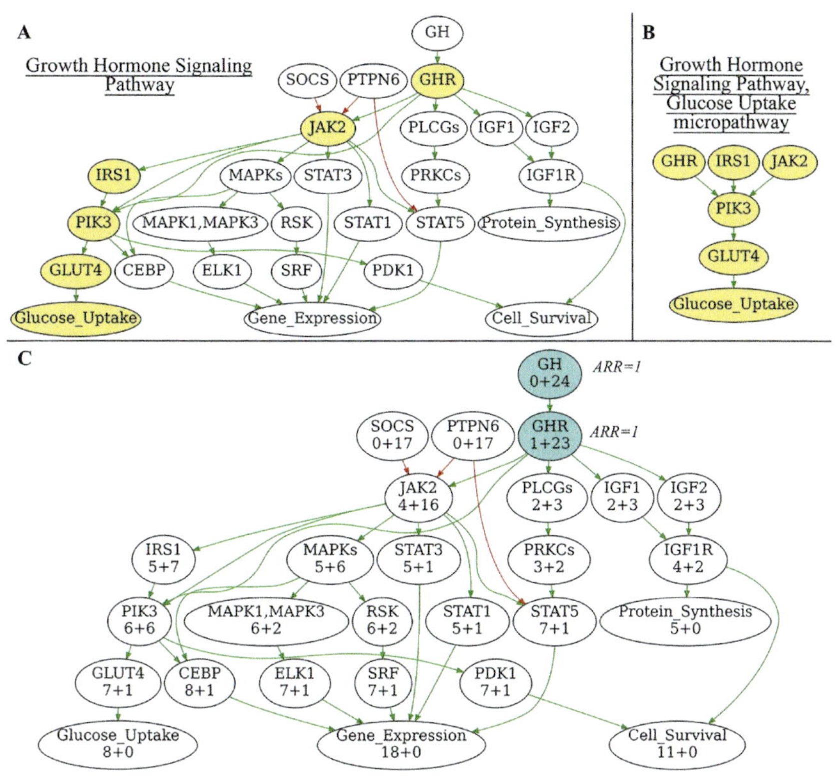

Fig. 3 (A) Growth Hormone Signaling Pathway with highlighted Glucose Uptake micropathway. (B) Separate graph representing Glucose Uptake micropathway. (C) N+M values for all vertices of Growth Hormone Signaling Pathway graph. The vertices with maximal N+M values (highlighted) are equal major node candidates and get ARR=1. Edge color indicates interaction type: "activation" (green) or "inhibition" (red).

automatically annotate gene products with ARR values in each individual molecular pathway (Borisov et al., 2020; Sorokin et al., 2021).

This algorithm is based on the machine reading of gene product interaction graph for each pathway (Fig. 3). This pathway graph is then built manually or using any molecular pathway database (Sorokin et al., 2021). For the calculation of ARR values, the pathway graph must be connected. If the pathway graph meets this criterion, then ARR values can be algorithmically assigned to the participating gene products.

Nodes on the graph correspond to gene products, and the ribs between every two proximate nodes show molecular interaction between the corresponding gene products. On the graph, each rib has direction and is characterized by an activating or repressor nature of its molecular interaction. For the biochemical pathways, enzyme gene names correspond to the pathway nodes, and the interaction ribs show directions of the reactions catalyzed by them.

The algorithm includes the following main steps.

(i) Initialization. At this step, a major node is algorithmically found to occupy the central place on the pathway graph (Fig. 3). The major node is then used as the gauge of pathway function with ARR = 1. The central node is identified as follows. For every node (say V) two parameters N and M are calculated where N is the number of other nodes that can be reached if moving directly from V, and M is the number of upstream nodes from which the node V can be reached. Consequently, $N+M$ is the number of nodes that are directly connected with V. The central node Vmax is then selected from all nodes, where $N+M$ reaches the maximum value. If two or several nodes have the same maximal $N+M$ value, then the central node for a pathway is selected randomly among those candidates. The central node then serves as the starting point with ARR = 1 for further recursive assignment of ARR for the remaining nodes. Thus, the algorithm is applicable also for cyclic pathways, where all many nodes may have the same $N+M$ value.

(ii) Recursion. For every node V, all connected nodes P_i under ARR annotation may have ribs either directed toward V ($P_i \rightarrow V$) or outward V ($P_i \leftarrow V$) on the graph. During recursion, each rib is considered only one time to prevent endless recursion for the case of cyclic interactions. For the "activator" rib, temporary ARRtemp = 1 is assigned to the node P_i. In contrast, for the "inhibitor" rib, P_i receives ARRtemp = −1. This means that all gene products included in this node will be assigned with the same ARRtemp values.

For gene product GPi that belongs to node P_i: if GPi wasn't previously assessed in the graph traversal, then ARR = ARRtemp(P_i) for the node P_i will be assigned to GPi. If GP_i was previously assessed in the graph traversal and received ARR equal to the current ARRtemp(P_i), then ARR = ARRtemp(P_i) will be assigned to GP_i. If GP_i was previously considered and received different ARR not equal to ARRtemp(P_i), then ARR is assigned to GP_i following the conflict resolution rule.

If GP_i with previously assigned ARR(s) is currently considered in the graph traversal but its previously assigned ARR(s) contradict(s) with the ARRtemp(P_i), then it should be resolved as follows:

- when the signs of ARRtemp(P_i) and the previous ARR coefficient(s) are different, then the resulting ARRfinal(P_i) = 0;

- when the difference between ARRtemp(P_i) and any previous ARR(GP_i) doesn't exceed 0.5, and at least one of those ARRs is positive, then the resulting ARRfinal(P_i) = 0.5;
- when the difference between ARRtemp(P_i) and any previous ARRs(GP_i) doesn't exceed 0.5 and at least one of those ARRs is negative, the resulting ARRfinal(P_i) = − 0.5.

Then the recursion is initiated starting from the nodes proximate to the central node V for every node P_i and all of its gene products. In this way, the algorithm will assign ARR values to all connected nodes on the graph and annotate all the corresponding gene products.

This recursion will be stopped when a vertex with 0, 0.5 or −0.5 ARR is encountered during the traversal of the graph, because otherwise all vertices will have ARR 0, 0.5 or −0.5 in case of only one ARR inconsistency identified. Thus, the gene products will receive ARR values reflecting their functional significances in a given molecular pathway. This operation enables further PAL calculations for the annotated molecular pathway (Borisov et al., 2020).

We applied this algorithm to annotate 3044 human molecular pathways from various databases, collectively covering ~9000 gene products. To this end we extracted structures of molecular pathways from the National Cancer Institute (NCI) (Schaefer et al., 2009), Biocarta (Bindea et al., 2009), Qiagen Pathway Central, HumanCyC (Romero et al., 2004), Reactome (Croft et al., 2014), and Kyoto Encyclopedia of Genes and Genomes (KEGG) (Kanehisa, Goto, Furumichi, Tanabe, & Hirakawa, 2010) databases (Table 1).

Annotations of similar pathways can be different between the source data collections. For example, EGFR signaling pathway is given in Qiagen SABiosciences database as "EGF_Pathway," in Biocarta collection—as "biocarta_egf_signaling_Main_Pathway," and in Reactome collection—as "reactome_Signaling_by_EGFR_Main_Pathway." Yet conceptually similar, these three pathways have different names, gene and edge compositions.

Many full-size pathways had two or more terminal branches that with different functional impact(s). In addition to the extracted full-size pathways we also generated a number of subsequent *micropathways* that were derivatives of the complete pathways to characterize molecular processes in more detail by separately analyzing different terminal branches of the pathways (Sorokin et al., 2021). In this case, micropathways were sub-graphs containing terminal "molecular function" node and nodes from all possible paths of length 3. Totally, we processed 2018 full-size and 1026

Table 1 Statistics of the automatically annotated pathway databases.

Pathway database	References	Number of Core pathways	All pathways	Unique genes	Pathway graph generation method
Biocarta	Nishimura (2001)	198	337	1082	Automated
Reactome	Croft et al. (2014)	945	945	6105	Automated
KEGG	Kanehisa et al. (2010)	288	288	5593	Automated
Qiagen	(Pathway Map Reference Guide)	57	380	2493	Manual
NCI	(Schaefer et al., 2009)	211	775	2214	Automated
HumanCYC	Romero et al. (2004)	319	319	1038	Automated
Total data collection	Sorokin et al. (2021)	2018	3044	9022	Mixed

"*All pathways*" includes *core pathways* and *micropathways*. Number of unique gene products participating in pathways from the corresponding databases is shown. For total data collection, the amount of unique gene products from all curated pathways is shown.

micropathways (Table 1). The number of pathway nodes was smaller than the number of genes involved in a pathway because one node could correspond to several gene products (Sorokin et al., 2021). The resulting knowledgebase is freely accessible (Sorokin et al., 2021) and can be applied for direct PAL calculations to assess signaling, metabolic and DNA repair pathway regulation using high throughput gene expression data.

4. Applications in life sciences

Functionally annotated molecular pathways can be used for further calculations of PALs using high-throughput gene expression data like quantitative proteomic and RNA sequencing profiles (Buzdin et al., 2018). To this end, several previously published bioinformatic algorithms can be utilized (Borisov et al., 2020; Buzdin, Zhavoronkov, Korzinkin, Roumiantsev, et al., 2014; Ozerov et al., 2016). The obtained PAL values can be used for many applications including fundamental research (Pasteuning-Vuhman et al., 2017), therapeutic research and development (Aliper, Jellen, et al., 2017; Aliper, Korzinkin, et al., 2017; Bakula et al., 2019; Ravi et al., 2018) and personalized diagnostics (Moisseev, Albert, Lubarsky, Schroeder, & Clark, 2020; Poddubskaya et al., 2019). Technically, PAL values can serve as the valuable next-generation molecular biomarkers (Aliper, Jellen, et al., 2017; Aliper, Korzinkin, et al., 2017; Borisov et al., 2017; Sorokin, Ignatev, Barbara, et al., 2020; Sorokin, Ignatev, Poddubskaya, et al., 2020; Sorokin, Kholodenko, Kalinovsky, et al., 2020; Sorokin, Poddubskaya, Baranova, et al., 2020) or as the substrates for various artificial intelligence applications like building of classifiers (Borisov et al., 2018; Borisov & Buzdin, 2019; Buzdin, Sorokin, Poddubskaya, et al., 2019; Tkachev et al., 2019; Tkachev, Sorokin, Borisov, et al., 2020).

PAL is the numeric characteristic that can be used for all types of comparisons in all objects and organisms, where pathway architecture is known. Compared to individual gene expression levels, PALs were found to be superior biomarkers of specific cancer types (Borisov et al., 2014; Lezhnina et al., 2014). Many PALs were characteristic for cancer drug response (Zhu et al., 2015), sensitivity to X-ray irradiation (Sorokin et al., 2018), for chronic diseases like Hutchinson-Gilford progeria (Aliper et al., 2015), asthma (Alexandrova et al., 2016), fibrosis (Makarev et al., 2016), macular degeneration (Makarev et al., 2014), for aging (Aliper et al., 2016) and viral infections (Buzdin et al., 2016). Algorithms were developed to convert knowledge of pathway activation levels into the optimized selection of cancer drugs (Artemov et al., 2015; Tkachev, Sorokin, Garazha, et al., 2020), which already had several successful clinical applications (; Poddubskaya, Baranova, et al., 2019; Poddubskaya et al., 2018; Poddubskaya, Bondarenko, et al., 2019; Sorokin, Poddubskaya, Baranova, et al., 2020).

From the point of view of biomarker's accuracy, molecular pathway activation is significantly better characteristic compared to individual gene expression profiles (Borisov et al., 2014; Buzdin et al., 2018; Zolotovskaia, Sorokin, Roumiantsev, et al., 2019). The same approach was used to calculate pathway mutation burden using whole-exome sequencing data (Zolotovskaia, Tkachev, Seryakov, et al., 2020). The relative number of high-quality biomarkers was up to three orders of magnitude higher for the pathway-based biomarkers compared to single gene mutations (Zolotovskaia, Sorokin, Emelianova, et al., 2019; Zolotovskaia, Sorokin, Roumiantsev, et al., 2019).

The same rationale can be used also for interrogating pathway regulation based on the micro RNA (miR) expression data. The PAL-based technique termed MiRImpact enables to quantitate activation of molecular pathways using total miR expression profiles (Artcibasova et al., 2016; Buzdin et al., 2016; Knyazeva et al., 2020). Furthermore, high-throughput epigenomic data like frequencies of chromatin immunoprecipitation-measured transcription factor binding sites in gene promoters (Nikitin et al., 2019b, 2018), or gene-specific profiles of histone modifications (Igolkina et al., 2019; Nikitin, Kolosov, et al., 2019) can all serve as the input for PAL-related quantitative assessment of pathway regulation. For these purposes, specific modifications are implemented for the basic PAL calculation formula (Borisov et al., 2020).

For *miR regulation*, a concept is employed that miRs affect target gene expressions via specific inhibition of target mRNAs (Zamore & Haley, 2005). The increased concentration of miR, therefore, leads to lower adjusted expression levels of the respective target mRNAs, and vice versa. The analysis is based on the knowledgebase of gene products and their mRNAs which are known molecular targets of individual miRs (Artcibasova et al., 2016; Buzdin et al., 2016). Wherein, each miR may have multiple gene targets, and each gene product may have multiple regulatory miRs (Artcibasova et al., 2016). In this case, the basic PAL formula is modified so that ln CNR_n member is calculated as follows:

$$\ln \mathrm{CNR_n} = -\sum\nolimits_j \ln \mathrm{miCNR_i} \cdot \mathrm{miII_{i,n}},$$

where "−1" coefficient in the beginning reflects the inhibitor role of miR for its target mRNAs; *n*—individual gene product under analysis; *j*—total number of miRs considered; *i*—individual miR under analysis. $miII_{i,n}$ (miR involvement index) is Boolean variable that indicates if gene product *n* is molecular target for miR *i*. It is equal to 1 when *n* is molecular target of *i* and equals to 0 when this is not the case. $miCNR_i$ is ratio of miR *i* expression level in the sample of interest to its expression level in the control group (Artcibasova et al., 2016).

For pathway regulation analysis using *epigenomic marks*, several additional parameters are defined. In pathway analysis using *transcription factor binding site (TFBS)* data, a consensus transcriptional start site (TSS) is determined for every gene (Nikitin et al., 2019a, 2019b, 2018). For each gene TSS, the 5-kb neighborhood is then determined, and in this neighborhood the number of mapped TFBS is identified. Then GRES (Gene Record Enrichment Score) value is calculated for every gene *n*:

$$\mathrm{GRES_n} = \mathrm{m} \cdot \mathrm{TES_n} / \sum\nolimits_{i=1}^{m} \mathrm{TES_i}$$

where m is total number of interrogating genes for a sample under analysis; TES_n—number of mapped TFBS in the neighborhood of n; i—gene identifier index; TES_i—total number of mapped TFBS in the neighborhood of m genes. For every gene, GRES ranks the saturation level of TFBS under analysis. For example, GRES =1 means average saturation level among all genes; GRES >1 means greater saturation; GRES <1 means lower saturation relatively to the average for all genes. Finally, analog of ln CNR_n from the core PAL formula is calculated as follows:

$$\ln CNR_n = \ln CNR(GRES_n)$$

where $CNR(GRES_n)$ is ratio of GRES for gene n in a sample under analysis to average GRES in the control group (Nikitin et al., 2018).

For the assessment of pathway regulation by the *histone modification marks* like *H3K4me1, H3K4me3, H3K9ac, H3K27ac, H3K27me3* and *H3K9me3* as in the two recently published examples (Igolkina et al., 2019; Nikitin, Kolosov, et al., 2019), the same values as above are calculated including GRES, TES, etc. by using the respective histone modification counts instead of TFBS frequencies (Igolkina et al., 2019; Nikitin, Kolosov, et al., 2019).

Finally, when using *genome-wide mutation data* for pathway analysis, *Pathway instability* (*PI*) scores can be obtained for all the pathways under analysis that depend on the mutation frequencies of the genes involved (Zolotovskaia, Sorokin, Emelianova, et al., 2019; Zolotovskaia, Sorokin, Garazha, et al., 2020; Zolotovskaia, Sorokin, Petrov, et al., 2020; Zolotovskaia, Sorokin, Roumiantsev, et al., 2019; Zolotovskaia, Tkachev, Seryakov, et al., 2020). First, mutation burdens of individual genes are calculated by using *Mutation rate* (*MR*) metric that is calculated as follows:

$$MR_n = \frac{N\ mut(n,g)}{N\ samples\ (g)},$$

where n is gene under analysis; $N\ mut(n,g)$ is the total number of mutations identified for gene n in a group of samples g; $N\ samples\ (g)$ is the number of samples in group g. However, MR strongly correlates with the length of gene coding DNA sequence (CDS), Spearman correlation ~0.8 (Zolotovskaia, Sorokin, Emelianova, et al., 2019; Zolotovskaia, Sorokin, Garazha, et al., 2020; Zolotovskaia, Sorokin, Petrov, et al., 2020; Zolotovskaia, Sorokin, Roumiantsev, et al., 2019; Zolotovskaia, Tkachev, Seryakov, et al., 2020), e.g., because longer CDS are more likely to accumulate mutations. In turn, this bias can be avoided by using a CDS length-normalized value termed *Normalized mutation rate* (*nMR*) that is calculated as follows:

$$nMR_n = \frac{1000 * MR_n}{Length\ CDS\ (n)},$$

where n is gene under analysis; MR_n is *Mutation rate* of gene n calculated as above; *Length CDS(n)* is CDS length of gene n in nucleotides. Finally, *Pathway instability* (*PI*) value can be calculated for every pathway to quantify relative enrichments by mutations. *PI* is expressed by the formula:

$$PI_p = \frac{\sum_n nMR_n PG_{p,n}}{N_p}$$

where p is pathway under analysis; nMR_n is the *Normalized mutation rate* of gene n;$PG_{p,\ n}$ is Boolean pathway-gene indicator that equals to 1 when n is included in pathway p, or equals to 0 otherwise; N_p is total number of gene products in pathway p (Zolotovskaia, Sorokin, Garazha, et al., 2020).

Those modifications allow using PAL scoring rationale for multitude of objects and tasks in the contemporary life sciences, where large scale analysis of proteomic, transcriptomic, mirnaomic, epigenomic, and mutational data is needed. In the next section, we concentrate on the clinical applications of PAL approach.

5. Applications in medicine

PAL methods were effective for finding next-generation biomarkers in a plethora of applications, including many fields in medicine and drug research (Alexandrova et al., 2016; Aliper et al., 2016, 2015; Aliper, Frieden-Korovkina, Buzdin, Roumiantsev, & Zhavoronkov, 2014; Borger et al., 2020, 2019; Borisov et al., 2018; Buzdin et al., 2016; Buzdin, Sorokin, Garazha, et al., 2019; Buzdin, Sorokin, Poddubskaya, et al., 2019; Comunanza et al., 2017; Emelianova et al., 2018; Jellen, Aliper, Buzdin, Zhavoronkov, & Zhavoronkov, 2015; Jovčevska et al., 2017; Kalasauskas et al., 2020; Kim et al., 2020; Kurz et al., 2017; Larkin et al., 2017; Lezhnina et al., 2014; Makarev et al., 2014, 2016; Marggraf et al., 2018; Negro et al., 2020; Petrov et al., 2017, 2016; Shepelin et al., 2016; Shtam et al., 2018, 2019; Sorokin, Ignatev, Barbara, et al., 2020; Sorokin, Ignatev, Poddubskaya, et al., 2020; Sorokin, Kholodenko, Grekhova, et al., 2018; Sorokin, Kholodenko, Kalinovsky, et al., 2020;

Sorokin, Kholodenko, Suntsova, et al., 2018; Sorokin, Poddubskaya, Baranova, et al., 2020; Spirin et al., 2014, 2017; Venkova et al., 2015; Wirsching et al., 2017; Zhavoronkov et al., 2014; Zhu et al., 2015; Zolotovskaia, Sorokin, Emelianova, et al., 2019; Zolotovskaia, Sorokin, Roumiantsev, et al., 2019). Overall, the pathway-based methods can use PAL scores per se as the putative biomarkers or for in-depth explaining complex molecular processes (Buzdin et al., 2018). Alternatively, PALs can be the intermediates for further metrics assessing various physiological impacts such as drug response (e.g., Kim et al., 2020; Tkachev, Sorokin, Garazha, et al., 2020; Zolotovskaia, Sorokin, Emelianova, et al., 2019), finding new effective medicines (Aliper et al., 2016; Emelianova et al., 2018; Marggraf et al., 2018; Sorokin, Kholodenko, Suntsova, et al., 2018; Spirin et al., 2017; Zhavoronkov et al., 2014) and their personalization (Sorokin, Poddubskaya, Baranova, et al., 2020; Tkachev, Sorokin, Borisov, et al., 2020; Tkachev, Sorokin, Garazha, et al., 2020; Zolotovskaia, Sorokin, Garazha, et al., 2020; Zolotovskaia, Sorokin, Petrov, et al., 2020; Zolotovskaia, Tkachev, Seryakov, et al., 2020). These methods may be classified in two main groups: (*i*) those using PALs as statistical coincidence biomarkers, and (*ii*) those using PALs in the context of drug specific molecular mechanisms (Buzdin, Sorokin, Garazha, et al., 2019).

The latter group of methods utilizes knowledge of specific molecular mechanisms for predicting drug responses in several human chronic diseases (Artemov et al., 2015; Tkachev, Sorokin, Garazha, et al., 2020) and even in anti-aging studies (Aliper et al., 2016). They combine profiling of molecular pathways and expression levels of drug target molecules to personalizing selection of drugs in case of individual comparisons (Moisseev et al., 2020; Poddubskaya, Baranova, et al., 2019; Poddubskaya et al., 2018; Poddubskaya, Bondarenko, et al., 2019; Sorokin, Poddubskaya, Baranova, et al., 2020), or for finding novel drug targets (Aliper et al., 2016; Spirin et al., 2014, 2017), new effective drug combinations (Comunanza et al., 2017; Sorokin, Kholodenko, Suntsova, et al., 2018; Venkova et al., 2015) and drugs repurposing (Aliper et al., 2016; Jellen et al., 2015; Levin et al., 2020). Presumably, it can be applied only for the drugs with established molecular specificities and known mechanism of action.

These techniques showed significant advances in the field of oncology (Buzdin, Sorokin, Garazha, et al., 2019). Carcinogenesis is well known for massive changes in regulation of gene-wide networks (Gyurkó et al., 2013; Zhou, Liu, Zhu, & Zhang, 2020). In a recent study of paired RNA sequencing and whole-exome sequencing profiles of cancer tissues from ~5000 patients with 13 cancer types, involvement of 278 signaling, 72 metabolic, 48 DNA repair and 47 cytoskeleton molecular pathways was interrogated (Zolotovskaia, Tkachev, Seryakov, et al., 2020). Cancer expression profiles were compared accordingly with the 655 tumor-matched adjacent pathologically normal human tissues. It was found that pathway activation profiles were largely congruent among the different cancer types, but there was no correlation between pathway mutation enrichment and expression changes both at the gene and the pathway levels (Zolotovskaia, Tkachev, Seryakov, et al., 2020). Overall, median activation levels of DNA repair pathways in cancer were positive, whereas the median levels for the other types of pathways were negative. The DNA repair pathways also demonstrated the highest values of mutation enrichment. In turn, the signaling and cytoskeleton pathways had the biggest proportions of genes which were outstandingly frequently mutated thus suggesting their initiator roles in carcinogenesis (Zolotovskaia, Tkachev, Seryakov, et al., 2020).

These findings are in line with the somatic mutation theory, which states that DNA damage is the key point in cancer transformation (Erenpreisa, Salmina, Anatskaya, & Cragg, 2021; Wu & Starr, 2014). Cancer cells accumulate mutations much faster than the normal cells (Antontseva et al., 2015), which can be connected with dysregulation of DNA repair pathways (Fouad & Aanei, 2017; Kulikov, Luchkina, Gogvadze, & Zhivotovsky, 2017; Zolotovskaia, Tkachev, Seryakov, et al., 2020). This may lead to epigenomic changes that can enhance malignization by repressing tumor suppressor genes (Sharma, Kelly, & Jones, 2010).

The signaling pathways mediate inter- and intracellular signal transduction and control all major biological processes in cancer cell such as migration, differentiation, cell growth, proliferation, and death (Dreesen & Brivanlou, 2007; Petrov et al., 2017; Whittaker, Marais, & Zhu, 2010). It is well known since 1920s (Xu et al., 2015) that metabolic pathways are strongly biased in cancer due to increased energy consumption that is required for forced proliferation (Sciacovelli, Gaude, Hilvo, & Frezza, 2014; Xu et al., 2015), especially in hypoxic conditions (Ilkhani et al., 2020; Samec et al., 2020). On the other hand, cancer cells partly or completely dedifferentiate and thus lose complexity of their metabolic patterns (Dang, 2012). In turn, affected cytoskeleton pathways in cancer can mediate oncogenic, invasive and metastatic alterations in mitosis, cell motility, morphogenesis, and intercellular contacts (Hall, 2009; Shtam et al., 2019). Finally, changes in DNA repair pathways is the key mechanism of the emergence and development of malignant tumors (Jeggo, Pearl, & Carr, 2016). They can induce multiplicative tumorigenic effects by promoting mutations and initiating accelerated molecular evolution of cancer cells (Dong et al., 2018; Galanos et al., 2018). There are also many functional links between the different types of molecular pathways in cancer (Cui et al., 2019; Turgeon, Perry, & Poulogiannis, 2018).

In the high-throughput study of 419 pathways in ~5000 cancers (Zolotovskaia, Tkachev, Seryakov, et al., 2020), all tumor samples had mutated genes. Among them, 94% of samples had at least one mutated gene from signaling, 78%—from cytoskeleton, 58%—from metabolic, and 53%—from DNA repair pathways (Zolotovskaia, Tkachev, Seryakov, et al., 2020). Furthermore, about half of all tumors had at least 9 mutated genes from signaling, 3 from cytoskeleton, 1 from metabolic, and 1 from DNA repair pathways (Zolotovskaia, Tkachev, Seryakov, et al., 2020). In the sampling investigated, 32.3% of signaling pathways had mutations in their genes in at least one tumor. This proportion was 30.4% for cytoskeleton, 29.2% for DNA repair, and 9.6% for metabolic pathways (Zolotovskaia, Tkachev, Seryakov, et al., 2020). Overall, the DNA repair pathways had the highest relative mutation rates per gene, the signaling pathways occupied intermediate position, and the metabolic and cytoskeleton pathways had the lowest detected relative mutation levels (Zolotovskaia, Tkachev, Seryakov, et al., 2020).

Structures, architecture and hierarchy of the most strongly mutated pathways can be significantly altered in cancer cells compared to the normal cells; the least mutated pathways may seem intact from the structural point of view (Zolotovskaia, Sorokin, Roumiantsev, et al., 2019). *The most strongly mutated* pathways dealt with (DNA repair) cellular response to stress and chromosome organization; (signaling) cell surface receptor pathways, phosphatidyl inositol-mediated regulation, growth, proliferation, regulation of calcium transport, RNA polymerase II transcription, and regulation of activation of immune cells; (cytoskeleton) cellular adhesion, exocytosis, cell communication and locomotion, cell cycle progression, processing and presentation of antigens using MHC class II machinery; (metabolic) drug response and catabolism, secretion of arachidonic acid, fatty acid derivative pathways, glycosylation, sulfur and benzene-containing compound pathways, production of mitochondrial RNA, and aminoacetylation of tRNAs.

The least mutated pathways were responsible for (DNA repair) DNA quality control and associated repair, and rearrangement of immunoglobulin loci; (signaling) negative regulation of glucocorticoid receptor pathway, regulation of hormone levels, circadian rhythms, H3 histone deacetylation, and catabolism of proteins; (cytoskeleton) membrane organization, receptor-mediated endocytosis, actin filament binding and locomotion, calcium-independent intercellular adhesion via plasma membrane; (metabolic) biosynthesis of nucleotides and DNA replication (Zolotovskaia, Tkachev, Seryakov, et al., 2020).

This picture was different from what was observed at the level of PAL values, which reflected activation or suppression of the pathways. Most of the pathways showed uniform, cancer type-independent regulation pattern in tumor samples (Zolotovskaia, Tkachev, Seryakov, et al., 2020). Most of the pathways were downregulated in cancer, except for the DNA repair pathways which were mostly upregulated (Fig. 4). The highest mean downregulation was observed for the cytoskeleton pathways. Overall, averaged PALs indicated cancer-specific activation of ~60% DNA repair, 45% metabolic, 43% signaling and 38% cytoskeleton pathways. Most of DNA repair pathways were activated in the majority of samples, whereas cytoskeleton, metabolic and signaling pathways showed mosaic patterns (Zolotovskaia, Tkachev, Seryakov, et al., 2020), Fig. 5.

The most strongly upregulated pathways dealt with (DNA repair) response to DNA-templated transcription and elongation, 7-methylguanosine RNA capping, protein localization on chromosomes and chromosomal organization, rearrangement of immune receptor loci; (signaling) extracellular matrix organization, cell migration and adhesion, transmembrane transport of calcium, regulation of cell death, maintaining mRNA stability, antigen processing and presentation via MHC class I machinery, and activation of immune cells; (cytoskeleton) there were no strongly activated cytoskeleton pathways; (metabolic) oligosaccharide and nucleotide biosynthesis, and DNA replication.

The most cancer-downregulated pathways were enriched in the following terms (DNA repair) there were no strongly cancer-downregulated DNA repair pathways; (signaling) regulation of transcription by RNA polymerase II, interleukin-7 pathway; (cytoskeleton) cell locomotion and motility of subcellular components, organs development and cell proliferation; (metabolic) cellular ketone body metabolism, catabolism of neurotransmitters, sulfur metabolism, drug response pathways (Zolotovskaia, Tkachev, Seryakov, et al., 2020). There was also detected that intertumoral similarities are much higher at the pathway level compared to the level of individual

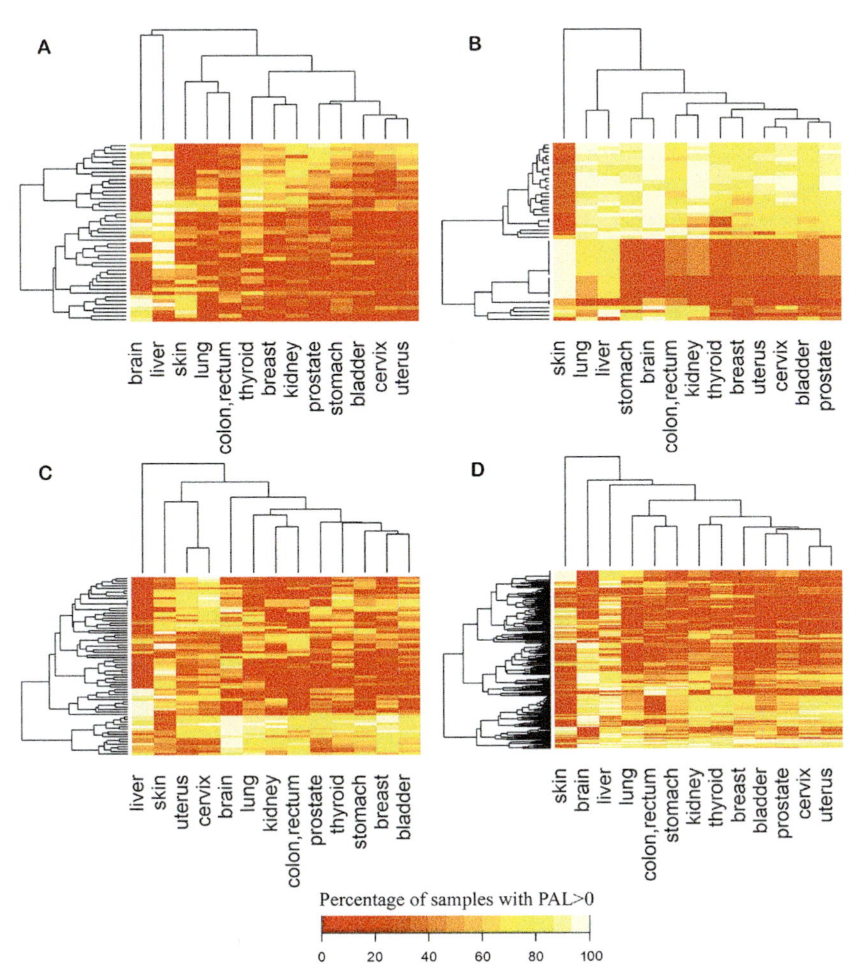

Fig. 4 Percentage of cancer samples with positive pathway activation levels (PAL > 0) in 13 human cancer types calculated for: (A) *cytoskeleton*, (B) *DNA repair*, (C) *metabolic*, (D) *signaling* pathways under study. Color scale represents percentage of samples with PAL > 0 for each pathway.

genes, and DNA repair pathways demonstrated the most congruent activation and mutation patterns among the tumors (Zolotovskaia, Tkachev, Seryakov, et al., 2020).

Thus, this type of analysis poses a challenge of analyzing whole disease interactome models.

In cancer studies, the PAS-based techniques have an advantage of ranking all possible targeted drugs by modeling their activities using a single

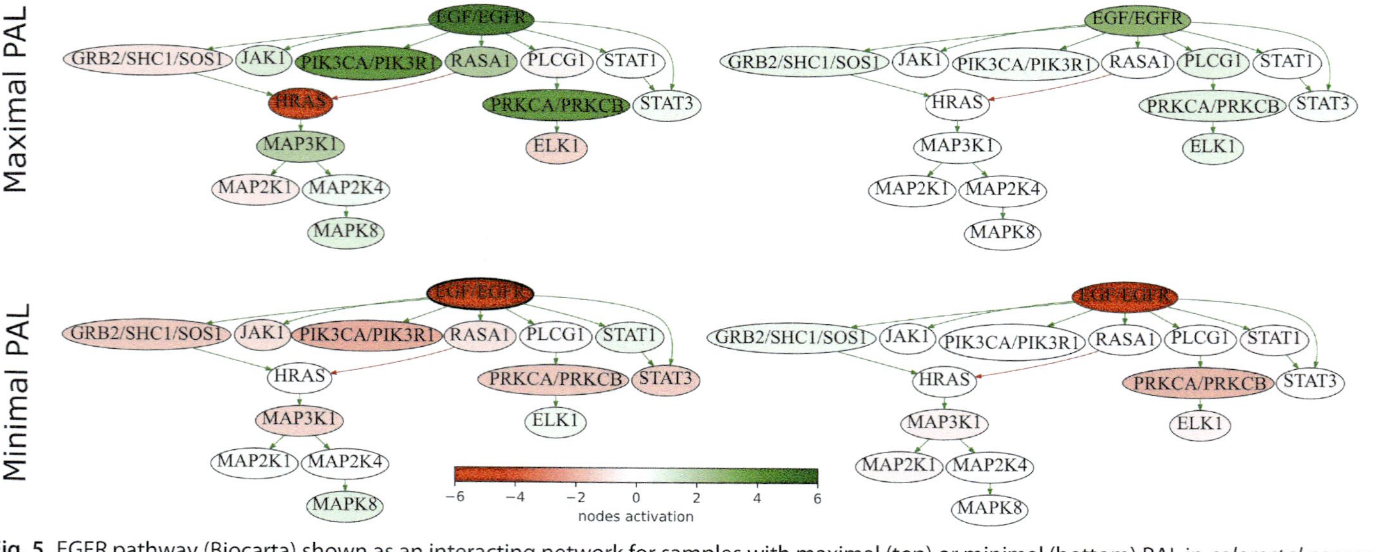

Fig. 5 EGFR pathway (Biocarta) shown as an interacting network for samples with maximal (top) or minimal (bottom) PAL in *colorectal cancer* (left) and *normal colon* (right). Green/red arrows indicate activation/inhibition interactions, respectively. Color depth of transcript nodes reflects the extent of node activation. The figure illustrates mosaic activation pattern of this signaling pathway in colorectal cancers and normal colon tissues. In both tissue types there are fractions of samples showing upregulated (PAL > 0) and downregulated (PAL < 0) values.

high throughput tumor expression profile. To this end, a Drug Score (DS) value is introduced that measures effectiveness of each targeted drug for a tumor. This type of drug scoring is based on the rationale that an effective drug must compensate pathological changes in IMP activations; simultaneously, the specific molecular target(s) of an effective drug must be expressed at the relatively high level (Aliper et al., 2016; Artemov et al., 2015; Tkachev, Sorokin, Garazha, et al., 2020; Zhavoronkov et al., 2014).

Different types of pathological tissues can be used, including fresh biopsy or formalin fixed, paraffin-embedded (FFPE) tumor tissues (Sorokin, Ignatev, Barbara, et al., 2020; Sorokin, Ignatev, Poddubskaya, et al., 2020; Sorokin, Poddubskaya, Baranova, et al., 2020). Theoretical considerations and clinical results suggest that for such investigations it is important to use possibly recent tumor biopsy specimens, and to have minimal possible number of lines of therapy in the interval between taking biopsy and molecular test (Poddubskaya et al., 2020; Suntsova & Buzdin, 2020). Additional lines of treatment can dramatically change molecular tumor landscape by influencing its heterogeneity and/or intensifying purifying selection of certain treatment-resistant clones (Kim et al., 2020; Zolotovskaia, Sorokin, Garazha, et al., 2020; Zolotovskaia, Sorokin, Petrov, et al., 2020).

For example, the *Oncobox* system of drug scoring interrogates concentrations of cancer drug targets by comparing them with the sets of control (normal) expression profiles (Tkachev, Sorokin, Garazha, et al., 2020). It also distinguishes the nature of different classes of targeted cancer drugs by their modes of action when applying drug score calculation algorithms (Tkachev, Sorokin, Garazha, et al., 2020). The capacity of this drug scoring algorithm was tested both preclinically and clinically by using different sets of gene expression data from different experimental platforms (Tkachev, Sorokin, Garazha, et al., 2020). For example, it was found efficient for discriminating good and poor responders on targeted cancer treatment in both retrospective and prospective investigations (Buzdin, Sorokin, Garazha, et al., 2019; Sorokin, Poddubskaya, Baranova, et al., 2020).

So far, there were several published attempts of translating these approaches to the field of clinical oncology. For example, recently published successful cases of Oncobox drug score-based prescriptions of targeted tyrosine kinase inhibitors for patients with metastatic cholangiocarcinoma (Poddubskaya et al., 2018), metastatic lung cancer (Poddubskaya, Bondarenko, et al., 2019), and recurrent ovarian cancer (Poddubskaya, Baranova, et al., 2019), and of anti-PD-1 immunotherapy to the recurrent and metastatic gastric cancer case (Moisseev et al., 2020). Importantly, in all these four cases the drugs were successfully prescribed in off-label mode as the experimental therapy, outside the accepted standards of care.

In another recent study, a group of retrospective patients with gastric cancer showed good agreement between the algorithmically predicted and actually observed efficacies of ramucirumab (a VEGF receptor-specific therapeutic antibody) (Sorokin, Poddubskaya, Baranova, et al., 2020).

In another pilot study, a group of 23 prospective patients with recurrent and/or metastatic solid cancers was prescribed with targeted drugs based on the results of microarray gene expression profiling (Buzdin, Prassolov, et al., 2017; Buzdin, Sorokin, et al., 2017). The objective response rate (complete + partial response according to RECIST criteria) for prescriptions made in accordance with the Oncobox algorithm recommendations was 61%. In the recent prospective study OMICSGLIOMA, Oncobox algorithm could effectively predict sensitivity of 160 glioblastomas to the treatment with temozolomide, a MGMT-specific cancer therapeutic (Kim et al., 2020), Fig. 6.

In 2018, a new multi-center clinical trial (NCT03724097) was launched to assess effectiveness of Oncobox method to identify lists of effective drugs using RNA sequencing data. The study included 239 patients with recurrent and/or metastatic cancer, biomaterials were paraffined samples of surgically removed tumor tissue or of tumor biopsies. The algorithmic annotation resulted in a rating of expected efficacy for 130 anticancer drugs. The treatments were prescribed by the treating doctors, and the patients were naturally divided into the three groups: (*i*) patients who received therapy in agreement with the Oncobox rating (monotherapy or combination therapy); (*ii*) who received drugs that were not consistent with the Oncobox rating; (*iii*) who received only palliative care. The preliminary study results suggest that the follow-up clinical response information was obtained for 144 patients, among them 25 (17%) died before therapeutic interventions, 19% received palliative care (group *iii*), 39% received therapy as recommended by Oncobox (group *i*), and 25% received other therapy (group *ii*). The control-over disease rate (complete response + partial response + stable disease) for the group (*i*) was 71% vs 44% for the group (*ii*), thus suggesting efficacy of PAL-based algorithmic personalization of cancer drug prescriptions (Poddubskaya et al., 2019). No clinical records could be found for any other possible PAL-based drug scoring algorithms.

On the different side, activities of molecular pathways can be associated with the susceptibility to cancer drugs using agnostic rationale, in statistical association studies. For example, PAL of "cAMP Pathway

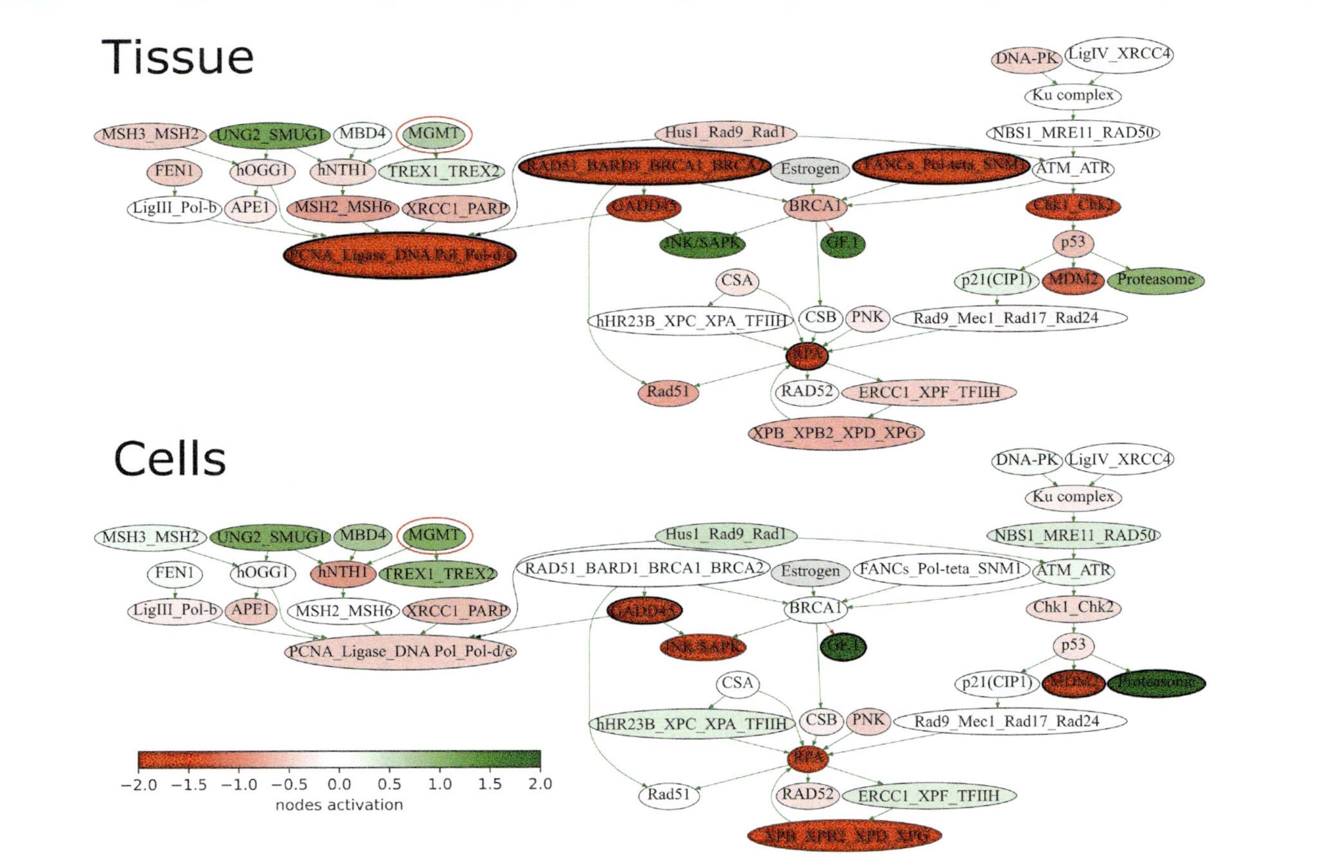

Fig. 6 Pathway activation profile of the DNA damage response pathway in recurrent glioblastoma samples non-responsive to treatment with temozolomide (TMZ) normalized on newly diagnosed glioblastoma samples that respond on TMZ. *MGMT* gene product (circled) counteracts with TMZ activity. The data are shown separately for glioblastoma *tissues* and for *glioma stem cells* that were isolated from the fresh glioblastoma biosamples and cultured in selective conditions. Green color denotes pathway node activation, purple—downregulation, scale is given on the bottom.

(Protein Retention)" pathway in a clinical trial was found to be connected with the efficacy of trastuzumab, a HER2-specific therapeutic antibody, in the metastatic or recurrent breast cancer patients. This pathway also predicted time to recurrence of the disease in these patients with Log-rank *p*-value 0.041 (Sorokin, Ignatev, Barbara, et al., 2020; Sorokin, Ignatev, Poddubskaya, et al., 2020; Sorokin, Kholodenko, Kalinovsky, et al., 2020; Sorokin, Poddubskaya, Baranova, et al., 2020). PALs for several other pathways were statistically associated with the responsiveness of *KRAS*-wild type colorectal tumors on the treatment with cetuximab, an EGFR-specific therapeutic antibody (Zhu et al., 2015). PAL values of some other pathways could serve as the biomarkers of response on chemotherapy treatment in kidney and ovarian cancers (Buzdin et al., 2018; Ozerov et al., 2016; Tkachev, Sorokin, Garazha, et al., 2020).

6. Software for PAL calculation

PAL values can be calculated using Oncofinder (Buzdin, Zhavoronkov, Korzinkin, Venkova, et al., 2014), iPANDA (Ozerov et al., 2016) and Oncobox (Tkachev, Sorokin, Garazha, et al., 2020) methods. We developed is a freely accessible software implemented as a Python library using annotated 3044-molecular pathways database and Oncobox algorithm for PAL calculations, available at https://pypi.org/project/oncoboxlib/. As an input, it takes gene expression data normalized by DESeq2, quantile normalization or any other method. Gene names must be given according to HGNC (genenames.org) rules. At least two groups of samples are required: cases and controls, and each group must be represented by at least one sample. Sample names should contain "Norm_" (for controls) or "Tumour_" (for cases). The program returns PAL values for each pathway in each sample. The annotated 3044-pathway dataset can be downloaded and used for PAL calculation using the same link. There is an available example of PAL calculation using real-world RNA sequencing data for gastric cancer samples ($n=16$) (Sorokin, Poddubskaya, Baranova, et al., 2020) in comparison with the healthy stomach tissue samples ($n=7$) from the ANTE collection (Suntsova et al., 2019). Both datasets under comparison were obtained by profiling with the same equipment, reagents and protocols, and are technically compatible. Cancer samples were marked as "Tumor_" and normal samples—as "Norm_". Then 3044 PAL values were calculated for all molecular profiles, and an output file "pal.csv" was obtained.

7. Concluding remarks

Current success in bioinformatics makes it possible to largely automatize the analysis of molecular profiles obtained for the biosamples of interest. It also enables to do multiple comparisons of different groups of biosamples and to identify altered molecular processes. The molecular data for such analyses must be provided by reliable high-throughput methods of gene expression quantization interrogating transcriptomic or proteomic data. Recent technological advances in proteomics and transcriptomics allow using both types of data for genome-wide profiling of complex molecular interaction networks, such as the human interactome. Furthermore, quantitative molecular pathway analysis can be useful not only for fundamental studies, but also for medicine and biotechnology. It can accelerate drug discovery and validation processes and can serve to improve molecular diagnostics and personalized medicine. Strong scientific line of evidence and ease of use make pathway analysis a method of choice for a wealth of tasks that require interpretation of big molecular data.

Acknowledgments

This study was supported by the Russian Science Foundation grant No. 20-75-10071 (Maria Suntsova), and by the Russian Foundation for Basic Research grant 19-29-01108 (Anton Buzdin).

References

Aebersold, R., & Mann, M. (2016). Mass-spectrometric exploration of proteome structure and function. *Nature*, *537*, 347–355. Nature Publishing Group https://doi.org/10.1038/nature19949.

Alexandrova, E., Nassa, G., Corleone, G., Buzdin, A., Aliper, A. M., Terekhanova, N., et al. (2016). Large-scale profiling of signalling pathways reveals an asthma specific signature in bronchial smooth muscle cells. *Oncotarget*, 7(18), 25150–25161. https://doi.org/10.18632/oncotarget.7209.

Aliper, A. M., Frieden-Korovkina, V. P., Buzdin, A., Roumiantsev, S. A., & Zhavoronkov, A. (2014). Interactome analysis of myeloid-derived suppressor cells in murine models of colon and breast cancer. *Oncotarget*, *5*(22), 11345–11353. https://doi.org/10.18632/oncotarget.2489.

Aliper, A. M., Csoka, A. B., Buzdin, A., Jetka, T., Roumiantsev, S., Moskalev, A., et al. (2015). Signaling pathway activation drift during aging: Hutchinson-Gilford Progeria Syndrome fibroblasts are comparable to normal middle-age and old-age cells. *Aging*, 7(1), 26–37. https://doi.org/10.18632/aging.100717.

Aliper, A., Belikov, A. V., Garazha, A., Jellen, L., Artemov, A., Suntsova, M., et al. (2016). In search for geroprotectors: In silico screening and in vitro validation of signalome-level mimetics of young healthy state. *Aging*, *8*(9), 2127–2152. https://doi.org/10.18632/aging.101047.

Aliper, A. M., Korzinkin, M. B., Kuzmina, N. B., Zenin, A. A., Venkova, L. S., Smirnov, P. Y., et al. (2017). Mathematical justification of expression-based pathway activation scoring (PAS). *Methods in Molecular Biology (Clifton, N.J.)*, *1613*, 31–51. https://doi.org/10.1007/978-1-4939-7027-8_3.

Aliper, A., Jellen, L., Cortese, F., Artemov, A., Karpinsky-Semper, D., Moskalev, A., et al. (2017). Towards natural mimetics of metformin and rapamycin. *Aging*, *9*(11), 2245–2268. https://doi.org/10.18632/aging.101319.

Anders, S., & Huber, W. (2010). Differential expression analysis for sequence count data. *Genome Biology*, *11*(10), R106. https://doi.org/10.1186/gb-2010-11-10-r106.

Andreev, D. E., O'Connor, P. B. F., Loughran, G., Dmitriev, S. E., Baranov, P. V., & Shatsky, I. N. (2017). Insights into the mechanisms of eukaryotic translation gained with ribosome profiling. *Nucleic Acids Research*, *45*, 513–526. Oxford University Press https://doi.org/10.1093/nar/gkw1190.

Anisimova, A. S., Meerson, M. B., Gerashchenko, M. V., Kulakovskiy, I. V., Dmitriev, S. E., & Gladyshev, V. N. (2020). Multifaceted deregulation of gene expression and protein synthesis with age. *Proceedings of the National Academy of Sciences of the United States of America*, *117*(27), 15581–15590. https://doi.org/10.1073/pnas.2001788117.

Antontseva, E. V., Matveeva, M. Y., Bondar, N. P., Kashina, E. V., Leberfarb, E. Y., Bryzgalov, L. O., et al. (2015). Regulatory single nucleotide polymorphisms at the beginning of intron 2 of the human KRAS gene. *Journal of Biosciences*, *40*(5), 873–883. https://doi.org/10.1007/s12038-015-9567-8.

Apweiler, R., Martin, M. J., O'Donovan, C., Magrane, M., Alam-Faruque, Y., Antunes, R., et al. (2011). Ongoing and future developments at the Universal Protein Resource. *Nucleic Acids Research*, *39*(Suppl. 1), D214–D219. https://doi.org/10.1093/nar/gkq1020.

Artcibasova, A. V., Korzinkin, M. B., Sorokin, M. I., Shegay, P. V., Zhavoronkov, A. A., Gaifullin, N., et al. (2016). MiRImpact, a new bioinformatic method using complete microRNA expression profiles to assess their overall influence on the activity of intracellular molecular pathways. *Cell Cycle*, *15*(5), 689–698. https://doi.org/10.1080/15384101.2016.1147633.

Artemov, A., Aliper, A., Korzinkin, M., Lezhnina, K., Jellen, L., Zhukov, N., et al. (2015). A method for predicting target drug efficiency in cancer based on the analysis of signaling pathway activation. *Oncotarget*, *6*(30), 29347–29356. https://doi.org/10.18632/oncotarget.5119.

Bakula, D., Ablasser, A., Aguzzi, A., Antebi, A., Barzilai, N., Bittner, M. I., et al. (2019). Latest advances in aging research and drug discovery. *Aging*, *11*(22), 9971–9981. https://doi.org/10.18632/aging.102487.

Barry, K. C., Ingolia, N. T., & Vance, R. E. (2017). Global analysis of gene expression reveals mRNA superinduction is required for the inducible immune response to a bacterial pathogen. *eLife*, *6*, e22707. https://doi.org/10.7554/eLife.22707.

Bartholomäus, A., Del Campo, C., & Ignatova, Z. (2016). Mapping the non-standardized biases of ribosome profiling. *Biological Chemistry*, *397*, 23–35. Walter de Gruyter GmbH https://doi.org/10.1515/hsz-2015-0197.

Bauer-Mehren, A., Furlong, L. I., & Sanz, F. (2009). Pathway databases and tools for their exploitation: Benefits, current limitations and challenges. *Molecular Systems Biology*, *5*, 290. https://doi.org/10.1038/msb.2009.47.

Betancourt, L. H., Pawłowski, K., Eriksson, J., Szasz, A. M., Mitra, S., Pla, I., et al. (2019). Improved survival prognostication of node-positive malignant melanoma patients utilizing shotgun proteomics guided by histopathological characterization and genomic data. *Scientific Reports*, *9*(1), 5154. https://doi.org/10.1038/s41598-019-41625-z.

Bindea, G., Mlecnik, B., Hackl, H., Charoentong, P., Tosolini, M., Kirilovsky, A., et al. (2009). ClueGO: A cytoscape plug-in to decipher functionally grouped gene ontology and pathway annotation networks. *Bioinformatics*, *25*(8), 1091–1093. https://doi.org/10.1093/bioinformatics/btp101.

Bolstad, B. M., Irizarry, R. A., Åstrand, M., & Speed, T. P. (2003). A comparison of normalization methods for high density oligonucleotide array data based on variance and bias. *Bioinformatics*, *19*(2), 185–193. https://doi.org/10.1093/bioinformatics/19.2.185.

Borger, P., Schneider, M., Frick, L., Langiewicz, M., Sorokin, M., Buzdin, A., et al. (2019). Exploration of the transcriptional landscape of ALPPS reveals the pathways of accelerated liver regeneration. *Frontiers in Oncology*, *9*, 1206. https://doi.org/10.3389/fonc.2019.01206.

Borger, P., Buzdin, A., Sorokin, M., Kachaylo, E., Humar, B., Graf, R., et al. (2020). Large-scale profiling of signaling pathways reveals a distinct demarcation between normal and extended liver resection. *Cell*, *9*(5), 1149. https://doi.org/10.3390/cells9051149.

Borisov, N., & Buzdin, A. (2019). New paradigm of machine learning (ML) in personalized oncology: Data trimming for squeezing more biomarkers from clinical datasets. *Frontiers in Oncology*, *9*, 658. https://doi.org/10.3389/fonc.2019.00658.

Borisov, N., Aksamitiene, E., Kiyatkin, A., Legewie, S., Berkhout, J., Maiwald, T., et al. (2009). Systems-level interactions between insulin-EGF networks amplify mitogenic signaling. *Molecular Systems Biology*, *5*, 256. https://doi.org/10.1038/msb.2009.19.

Borisov, N. M., Terekhanova, N. V., Aliper, A. M., Venkova, L. S., Smirnov, P. Y., Roumiantsev, S., et al. (2014). Signaling pathways activation profiles make better markers of cancer than expression of individual genes. *Oncotarget*, *5*(20), 10198–10205. https://doi.org/10.18632/oncotarget.2548.

Borisov, N., Suntsova, M., Sorokin, M., Garazha, A., Kovalchuk, O., Aliper, A., et al. (2017). Data aggregation at the level of molecular pathways improves stability of experimental transcriptomic and proteomic data. *Cell Cycle*, *16*(19), 1810–1823. https://doi.org/10.1080/15384101.2017.1361068.

Borisov, N., Tkachev, V., Suntsova, M., Kovalchuk, O., Zhavoronkov, A., Muchnik, I., et al. (2018). A method of gene expression data transfer from cell lines to cancer patients for machine-learning prediction of drug efficiency. *Cell Cycle*, *17*(4), 486–491. https://doi.org/10.1080/15384101.2017.1417706.

Borisov, N., Shabalina, I., Tkachev, V., Sorokin, M., Garazha, A., Pulin, A., et al. (2019). Shambhala: A platform-agnostic data harmonizer for gene expression data. *BMC Bioinformatics*, *20*(1), 66. https://doi.org/10.1186/s12859-019-2641-8.

Borisov, N. I., Sorokin, M., Garazha, A., & Buzdin, A. (2020). Quantitation of molecular pathway activation using RNA sequencing data. *Methods in Molecular Biology (Clifton, N.J.)*, *2063*, 189–206. https://doi.org/10.1007/978-1-0716-0138-9_15.

Bossel Ben-Moshe, N., Gilad, S., Perry, G., Benjamin, S., Balint-Lahat, N., Pavlovsky, A., et al. (2018). mRNA-seq whole transcriptome profiling of fresh frozen versus archived fixed tissues. *BMC Genomics*, *19*(1), 419. https://doi.org/10.1186/s12864-018-4761-3.

Branzei, D., & Foiani, M. (2008). Regulation of DNA repair throughout the cell cycle. *Nature Reviews Molecular Cell Biology*, *9*, 297–308. https://doi.org/10.1038/nrm2351.

Buzdin, A. A., Zhavoronkov, A. A., Korzinkin, M. B., Roumiantsev, S. A., Aliper, A. M., Venkova, L. S., et al. (2014). The OncoFinder algorithm for minimizing the errors introduced by the high-throughput methods of transcriptome analysis. *Frontiers in Molecular Biosciences*, *1*(AUG), 8. https://doi.org/10.3389/fmolb.2014.00008.

Buzdin, A. A., Zhavoronkov, A. A., Korzinkin, M. B., Venkova, L. S., Zenin, A. A., Smirnov, P. Y., et al. (2014). Oncofinder, a new method for the analysis of intracellular signaling pathway activation using transcriptomic data. *Frontiers in Genetics*, *5*(MAR), 55. https://doi.org/10.3389/fgene.2014.00055.

Buzdin, A. A., Artcibasova, A. V., Fedorova, N. F., Suntsova, M. V., Garazha, A. V., Sorokin, M. I., et al. (2016). Early stage of cytomegalovirus infection suppresses host microRNA expression regulation in human fibroblasts. *Cell Cycle (Georgetown, Texas)*, *15*(24), 3378–3389. https://doi.org/10.1080/15384101.2016.1241928.

Buzdin, A. A., Prassolov, V., Zhavoronkov, A. A., & Borisov, N. M. (2017). Bioinformatics meets biomedicine: Oncofinder, a quantitative approach for interrogating molecular pathways using gene expression data. *Methods in Molecular Biology (Clifton, N.J.)*, *1613*, 53–83. https://doi.org/10.1007/978-1-4939-7027-8_4.

Buzdin, A., Sorokin, M., Glusker, A., Garazha, A., Poddubskaya, E., Shirokorad, V., et al. (2017). Activation of intracellular signaling pathways as a new type of biomarkers for selection of target anticancer drugs. *Journal of Clinical Oncology*, *35*(15_suppl), e23142. https://doi.org/10.1200/JCO.2017.35.15_suppl.e23142.

Buzdin, A., Sorokin, M., Garazha, A., Sekacheva, M., Kim, E., Zhukov, N., et al. (2018). Molecular pathway activation—New type of biomarkers for tumor morphology and personalized selection of target drugs. *Seminars in Cancer Biology*, *53*, 110–124. https://doi.org/10.1016/j.semcancer.2018.06.003.

Buzdin, A., Sorokin, M., Garazha, A., Glusker, A., Aleshin, A., Poddubskaya, E., et al. (2019). RNA sequencing for research and diagnostics in clinical oncology. *Seminars in Cancer Biology*, *60*, 311–323. Academic Press https://doi.org/10.1016/j.semcancer.2019.07.010.

Buzdin, A., Sorokin, M., Poddubskaya, E., & Borisov, N. (2019). High-throughput mutation data now complement transcriptomic profiling: Advances in molecular pathway activation analysis approach in cancer biology. *Cancer Informatics*, *18*, 1176935119838844. https://doi.org/10.1177/1176935119838844.

Buzdin, A., Skvortsova, I. I., Li, X., & Wang, Y. (2021). Editorial: Next generation sequencing based diagnostic approaches in clinical oncology. *Frontiers in Oncology*, *10*, 3276. https://doi.org/10.3389/fonc.2020.635555.

Byron, S. A., Van Keuren-Jensen, K. R., Engelthaler, D. M., Carpten, J. D., & Craig, D. W. (2016). Translating RNA sequencing into clinical diagnostics: Opportunities and challenges. *Nature Reviews Genetics*, *17*, 257–271. Nature Publishing Group https://doi.org/10.1038/nrg.2016.10.

Carlson, C. S., Emerson, R. O., Sherwood, A. M., Desmarais, C., Chung, M. W., Parsons, J. M., et al. (2013). Using synthetic templates to design an unbiased multiplex PCR assay. *Nature Communications*, *4*, 2680. https://doi.org/10.1038/ncomms3680.

Casbas-Hernandez, P., Sun, X., Roman-Perez, E., D'Arcy, M., Sandhu, R., Hishida, A., et al. (2015). Tumor intrinsic subtype is reflected in cancer-adjacent tissue. *Cancer Epidemiology, Biomarkers & Prevention*, *24*(2), 406–414. https://doi.org/10.1158/1055-9965.EPI-14-0934.

Caspi, R., Billington, R., Keseler, I. M., Kothari, A., Krummenacker, M., Midford, P. E., et al. (2020). The MetaCyc database of metabolic pathways and enzymes-a 2019 update. *Nucleic Acids Research*, *48*(D1), D445–D453. https://doi.org/10.1093/nar/gkz862.

Castillo, D., Gálvez, J. M., Herrera, L. J., Román, B. S., Rojas, F., & Rojas, I. (2017). Integration of RNA-Seq data with heterogeneous microarray data for breast cancer profiling. *BMC Bioinformatics*, *18*(1), 506. https://doi.org/10.1186/s12859-017-1925-0.

Chang, T.-W. (1983). Binding of cells to matrixes of distinct antibodies coated on solid surface. *Journal of Immunological Methods*, *65*(1–2), 217–223. https://doi.org/10.1016/0022-1759(83)90318-6.

Comunanza, V., Corà, D., Orso, F., Consonni, F. M., Middonti, E., Di Nicolantonio, F., et al. (2017). VEGF blockade enhances the antitumor effect of BRAF V 600E inhibition. *EMBO Molecular Medicine*, *9*(2), 219–237. https://doi.org/10.15252/emmm.201505774.

Coscia, F., Watters, K. M., Curtis, M., Eckert, M. A., Chiang, C. Y., Tyanova, S., et al. (2016). Integrative proteomic profiling of ovarian cancer cell lines reveals precursor cell associated proteins and functional status. *Nature Communications*, 7, 12645. https://doi.org/10.1038/ncomms12645.

Croft, D., Mundo, A. F., Haw, R., Milacic, M., Weiser, J., Wu, G., et al. (2014). The reactome pathway knowledgebase. *Nucleic Acids Research*, *42*(D1), D472–D477. https://doi.org/10.1093/nar/gkt1102.

Cui, J., Qu, Z., Harata-Lee, Y., Nwe Aung, T., Shen, H., Wang, W., et al. (2019). Cell cycle, energy metabolism and DNA repair pathways in cancer cells are suppressed by Compound Kushen Injection. *BMC Cancer*, *19*(1), 103. https://doi.org/10.1186/s12885-018-5230-8.

Dalma-Weiszhausz, D. D., Warrington, J., Tanimoto, E. Y., & Miyada, C. G. (2006). The affymetrix GeneChip platform: An overview. *Methods in Enzymology*, *410*, 3–28. https://doi.org/10.1016/S0076-6879(06)10001-4.

Dang, C. V. (2012). Links between metabolism and cancer. *Genes & Development*, *26*(9), 877–890. https://doi.org/10.1101/gad.189365.112.

Davis, C. A., Hitz, B. C., Sloan, C. A., Chan, E. T., Davidson, J. M., Gabdank, I., et al. (2018). The Encyclopedia of DNA elements (ENCODE): Data portal update. *Nucleic Acids Research*, *46*(D1), D794–D801. https://doi.org/10.1093/nar/gkx1081.

De Klerk, E., Fokkema, I. F. A. C., Thiadens, K. A. M. H., Goeman, J. J., Palmblad, M., Den Dunnen, J. T., et al. (2015). Assessing the translational landscape of myogenic differentiation by ribosome profiling. *Nucleic Acids Research*, *43*(9), 4408–4428. https://doi.org/10.1093/nar/gkv281.

Denis, J. A., Nectoux, J., Lamy, P. J., Sciellour Le, C. R., Guermouche, H., Alary, A. S., et al. (2018). Development of digital PCR molecular tests for clinical practice: Principles, practical implementation and recommendations. *Annales de Biologie Clinique*, *76*(5), 505–523. https://doi.org/10.1684/abc.2018.1372.

Disanza, A., Frittoli, E., Palamidessi, A., & Scita, G. (2009). Endocytosis and spatial restriction of cell signaling. *Molecular Oncology*, *3*, 280–296. https://doi.org/10.1016/j.molonc.2009.05.008.

Dong, L.-Q. Q., Shi, Y., Ma, L.-J. J., Yang, L.-X. X., Wang, X.-Y. Y., Zhang, S., et al. (2018). Spatial and temporal clonal evolution of intrahepatic cholangiocarcinoma. *Journal of Hepatology*, *69*(1), 89–98. https://doi.org/10.1016/j.jhep.2018.02.029.

Draghici, S., Khatri, P., Tarca, A. L., Amin, K., Done, A., Voichita, C., et al. (2007). A systems biology approach for pathway level analysis. *Genome Research*, *17*(10), 1537–1545. https://doi.org/10.1101/gr.6202607.

Dreesen, O., & Brivanlou, A. H. (2007). Signaling pathways in cancer and embryonic stem cells. *Stem Cell Reviews*, *3*(1), 7–17. https://doi.org/10.1007/s12015-007-0004-8.

Duarte, J. G., & Blackburn, J. M. (2017). Advances in the development of human protein microarrays. *Expert Review of Proteomics*, *14*, 627–641. Taylor and Francis Ltd https://doi.org/10.1080/14789450.2017.1347042.

Dubovenko, A., Nikolsky, Y., Rakhmatulin, E., & Nikolskaya, T. (2017). Functional analysis of OMICs data and small molecule compounds in an integrated "knowledge-based" platform. In *Vol. 1613. Methods in molecular biology* (pp. 101–124). Humana Press Inc. https://doi.org/10.1007/978-1-4939-7027-8_6.

Dunn, J. G., Foo, C. K., Belletier, N. G., Gavis, E. R., & Weissman, J. S. (2013). Ribosome profiling reveals pervasive and regulated stop codon readthrough in Drosophila melanogaster. *eLife*, *2013*(2), e01179. https://doi.org/10.7554/eLife.01179.

Duo, J., Chiriac, C., Huang, R. Y. C., Mehl, J., Chen, G., Tymiak, A., et al. (2018). Slow off-rate modified aptamer (SOMAmer) as a novel reagent in immunoassay development for accurate soluble glypican-3 quantification in clinical samples. *Analytical Chemistry*, *90*(8), 5162–5170. https://doi.org/10.1021/acs.analchem.7b05277.

Eastel, J. M., Lam, K. W., Lee, N. L., Lok, W. Y., Tsang, A. H. F., Pei, X. M., et al. (2019). Application of NanoString technologies in companion diagnostic development. *Expert Review of Molecular Diagnostics*, *19*, 591–598. Taylor and Francis Ltd https://doi.org/10.1080/14737159.2019.1623672.

Eden, E., Navon, R., Steinfeld, I., Lipson, D., & Yakhini, Z. (2009). GOrilla: A tool for discovery and visualization of enriched GO terms in ranked gene lists. *BMC Bioinformatics*, *10*, 48. https://doi.org/10.1186/1471-2105-10-48.

Edwards, N. J., Oberti, M., Thangudu, R. R., Cai, S., McGarvey, P. B., Jacob, S., et al. (2015). The CPTAC data portal: A resource for cancer proteomics research. *Journal of Proteome Research*, *14*(6), 2707–2713. https://doi.org/10.1021/pr501254j.

von Eichborn, J., Dunkel, M., Gohlke, B. O., Preissner, S. C., Hoffmann, M. F., Bauer, J. M. J., et al. (2012). SynSysNet: Integration of experimental data on synaptic protein–protein interactions with drug-target relations. *Nucleic Acids Research*, *41*(D1), D834–D840. https://doi.org/10.1093/nar/gks1040.

Ekins, S., Nikolsky, Y., Bugrim, A., Kirillov, E., & Nikolskaya, T. (2007). Pathway mapping tools for analysis of high content data. *Methods in Molecular Biology (Clifton, N.J.)*, *356*, 319–350. https://doi.org/10.1385/1-59745-217-3:319.

Elkon, R., Vesterman, R., Amit, N., Ulitsky, I., Zohar, I., Weisz, M., et al. (2008). SPIKE—A database, visualization and analysis tool of cellular signaling pathways. *BMC Bioinformatics*, *9*, 110. https://doi.org/10.1186/1471-2105-9-110.

Emelianova, A. A., Kuzmin, D. V., Panteleev, P. V., Sorokin, M., Buzdin, A. A., & Ovchinnikova, T. V. (2018). Anticancer activity of the goat antimicrobial peptide ChMAP-28. *Frontiers in Pharmacology*, *9*, 1501. https://doi.org/10.3389/fphar.2018.01501.

Erenpreisa, J., Salmina, K., Anatskaya, O., & Cragg, M. S. (2021). Paradoxes of cancer: Survival at the brink. *Seminars in Cancer Biology*. Academic Press https://doi.org/10.1016/j.semcancer.2020.12.009.

Fidler, M. J., Fhied, C. L., Roder, J., Basu, S., Sayidine, S., Fughhi, I., et al. (2018). The serum-based VeriStrat® test is associated with proinflammatory reactants and clinical outcome in non-small cell lung cancer patients. *BMC Cancer*, *18*(1), 310. https://doi.org/10.1186/s12885-018-4193-0.

Filteau, M., Diss, G., Torres-Quiroz, F., Dubé, A. K., Schraffl, A., Bachmann, V. A., et al. (2015). Systematic identification of signal integration by protein kinase A. *Proceedings of the National Academy of Sciences of the United States of America*, *112*(14), 4501–4506. https://doi.org/10.1073/pnas.1409938112.

Fouad, Y. A., & Aanei, C. (2017). Revisiting the hallmarks of cancer. *American Journal of Cancer Research*, 7(5), 1016–1036. Retrieved from http://www.ncbi.nlm.nih.gov/pubmed/28560055.

Galanos, P., Pappas, G., Polyzos, A., Kotsinas, A., Svolaki, I., Giakoumakis, N. N., et al. (2018). Mutational signatures reveal the role of RAD52 in p53-independent p21-driven genomic instability. *Genome Biology*, *19*(1), 37. https://doi.org/10.1186/s13059-018-1401-9.

Gao, S., & Wang, X. (2007). TAPPA: topological analysis of pathway phenotype association. *Bioinformatics*, *23*(22), 3100–3102. https://doi.org/10.1093/bioinformatics/btm460.

Gyurkó, D. M., Veres, D. V., Módos, D., Lenti, K., Korcsmáros, T., & Csermely, P. (2013). Adaptation and learning of molecular networks as a description of cancer development at the systems-level: Potential use in anti-cancer therapies. *Seminars in Cancer Biology*, *23*, 262–269. https://doi.org/10.1016/j.semcancer.2013.06.005.

Hall, A. (2009). The cytoskeleton and cancer. *Cancer and Metastasis Reviews*, *28*(1–2), 5–14. https://doi.org/10.1007/s10555-008-9166-3.

Huang, D. W., Sherman, B. T., & Lempicki, R. A. (2009a). Bioinformatics enrichment tools: paths toward the comprehensive functional analysis of large gene lists. *Nucleic Acids Research*, *37*(1), 1–13. https://doi.org/10.1093/nar/gkn923.

Huang, D. W., Sherman, B. T., & Lempicki, R. A. (2009b). Systematic and integrative analysis of large gene lists using DAVID bioinformatics resources. *Nature Protocols*, *4*(1), 44–57. https://doi.org/10.1038/nprot.2008.211.

Huang, X., Stern, D. F., & Zhao, H. (2016). Transcriptional profiles from paired normal samples offer complementary information on cancer patient survival—Evidence from TCGA pan-cancer data. *Scientific Reports*, *6*, 20567. https://doi.org/10.1038/srep20567.

Ibrahim, M. A.-H. H., Jassim, S., Cawthorne, M. A., & Langlands, K. (2012). A topology-based score for pathway enrichment. *Journal of Computational Biology*, *19*(5), 563–573. https://doi.org/10.1089/cmb.2011.0182.

Igolkina, A. A., Zinkevich, A., Karandasheva, K. O., Popov, A. A., Selifanova, M. V., Nikolaeva, D., et al. (2019). H3K4me3, H3K9ac, H3K27ac, H3K27me3 and H3K9me3 histone tags suggest distinct regulatory evolution of open and condensed chromatin landmarks. *Cells*, *8*(9), 1034. https://doi.org/10.3390/cells8091034.

Ilkhani, K., Bastami, M., Delgir, S., Safi, A., Talebian, S., & Alivand, M.-R. (2020). The engaged role of tumor microenvironment in cancer metabolism: Focusing on cancer-associated fibroblast and exosome mediators. *Anti-Cancer Agents in Medicinal Chemistry*, *21*(2), 254–266. https://doi.org/10.2174/1871520620666200910123428.

Ingolia, N. T., Ghaemmaghami, S., Newman, J. R. S., & Weissman, J. S. (2009). Genome-wide analysis in vivo of translation with nucleotide resolution using ribosome profiling. *Science*, *324*(5924), 218–223. https://doi.org/10.1126/science.1168978.

Jeggo, P. A., Pearl, L. H., & Carr, A. M. (2016). DNA repair, genome stability and cancer: A historical perspective. *Nature Reviews Cancer*, *16*(1), 35–42. https://doi.org/10.1038/nrc.2015.4.

Jellen, L. C., Aliper, A., Buzdin, A., Zhavoronkov, A., & Zhavoronkov, A. (2015). Screening and personalizing nootropic drugs and cognitive modulator regimens in silico. *Frontiers in Systems Neuroscience*, *9*(FEB), 4. https://doi.org/10.3389/fnsys.2015.00004.

Jones, A. C., Antillon, K. S., Jenkins, S. M., Janos, S. N., Overton, H. N., Shoshan, D. S., et al. (2015). Prostate field cancerization: Deregulated expression of macrophage inhibitory cytokine 1 (MIC-1) and platelet derived growth factor a (PDGF-A) in tumor adjacent tissue. *PLoS One*, *10*(3), e0119314. https://doi.org/10.1371/journal.pone.0119314.

Jovčevska, I., Zupanec, N., Urlep, Ž., Vranic, A., Matos, B., Stokin, C. L., et al. (2017). Differentially expressed proteins in glioblastoma multiforme identified with a nanobody-based anti-proteome approach and confirmed by OncoFinder as possible tumor-class predictive biomarker candidates. *Oncotarget*, *8*(27), 44141–44158. https://doi.org/10.18632/oncotarget.17390.

Junaid, M., Akter, Y., Afrose, S. S., Tania, M., & Khan, M. A. (2020). Biological role of AKT, and regulation of AKT signaling pathway by thymoquinone: perspectives in cancer therapeutics. *Mini-Reviews in Medicinal Chemistry*, *20*. https://doi.org/10.2174/1389557520666201005143818.

Kalasauskas, D., Sorokin, M., Sprang, B., Elmasri, A., Viehweg, S., Salinas, G., et al. (2020). Diversity of clinically relevant outcomes resulting from hypofractionated radiation in human glioma stem cells mirrors distinct patterns of transcriptomic changes. *Cancers*, *12*(3), 570. https://doi.org/10.3390/cancers12030570.

Kanehisa, M., Goto, S., Furumichi, M., Tanabe, M., & Hirakawa, M. (2010). KEGG for representation and analysis of molecular networks involving diseases and drugs. *Nucleic Acids Research*, *38*(Suppl. 1), D355–D360. https://doi.org/10.1093/nar/gkp896.

Karpova, M. A., Moshkovskii, S. A., Toropygin, I. Y., & Archakov, A. I. (2010). Cancer-specific MALDI-TOF profiles of blood serum and plasma: Biological meaning and perspectives. *Journal of Proteomics*, *73*, 537–551. https://doi.org/10.1016/j.jprot.2009.09.011.

Khatri, P., & Drăghici, S. (2005). Ontological analysis of gene expression data: Current tools, limitations, and open problems. *Bioinformatics*, *21*, 3587–3595. https://doi.org/10.1093/bioinformatics/bti565.

Khatri, P., Sirota, M., & Butte, A. J. (2012). Ten years of pathway analysis: Current approaches and outstanding challenges. *PLoS Computational Biology*, *8*(2), e1002375. https://doi.org/10.1371/journal.pcbi.1002375.

Kholodenko, B. N., Demin, O. V., Moehren, G., & Hoek, J. B. (1999). Quantification of short term signaling by the epidermal growth factor receptor. *Journal of Biological Chemistry*, *274*(42), 30169–30181. https://doi.org/10.1074/jbc.274.42.30169.

Kim, E. L., Sorokin, M., Kantelhardt, S. R., Kalasauskas, D., Sprang, B., Fauss, J., et al. (2020). Intratumoral heterogeneity and longitudinal changes in gene expression predict differential drug sensitivity in newly diagnosed and recurrent glioblastoma. *Cancers*, *12*(2), 520. https://doi.org/10.3390/cancers12020520.

King, H. A., & Gerber, A. P. (2016). Translatome profiling: Methods for genome-scale analysis of mRNA translation. *Briefings in Functional Genomics*, *15*, 22–31. Oxford University Press https://doi.org/10.1093/bfgp/elu045.

Kiyatkin, A., Aksamitiene, E., Markevich, N. I., Borisov, N. M., Hoek, J. B., & Kholodenko, B. N. (2006). Scaffolding protein Grb2-associated binder 1 sustains epidermal growth factor-induced mitogenic and survival signaling by multiple positive feedback loops. *Journal of Biological Chemistry*, *281*(29), 19925–19938. https://doi.org/10.1074/jbc.M600482200.

Knyazeva, M., Korobkina, E., Karizky, A., Sorokin, M., Buzdin, A., Vorobyev, S., et al. (2020). Reciprocal dysregulation of mir-146b and mir-451 contributes in malignant phenotype of follicular thyroid tumor. *International Journal of Molecular Sciences*, *21*(17), 1–17. https://doi.org/10.3390/ijms21175950.

Kono, N., & Arakawa, K. (2019). Nanopore sequencing: Review of potential applications in functional genomics. *Development, Growth & Differentiation*, *61*, 316–326. Blackwell Publishing https://doi.org/10.1111/dgd.12608.

Kuenzi, B. M., Remsing Rix, L. L., Stewart, P. A., Fang, B., Kinose, F., Bryant, A. T., et al. (2017). Polypharmacology-based ceritinib repurposing using integrated functional proteomics. *Nature Chemical Biology*, *13*(12), 1222–1231. https://doi.org/10.1038/nchembio.2489.

Kulikov, A. V., Luchkina, E. A., Gogvadze, V., & Zhivotovsky, B. (2017). Mitophagy: Link to cancer development and therapy. *Biochemical and Biophysical Research Communications*, *482*(3), 432–439. https://doi.org/10.1016/j.bbrc.2016.10.088.

Kurz, S., Thieme, R., Amberg, R., Groth, M., Jahnke, H.-G., Pieroh, P., et al. (2017). The anti-tumorigenic activity of A2M-A lesson from the naked mole-rat. *PLoS One*, *12*(12), e0189514. https://doi.org/10.1371/journal.pone.0189514.

Lahens, N. F., Ricciotti, E., Smirnova, O., Toorens, E., Kim, E. J., Baruzzo, G., et al. (2017). A comparison of illumina and ion torrent sequencing platforms in the context of differential gene expression. *BMC Genomics*, *18*(1), 602. https://doi.org/10.1186/s12864-017-4011-0.

Larkin, B., Ilyukha, V., Sorokin, M., Buzdin, A., Vannier, E., & Poltorak, A. (2017). Cutting edge: Activation of STING in T cells induces type I IFN responses and cell death. *The Journal of Immunology*, *199*(2), 397–402. https://doi.org/10.4049/jimmunol.1601999.

Levin, J. M., Oprea, T. I., Davidovich, S., Clozel, T., Overington, J. P., Vanhaelen, Q., et al. (2020). Artificial intelligence, drug repurposing and peer review. *Nature Biotechnology*, *38*, 1127–1131. Nature Research https://doi.org/10.1038/s41587-020-0686-x.

Lezhnina, K., Kovalchuk, O., Zhavoronkov, A. A., Korzinkin, M. B., Zabolotneva, A. A., Shegay, P. V., et al. (2014). Novel robust biomarkers for human bladder cancer based on activation of intracellular signaling pathways. *Oncotarget*, *5*(19), 9022–9032. https://doi.org/10.18632/oncotarget.2493.

Li, J., & Zhan, X. (2021). Mass spectrometry-based proteomics analyses of post-translational modifications and proteoforms in human pituitary adenomas. *Biochimica et Biophysica Acta, Proteins and Proteomics*, *1869*, 140584. Elsevier B.V. https://doi.org/10.1016/j.bbapap.2020.140584.

Li, Q., Zhao, X., Zhang, W., Wang, L., Wang, J., Xu, D., et al. (2019). Reliable multiplex sequencing with rare index mis-assignment on DNB-based NGS platform. *BMC Genomics*, *20*(1), 215. https://doi.org/10.1186/s12864-019-5569-5.

Lin, B., Wang, J., & Cheng, Y. (2010). Recent patents and advances in the next-generation sequencing technologies. *Recent Patents on Biomedical Engineering*, *1*(1), 60–67. https://doi.org/10.2174/1874764710801010060.

Lin, S. H., Beane, L., Chasse, D., Zhu, K. W., Mathey-Prevot, B., & Chang, J. T. (2013). Cross-platform prediction of gene expression signatures. *PLoS One*, *8*(11), e79228. https://doi.org/10.1371/journal.pone.0079228.

Liu, J., François, J. M., & Capp, J. P. (2016). Use of noise in gene expression as an experimental parameter to test phenotypic effects. *Yeast*, *33*(6), 209–216. https://doi.org/10.1002/yea.3152.

Lonsdale, J., Thomas, J., Salvatore, M., Phillips, R., Lo, E., Shad, S., et al. (2013). The genotype-tissue expression (GTEx) project. *Nature Genetics*, *45*, 580–585.

Love, M. I., Huber, W., & Anders, S. (2014). Moderated estimation of fold change and dispersion for RNA-seq data with DESeq2. *Genome Biology*, *15*(12), 550. https://doi.org/10.1186/s13059-014-0550-8.

Ma, C.-Y., & Liao, C.-S. (2020). A review of protein-protein interaction network alignment: From pathway comparison to global alignment. *Computational and Structural Biotechnology Journal*, *18*, 2647–2656. https://doi.org/10.1016/j.csbj.2020.09.011.

Ma, L., Liang, Z., Zhou, H., & Qu, L. (2018). Applications of RNA indexes for precision oncology in breast cancer. *Genomics, Proteomics & Bioinformatics*, *16*, 108–119. Beijing Genomics Institute https://doi.org/10.1016/j.gpb.2018.03.002.

Makarev, E., Cantor, C., Zhavoronkov, A., Buzdin, A., Aliper, A., & Csoka, A. B. (2014). Pathway activation profiling reveals new insights into age-related macular degeneration and provides avenues for therapeutic interventions. *Aging*, *6*(12), 1064–1075. https://doi.org/10.18632/aging.100711.

Makarev, E., Izumchenko, E., Aihara, F., Wysocki, P. T., Zhu, Q., Buzdin, A., et al. (2016). Common pathway signature in lung and liver fibrosis. *Cell Cycle*, *15*, 1667–1673. Taylor and Francis Inc https://doi.org/10.1080/15384101.2016.1152435.

Malumbres, M., & Barbacid, M. (2009). Cell cycle, CDKs and cancer: A changing paradigm. *Nature Reviews Cancer*, *9*, 153–166. https://doi.org/10.1038/nrc2602.

Mann, M. (2016). Origins of mass spectrometry-based proteomics. *Nature Reviews Molecular Cell Biology*, *17*, 678. Nature Publishing Group https://doi.org/10.1038/nrm.2016.135.

Maouche, S., Poirier, O., Godefroy, T., Olaso, R., Gut, I., Collet, J. P., et al. (2008). Performance comparison of two microarray platforms to assess differential gene expression in human monocyte and macrophage cells. *BMC Genomics*, *9*, 302. https://doi.org/10.1186/1471-2164-9-302.

Marggraf, M. B., Panteleev, P. V., Emelianova, A. A., Sorokin, M. I., Bolosov, I. A., Buzdin, A. A., et al. (2018). Cytotoxic potential of the novel horseshoe crab peptide polyphemusin III. *Marine Drugs*, *16*(12), 466. https://doi.org/10.3390/md16120466.

Marshall, C. J. (1995). Specificity of receptor tyrosine kinase signaling: Transient versus sustained extracellular signal-regulated kinase activation. *Cell*, *80*, 179–185. https://doi.org/10.1016/0092-8674(95)90401-8.

Mathivanan, S., Periaswamy, B., Gandhi, T. K. B., Kandasamy, K., Suresh, S., Mohmood, R., et al. (2006). An evaluation of human protein-protein interaction data in the public domain. *BMC Bioinformatics*, 7(Suppl. 5), S19. https://doi.org/10.1186/1471-2105-7-S5-S19.

Michel, A. M., & Baranov, P. V. (2013). Ribosome profiling: A Hi-Def monitor for protein synthesis at the genome-wide scale. *Wiley Interdisciplinary Reviews: RNA*, *4*, 473–490. Blackwell Publishing Ltd https://doi.org/10.1002/wrna.1172.

Mirus, J. E., Zhang, Y., Hollingsworth, M. A., Solan, J. L., Lampe, P. D., & Hingorani, S. R. (2014). Spatiotemporal proteomic analyses during pancreas cancer progression identifies serine/threonine stress kinase 4 (STK4) as a novel candidate biomarker for early stage disease. *Molecular and Cellular Proteomics*, *13*(12), 3484–3496. https://doi.org/10.1074/mcp.M113.036517.

Mitrea, C., Taghavi, Z., Bokanizad, B., Hanoudi, S., Tagett, R., Donato, M., et al. (2013). Methods and approaches in the topology-based analysis of biological pathways. *Frontiers in Physiology*, *4*, 278. https://doi.org/10.3389/fphys.2013.00278.

Moisseev, A., Albert, E., Lubarsky, D., Schroeder, D., & Clark, J. (2020). Transcriptomic and genomic testing to guide individualized treatment in chemoresistant gastric cancer case. *Biomedicine*, *8*(3), 67. https://doi.org/10.3390/biomedicines8030067.

Moshkovskii, S. A., Vlasova, M. A., Pyatnitskiy, M. A., Tikhonova, O. V., Safarova, M. R., Makarov, O. V., et al. (2007). Acute phase serum amyloid A in ovarian cancer as an important component of proteome diagnostic profiling. *Proteomics - Clinical Applications*, *1*(1), 107–117. https://doi.org/10.1002/prca.200600229.

Nakaya, A., Katayama, T., Itoh, M., Hiranuka, K., Kawashima, S., Moriya, Y., et al. (2013). KEGG OC: A large-scale automatic construction of taxonomy-based ortholog clusters. *Nucleic Acids Research*, *41*, D353–D357. https://doi.org/10.1093/nar/gks1239.

Nault, R., Fader, K. A., & Zacharewski, T. (2015). RNA-Seq versus oligonucleotide array assessment of dose-dependent TCDD-elicited hepatic gene expression in mice. *BMC Genomics*, *16*(1), 373. https://doi.org/10.1186/s12864-015-1527-z.

Navajas, R., Corrales, F., & Paradela, A. (2021). Quantitative proteomics-based analyses performed on pre-eclampsia samples in the 2004–2020 period: A systematic review. *Clinical Proteomics*, *18*(1), 6. https://doi.org/10.1186/s12014-021-09313-1.

Negro, G., Aschenbrenner, B., Brezar, S. K., Cemazar, M., Coer, A., Gasljevic, G., et al. (2020). Molecular heterogeneity in breast carcinoma cells with increased invasive capacities. *Radiology and Oncology*, *54*(1), 103–118. https://doi.org/10.2478/raon-2020-0007.

Nieuwenhuis, T. O., Yang, S. Y., Verma, R. X., Pillalamarri, V., Arking, D. E., Rosenberg, A. Z., et al. (2020). Consistent RNA sequencing contamination in GTEx and other data sets. *Nature Communications*, *11*(1), 1933. https://doi.org/10.1038/s41467-020-15821-9.

Nikitin, A., Egorov, S., Daraselia, N., & Mazo, I. (2003). Pathway studio—The analysis and navigation of molecular networks. *Bioinformatics*, *19*(16), 2155–2157. https://doi.org/10.1093/bioinformatics/btg290.

Nikitin, D., Penzar, D., Garazha, A., Sorokin, M., Tkachev, V., Borisov, N., et al. (2018). Profiling of human molecular pathways affected by retrotransposons at the level of regulation by transcription factor proteins. *Frontiers in Immunology*, *9*(JAN), 30. https://doi.org/10.3389/fimmu.2018.00030.

Nikitin, D., Kolosov, N., Murzina, A., Pats, K., Zamyatin, A., Tkachev, V., et al. (2019). Retroelement-linked H3K4me1 histone tags uncover regulatory evolution trends of gene enhancers and feature quickly evolving molecular processes in human physiology. *Cell*, *8*(10), 1219. https://doi.org/10.3390/cells8101219.

Nikitin, D., Garazha, A., Sorokin, M., Penzar, D., Tkachev, V., Markov, A., et al. (2019a). Correction: Nikitin, D., et al. Retroelement—Linked transcription factor binding patterns point to quickly developing molecular pathways in human evolution. *Cells*, *8*, 130. 2019. Cells, 8(8), 832 https://doi.org/10.3390/cells8080832.

Nikitin, D., Garazha, A., Sorokin, M., Penzar, D., Tkachev, V., Markov, A., et al. (2019b). Retroelement—Linked transcription factor binding patterns point to quickly developing molecular pathways in human evolution. *Cells*, *8*(2), 130. https://doi.org/10.3390/cells8020130.

Nishimura, D. (2001). BioCarta. *Biotech Software & Internet Report*, *2*, 117–120. https://doi.org/10.1089/152791601750294344.

Otto, G. M., & Brar, G. A. (2018). Seq-ing answers: Uncovering the unexpected in global gene regulation. *Current Genetics*, *64*, 1183–1188. Springer Verlag https://doi.org/10.1007/s00294-018-0839-3.

Ozerov, I. V., Lezhnina, K. V., Izumchenko, E., Artemov, A. V., Medintsev, S., Vanhaelen, Q., et al. (2016). In silico pathway activation network decomposition analysis (iPANDA) as a method for biomarker development. *Nature Communications*, *7*, 13427. https://doi.org/10.1038/ncomms13427.

O'Neill, J. R., Pak, H. S., Pairo-Castineira, E., Save, V., Paterson-Brown, S., Nenutil, R., et al. (2017). Quantitative shotgun proteomics unveils candidate novel esophageal adenocarcinoma (EAC)-specific proteins. *Molecular and Cellular Proteomics*, *16*(6), 1138–1150. https://doi.org/10.1074/mcp.M116.065078.

Painter, J. T., Clayton, N. P., & Herbert, R. A. (2010). Useful immunohistochemical markers of tumor differentiation. *Toxicologic Pathology*, *38*, 131–141. https://doi.org/10.1177/0192623309356449.

Parkhitko, A. A., Filine, E., Mohr, S. E., Moskalev, A., & Perrimon, N. (2020). Targeting metabolic pathways for extension of lifespan and healthspan across multiple species. *Ageing Research Reviews*, *64*, 101188. https://doi.org/10.1016/j.arr.2020.101188.

Pasteuning-Vuhman, S., Boertje-Van Der Meulen, J. W., Van Putten, M., Overzier, M., Ten Dijke, P., Kiełbasa, S. M., et al. (2017). New function of the myostatin/activin type I receptor (ALK4) as a mediator of muscle atrophy and muscle regeneration. *FASEB Journal*, *31*(1), 238–255. https://doi.org/10.1096/fj.201600675R.

Petricoin, E. F., Ardekani, A. M., Hitt, B. A., Levine, P. J., Fusaro, V. A., Steinberg, S. M., et al. (2002). Use of proteomic patterns in serum to identify ovarian cancer. *Lancet*, *359*(9306), 572–577. https://doi.org/10.1016/S0140-6736(02)07746-2.

Petrov, I., Suntsova, M., Mutorova, O., Sorokin, M., Garazha, A., Ilnitskaya, E., et al. (2016). Molecular pathway activation features of pediatric acute myeloid leukemia (AML) and acute lymphoblast leukemia (ALL) cells. *Aging*, *8*(11), 2936–2947. https://doi.org/10.18632/aging.101102.

Petrov, I., Suntsova, M., Ilnitskaya, E., Roumiantsev, S., Sorokin, M., Garazha, A., et al. (2017). Gene expression and molecular pathway activation signatures of MYCN-amplified neuroblastomas. *Oncotarget*, *8*(48), 83768–83780. https://doi.org/10.18632/oncotarget.19662.

Poddubskaya, E. V., Baranova, M. P., Allina, D. O., Smirnov, P. Y., Albert, E. A., Kirilchev, A. P., et al. (2018). Personalized prescription of tyrosine kinase inhibitors in unresectable metastatic cholangiocarcinoma. *Experimental Hematology & Oncology*, 7(1), 21. https://doi.org/10.1186/s40164-018-0113-x.

Poddubskaya, E., Bondarenko, A., Boroda, A., Zotova, E., Glusker, A., Sletina, S., et al. (2019). Transcriptomics-guided personalized prescription of targeted therapeutics for metastatic ALK-positive lung cancer case following recurrence on ALK inhibitors. *Frontiers in Oncology*, *9*, 1026. https://doi.org/10.3389/fonc.2019.01026.

Poddubskaya, E., Buzdin, A., Garazha, A., Sorokin, M., Glusker, A., Aleshin, A., et al. (2019). Oncobox, gene expression-based second opinion system for predicting response to treatment in advanced solid tumors. *Journal of Clinical Oncology*, *37*(15_suppl), e13143. https://doi.org/10.1200/JCO.2019.37.15_suppl.e13143.

Poddubskaya, E. V., Baranova, M. P., Allina, D. O., Sekacheva, M. I., Makovskaia, L. A., Kamashev, D. E., et al. (2019). Personalized prescription of imatinib in recurrent granulosa cell tumor of the ovary: Case report. *Molecular Case Studies*, *5*(2), mcs.a003434. https://doi.org/10.1101/mcs.a003434.

Poddubskaya, E., Sorokin, M., Garazha, A., Glusker, A., Moisseev, A., Sekacheva, M., et al. (2020). Clinical use of RNA sequencing and oncobox analytics to predict personalized targeted therapeutic efficacy. *Journal of Clinical Oncology*, *38*(15_suppl), e13676. https://doi.org/10.1200/jco.2020.38.15_suppl.e13676.

Polyakova, A., Kuznetsova, K., & Moshkovskii, S. (2015). Proteogenomics meets cancer immunology: Mass spectrometric discovery and analysis of neoantigens. *Expert Review of Proteomics*, *12*, 533–541. Taylor and Francis Ltd https://doi.org/10.1586/14789450.2015.1070100.

Principe, S., Mejia-Guerrero, S., Ignatchenko, V., Sinha, A., Ignatchenko, A., Shi, W., et al. (2018). Proteomic analysis of cancer-associated fibroblasts reveals a paracrine role for MFAP5 in human oral tongue squamous cell carcinoma. *Journal of Proteome Research, 17*(6), 2045–2059. https://doi.org/10.1021/acs.jproteome.7b00925.

Rai, M. F., Tycksen, E. D., Sandell, L. J., & Brophy, R. H. (2018). Advantages of RNA-seq compared to RNA microarrays for transcriptome profiling of anterior cruciate ligament tears. *Journal of Orthopaedic Research, 36*(1), 484–497. https://doi.org/10.1002/jor.23661.

Ravi, R., Noonan, K. A., Pham, V., Bedi, R., Zhavoronkov, A., Ozerov, I. V., et al. (2018). Bifunctional immune checkpoint-targeted antibody-ligand traps that simultaneously disable TGFβ enhance the efficacy of cancer immunotherapy. *Nature Communications, 9*(1), 741. https://doi.org/10.1038/s41467-017-02696-6.

Reymond, M. A., & Schlegel, W. (2007). Proteomics in cancer. *Advances in Clinical Chemistry, 44*, 103–142. https://doi.org/10.1016/S0065-2423(07)44004-5.

Romero, P., Wagg, J., Green, M. L., Kaiser, D., Krummenacker, M., & Karp, P. D. (2004). Computational prediction of human metabolic pathways from the complete human genome. *Genome Biology, 6*(1), R2. https://doi.org/10.1186/gb-2004-6-1-r2.

Rosenberg, J. M., & Utz, P. J. (2015). Protein microarrays: A new tool for the study of autoantibodies in immunodeficiency. *Frontiers in Immunology, 6*, 138. Frontiers Media S.A https://doi.org/10.3389/fimmu.2015.00138.

Rudy, J., & Valafar, F. (2011). Empirical comparison of cross-platform normalization methods for gene expression data. *BMC Bioinformatics, 12*(1), 467. https://doi.org/10.1186/1471-2105-12-467.

Samec, M., Liskova, A., Koklesova, L., Samuel, S. M., Zhai, K., Buhrmann, C., et al. (2020). Flavonoids against the Warburg phenotype—Concepts of predictive, preventive and personalised medicine to cut the Gordian knot of cancer cell metabolism. *The EPMA Journal, 11*, 377–398. Springer https://doi.org/10.1007/s13167-020-00217-y.

Sanchez, A., Choubey, S., & Kondev, J. (2013). Regulation of noise in gene expression. *Annual Review of Biophysics, 42*(1), 469–491. https://doi.org/10.1146/annurev-biophys-083012-130401.

Schaefer, C. F., Anthony, K., Krupa, S., Buchoff, J., Day, M., Hannay, T., et al. (2009). PID: The pathway interaction database. *Nucleic Acids Research, 37*(Database issue), D674–D679. https://doi.org/10.1093/nar/gkn653.

Schena, M., Shalon, D., Davis, R. W., & Brown, P. O. (1995). Quantitative monitoring of gene expression patterns with a complementary DNA microarray. *Science, 270*(5235), 467–470. https://doi.org/10.1126/science.270.5235.467.

Schulze, A., & Downward, J. (2001). Navigating gene expression using microarrays—A technology review. *Nature Cell Biology, 3*, E190–E195. https://doi.org/10.1038/35087138.

Sciacovelli, M., Gaude, E., Hilvo, M., & Frezza, C. (2014). The metabolic alterations of cancer cells. In *Vol. 542. Methods in enzymology* (pp. 1–23). Academic Press Inc. https://doi.org/10.1016/B978-0-12-416618-9.00001-7.

Shabalin, A. A., Tjelmeland, H., Fan, C., Perou, C. M., & Nobel, A. B. (2008). Merging two gene-expression studies via cross-platform normalization. *Bioinformatics, 24*(9), 1154–1160. https://doi.org/10.1093/bioinformatics/btn083.

Sharma, S., Kelly, T. K., & Jones, P. A. (2010). Epigenetics in cancer. *Carcinogenesis, 31*(1), 27–36. https://doi.org/10.1093/carcin/bgp220.

Shepelin, D., Korzinkin, M., Vanyushina, A., Aliper, A., Borisov, N., Vasilov, R., et al. (2016). Molecular pathway activation features linked with transition from normal skin to primary and metastatic melanomas in human. *Oncotarget*, 7(1), 656–670. https://doi.org/10.18632/ONCOTARGET.6394.

Shih, W., Chetty, R., & Tsao, M. S. (2005). Expression profiling by microarrays in colorectal cancer (review). *Oncology Reports, 13*, 517–524. Spandidos Publications https://doi.org/10.3892/or.13.3.517.

Shtam, T., Naryzhny, S., Kopylov, A., Petrenko, E., Samsonov, R., Kamyshinsky, R., et al. (2018). Functional properties of circulating exosomes mediated by surface-attached plasma proteins. *Journal of Hematology*, 7(4), 149–153. https://doi.org/10.14740/jh412w.

Shtam, T., Naryzhny, S., Samsonov, R., Karasik, D., Mizgirev, I., Kopylov, A., et al. (2019). Plasma exosomes stimulate breast cancer metastasis through surface interactions and activation of FAK signaling. *Breast Cancer Research and Treatment, 174*(1), 129–141. https://doi.org/10.1007/s10549-018-5043-0.

Sorokin, M., Kholodenko, R., Grekhova, A., Suntsova, M., Pustovalova, M., Vorobyeva, N., et al. (2018). Acquired resistance to tyrosine kinase inhibitors may be linked with the decreased sensitivity to X-ray irradiation. *Oncotarget, 9*(4), 5111–5124. https://doi.org/10.18632/oncotarget.23700.

Sorokin, M., Kholodenko, R., Suntsova, M., Malakhova, G., Garazha, A., Kholodenko, I., et al. (2018). Oncobox bioinformatical platform for selecting potentially effective combinations of target cancer drugs using high-throughput gene expression data. *Cancers, 10*(10), 365. https://doi.org/10.3390/cancers10100365.

Sorokin, M., Poddubskaya, E., Baranova, M., Glusker, A., Kogoniya, L., Markarova, E., et al. (2020). RNA sequencing profiles and diagnostic signatures linked with response to ramucirumab in gastric cancer. *Cold Spring Harbor Molecular Case Studies, 6*(2), a004945. https://doi.org/10.1101/mcs.a004945.

Sorokin, M., Kholodenko, I., Kalinovsky, D., Shamanskaya, T., Doronin, I., Konovalov, D., et al. (2020). RNA sequencing-based identification of ganglioside GD2-positive cancer phenotype. *Biomedicine, 8*(6), 142. https://doi.org/10.3390/BIOMEDICINES8060142.

Sorokin, M., Ignatev, K., Poddubskaya, E., Vladimirova, U., Gaifullin, N., Lantsov, D., et al. (2020). RNA sequencing in comparison to immunohistochemistry for measuring cancer biomarkers in breast cancer and lung cancer specimens. *Biomedicine, 8*(5), 114. https://doi.org/10.3390/BIOMEDICINES8050114.

Sorokin, M., Ignatev, K., Barbara, V., Vladimirova, U., Muraveva, A., Suntsova, M., et al. (2020). Molecular pathway activation markers are associated with efficacy of trastuzumab therapy in metastatic HER2-positive breast cancer better than individual gene expression levels. *Biochemistry (Moscow), 85*(7), 758–772. https://doi.org/10.1134/S0006297920070044.

Sorokin, M., Borisov, N., Kuzmin, D., Gudkov, A., Zolotovskaia, M., Garazha, A., et al. (2021). Algorithmic annotation of functional roles for components of 3044 human molecular pathways. *Frontiers in Genetics, 12*, 139. https://doi.org/10.3389/FGENE.2021.617059.

Spirin, P. V., Lebedev, T. D., Orlova, N. N., Gornostaeva, A. S., Prokofjeva, M. M., Nikitenko, N. A., et al. (2014). Silencing AML1-ETO gene expression leads to simultaneous activation of both pro-apoptotic and proliferation signaling. *Leukemia, 28*(11), 2222–2228. https://doi.org/10.1038/leu.2014.130.

Spirin, P., Lebedev, T., Orlova, N., Morozov, A., Poymenova, N., Dmitriev, S. E., et al. (2017). Synergistic suppression of t(8;21)-positive leukemia cell growth by combining oridonin and MAPK1/ERK2 inhibitors. *Oncotarget, 8*(34), 56991–57002. https://doi.org/10.18632/oncotarget.18503.

Spisak, S., & Guttman, A. (2009). Biomedical applications of protein microarrays. *Current Medicinal Chemistry, 16*(22), 2806–2815. https://doi.org/10.2174/092986709788803141.

Sreekumar, A., Nyati, M. K., Varambally, S., Barrette, T. R., Ghosh, D., Lawrence, T. S., et al. (2001). Profiling of cancer cells using protein microarrays: Discovery of novel radiation-regulated proteins 1. In *Vol. 61. Cancer research*. Retrieved from www.microarrays.org.

Stephen, L. (2017). Multiplex immunoassay profiling. In *Vol. 1546. Methods in molecular biology* (pp. 169–176). Humana Press Inc. https://doi.org/10.1007/978-1-4939-6730-8_13.

Stephen, L., & Guest, P. C. (2018). Multiplex immunoassay profiling of hormones involved in metabolic regulation. In *Vol. 1735. Methods in molecular biology* (pp. 449–456). Humana Press Inc. https://doi.org/10.1007/978-1-4939-7614-0_32.

Stetson, L. C., Dazard, J. E., & Barnholtz-Sloan, J. S. (2016). Protein markers predict survival in glioma patients. *Molecular and Cellular Proteomics*, *15*(7), 2356–2365. https://doi.org/10.1074/mcp.M116.060657.

Su, Z., Łabaj, P. P., Li, S., Thierry-Mieg, J., Thierry-Mieg, D., Shi, W., et al. (2014). A comprehensive assessment of RNA-seq accuracy, reproducibility and information content by the sequencing quality control consortium. *Nature Biotechnology*, *32*(9), 903–914. https://doi.org/10.1038/nbt.2957.

Suntsova, M. V., & Buzdin, A. A. (2020). Differences between human and chimpanzee genomes and their implications in gene expression, protein functions and biochemical properties of the two species. *BMC Genomics*, *21*, 535. BioMed Central Ltd https://doi.org/10.1186/s12864-020-06962-8.

Suntsova, M., Gaifullin, N., Allina, D., Reshetun, A., Li, X., Mendeleeva, L., et al. (2019). Atlas of RNA sequencing profiles for normal human tissues. *Scientific Data*, *6*(1), 36. https://doi.org/10.1038/s41597-019-0043-4.

Sîrbu, A., Kerr, G., Crane, M., & Ruskin, H. J. (2012). RNA-Seq vs dual- and single-channel microarray data: Sensitivity analysis for differential expression and clustering. *PLoS One*, 7(12), e50986. https://doi.org/10.1371/journal.pone.0050986.

Tao, Z., Shi, A., Li, R., Wang, Y., Wang, X., & Zhao, J. (2017). Microarray bioinformatics in cancer—A review. *Journal of B.U.ON.*, *22*, 838–843. Zerbinis Publications. Retrieved from https://europepmc.org/article/med/29155508.

Tarca, A. L., Draghici, S., Khatri, P., Hassan, S. S., Mittal, P., Kim, J.-S. S., et al. (2009). A novel signaling pathway impact analysis. *Bioinformatics*, *25*(1), 75–82. https://doi.org/10.1093/bioinformatics/btn577.

Teumer, A., Schurmann, C., Schillert, A., Schramm, K., Ziegler, A., & Prokisch, H. (2016). Analyzing illumina gene expression microarray data obtained from human whole blood cell and blood monocyte samples. In *Vol. 1368. Methods in molecular biology* (pp. 85–97). Humana Press Inc. https://doi.org/10.1007/978-1-4939-3136-1_7.

Thomas, S., & Bonchev, D. (2010). A survey of current software for network analysis in molecular biology. *Human Genomics*, *4*(5), 353–360. https://doi.org/10.1186/1479-7364-4-5-353.

Tian, L., Greenberg, S. A., Kong, S. W., Altschuler, J., Kohane, I. S., & Park, P. J. (2005). Discovering statistically significant pathways in expression profiling studies. *Proceedings of the National Academy of Sciences of the United States of America*, *102*(38), 13544–13549. https://doi.org/10.1073/pnas.0506577102.

Tkachev, V., Sorokin, M., Mescheryakov, A., Simonov, A., Garazha, A., Buzdin, A., et al. (2018). FLOating-window projective separator (FloWPS): A data trimming tool for support vector machines (SVM) to improve robustness of the classifier. *Frontiers in Genetics*, *9*, 717. https://doi.org/10.3389/fgene.2018.00717.

Tkachev, V., Sorokin, M., Mescheryakov, A., Simonov, A., Garazha, A., Buzdin, A., et al. (2019). Floating-window projective separator (FLOWPS): A data trimming tool for support vector machines (SVM) to improve robustness of the classifier. *Frontiers in Genetics*, *10*(JAN), 717. https://doi.org/10.3389/fgene.2018.00717.

Tkachev, V., Sorokin, M., Borisov, C., Garazha, A., Buzdin, A., & Borisov, N. (2020). Flexible data trimming improves performance of global machine learning methods in omics- based personalized oncology. *International Journal of Molecular Sciences*, *21*(3), 713. https://doi.org/10.3390/ijms21030713.

Tkachev, V., Sorokin, M., Garazha, A., Borisov, N., & Buzdin, A. (2020). Oncobox method for scoring efficiencies of anticancer drugs based on gene expression data. *Methods in molecular biology, Vol. 2063*, 235–255. Humana Press Inc https://doi.org/10.1007/978-1-0716-0138-9_17.

Turgeon, M.-O., Perry, N. J. S., & Poulogiannis, G. (2018). DNA damage, repair, and cancer metabolism. *Frontiers in Oncology, 8*(FEB), 15. https://doi.org/10.3389/fonc.2018.00015.

Venkova, L., Aliper, A., Suntsova, M., Kholodenko, R., Shepelin, D., Borisov, N., et al. (2015). Combinatorial high-throughput experimental and bioinformatic approach identifies molecular pathways linked with the sensitivity to anticancer target drugs. *Oncotarget, 6*(29), 27227–27238. https://doi.org/10.18632/oncotarget.4507.

Vermeulen, K., Van Bockstaele, D. R., & Berneman, Z. N. (2003). The cell cycle: A review of regulation, deregulation and therapeutic targets in cancer. *Cell Proliferation, 36*, 131–149. https://doi.org/10.1046/j.1365-2184.2003.00266.x.

Vidal, M., Cusick, M. E., & Barabási, A.-L. (2011). Interactome networks and human disease. *Cell, 144*(6), 986–998. https://doi.org/10.1016/j.cell.2011.02.016.

Vladimirova, U., Rumiantsev, P., Zolotovskaia, M., Albert, E., Abrosimov, A., Slashchuk, K., et al. (2021). *DNA repair pathway activation features in follicular and papillary thyroid tumors, interrogated using 95 experimental RNA sequencing profiles*. Heliyon. in press. Retrieved from https://ya.ru/.

Wang, Y., Mashock, M., Tong, Z., Mu, X., Chen, H., Zhou, X., et al. (2020). Changing technologies of RNA sequencing and their applications in clinical oncology. *Frontiers in Oncology, 10*, 447. Frontiers Media S.A. https://doi.org/10.3389/fonc.2020.00447.

Watson, A., Mazumder, A., Stewart, M., & Balasubramanian, S. (1998). Technology for microarray analysis of gene expression. *Current Opinion in Biotechnology, 9*(6), 609–614. https://doi.org/10.1016/S0958-1669(98)80138-9.

Webber, J., Stone, T. C., Katilius, E., Smith, B. C., Gordon, B., Mason, M. D., et al. (2014). Proteomics analysis of cancer exosomes using a novel modified aptamer-based array (somascantm) platform. *Molecular and Cellular Proteomics, 13*(4), 1050–1064. https://doi.org/10.1074/mcp.M113.032136.

Weinstein, J. N., Collisson, E. A., Mills, G. B., Shaw, K. R. M. M., Ozenberger, B. A., Ellrott, K., et al. (2013). The cancer genome atlas pan-cancer analysis project. *Nature Genetics, 45*, 1113–1120. Nature Publishing Group.

Wen, Z., Wang, C., Shi, Q., Huang, Y., Su, Z., Hong, H., et al. (2010). Evaluation of gene expression data generated from expired Affymetrix GeneChip® microarrays using MAQC reference RNA samples. *BMC Bioinformatics, 11*(S6), S10. https://doi.org/10.1186/1471-2105-11-s6-s10.

Whittaker, S., Marais, R., & Zhu, A. X. (2010). The role of signaling pathways in the development and treatment of hepatocellular carcinoma. *Oncogene, 29*(36), 4989–5005. https://doi.org/10.1038/onc.2010.236.

Willier, S., Butt, E., & Grunewald, T. G. P. (2013). Lysophosphatidic acid (LPA) signalling in cell migration and cancer invasion: A focussed review and analysis of LPA receptor gene expression on the basis of more than 1700 cancer microarrays. *Biology of the Cell, 105*(8), 317–333. https://doi.org/10.1111/boc.201300011.

Wirsching, A., Melloul, E., Lezhnina, K., Buzdin, A. A., Ogunshola, O. O., Borger, P., et al. (2017). Temporary portal vein embolization is as efficient as permanent portal vein embolization in mice. *Surgery, 162*(1), 68–81. https://doi.org/10.1016/j.surg.2017.01.032.

Wishart, D. S., Mandal, R., Stanislaus, A., & Ramirez-Gaona, M. (2016). Cancer metabolomics and the human metabolome database. *Metabolites, 6*, 10. MDPI AG https://doi.org/10.3390/metabo6010010.

Wishart, D. S., Li, C., Marcu, A., Badran, H., Pon, A., Budinski, Z., et al. (2020). PathBank: A comprehensive pathway database for model organisms. *Nucleic Acids Research*, *48*(D1), D470–D478. https://doi.org/10.1093/nar/gkz861.

Witt, M., Walter, J.-G., & Stahl, F. (2015). Aptamer microarrays—Current status and future prospects. *Microarrays*, *4*(2), 115–132. https://doi.org/10.3390/microarrays4020115.

Wolber, P. K., Collins, P. J., Lucas, A. B., De Witte, A., & Shannon, K. W. (2006). The agilent in situ-synthesized microarray platform. *Methods in Enzymology*, *410*, 28–57. https://doi.org/10.1016/S0076-6879(06)10002-6.

Workman, R. E., Tang, A. D., Tang, P. S., Jain, M., Tyson, J. R., Razaghi, R., et al. (2019). Nanopore native RNA sequencing of a human poly(A) transcriptome. *Nature Methods*, *16*(12), 1297–1305. https://doi.org/10.1038/s41592-019-0617-2.

Wu, J., & Starr, S. (2014). Low-fidelity compensatory backup alternative DNA repair pathways may unify current carcinogenesis theories. *Future Oncology*, *10*, 1239–1253. Future Medicine Ltd https://doi.org/10.2217/fon.13.272.

Xu, X. D., Shao, S. X., Jiang, H. P., Cao, Y. W., Wang, Y. H., Yang, X. C., et al. (2015). Warburg effect or reverse Warburg effect? A review of cancer metabolism. *Oncology Research and Treatment*, *38*(3), 117–122. https://doi.org/10.1159/000375435.

Yang, W., Freeman, M. R., & Kyprianou, N. (2018). Personalization of prostate cancer therapy through phosphoproteomics. *Nature Reviews Urology*, *15*, 483–497. Nature Publishing Group https://doi.org/10.1038/s41585-018-0014-0.

Yang, K. C., Sathiyaseelan, P., Ho, C., & Gorski, S. M. (2018). Evolution of tools and methods for monitoring autophagic flux in mammalian cells. *Biochemical Society Transactions*, *46*, 97–110. Portland Press Ltd https://doi.org/10.1042/BST20170102.

Zamore, P. D., & Haley, B. (2005). Ribo-gnome: The big world of small RNAs. *Science*, *309*, 1519–1524. https://doi.org/10.1126/science.1111444.

Zhang, Z., Bast, R. C., Yu, Y., Li, J., Sokoll, L. J., Rai, A. J., et al. (2004). Three biomarkers identified from serum proteomic analysis for the detection of early stage ovarian cancer. *Cancer Research*, *64*(16), 5882–5890. https://doi.org/10.1158/0008-5472.CAN-04-0746.

Zhang, L., Zhang, J., Yang, G., Wu, D., Jiang, L., Wen, Z., et al. (2013). Investigating the concordance of Gene Ontology terms reveals the intra- and inter-platform reproducibility of enrichment analysis. *BMC Bioinformatics*, *14*, 143. https://doi.org/10.1186/1471-2105-14-143.

Zhang, W., Yu, Y., Hertwig, F., Thierry-Mieg, J., Zhang, W., Thierry-Mieg, D., et al. (2015). Comparison of RNA-seq and microarray-based models for clinical endpoint prediction. *Genome Biology*, *16*(1), 133. https://doi.org/10.1186/s13059-015-0694-1.

Zhang, B., Whiteaker, J. R., Hoofnagle, A. N., Baird, G. S., Rodland, K. D., & Paulovich, A. G. (2019). Clinical potential of mass spectrometry-based proteogenomics. *Nature Reviews. Clinical Oncology*, *16*, 256–268. Nature Publishing Group https://doi.org/10.1038/s41571-018-0135-7.

Zhao, Y., Yu, P., Wu, R., Ge, Y., Wu, J., Zhu, J., et al. (2013). Renal cell carcinoma-adjacent tissues enhance mobilization and recruitment of endothelial progenitor cells to promote the invasion of the neoplasm. *Biomedicine and Pharmacotherapy*, *67*(7), 643–649. https://doi.org/10.1016/j.biopha.2013.06.009.

Zhao, J., Qin, B., Nikolay, R., Spahn, C. M. T., & Zhang, G. (2019). Translatomics: The global view of translation. *International Journal of Molecular Sciences*, *20*, 212. MDPI AG https://doi.org/10.3390/ijms20010212.

Zhavoronkov, A., Buzdin, A. A., Garazha, A. V., Borisov, N. M., & Moskalev, A. A. (2014). Signaling pathway cloud regulation for in silico screening and ranking of the potential geroprotective drugs. *Frontiers in Genetics*, *5*(MAR), 49. https://doi.org/10.3389/fgene.2014.00049.

Zheng, J., Yu, H., Zhou, A., Wu, B., Liu, J., Jia, Y., et al. (2020). It takes two to tango: Coupling of Hippo pathway and redox signaling in biological process. *Cell Cycle*, 1–16. https://doi.org/10.1080/15384101.2020.1824448.

Zhou, L., Li, Q., Wang, J., Huang, C., & Nice, E. C. (2016). Oncoproteomics: Trials and tribulations. *Proteomics—Clinical Applications*, *10*, 516–531. Wiley-VCH Verlag https://doi.org/10.1002/prca.201500081.

Zhou, R., Liu, D., Zhu, J., & Zhang, T. (2020). Common gene signatures and key pathways in hypopharyngeal and esophageal squamous cell carcinoma: Evidence from bioinformatic analysis. *Medicine*, *99*(42), e22434. https://doi.org/10.1097/MD.0000000000022434.

Zhu, Q., Izumchenko, E., Aliper, A. M., Makarev, E., Paz, K., Buzdin, A. A., et al. (2015). Pathway activation strength is a novel independent prognostic biomarker for cetuximab sensitivity in colorectal cancer patients. *Human Genome Variation*, *2*(1), 15009. https://doi.org/10.1038/hgv.2015.9.

Zolotovskaia, M. A., Sorokin, M. I., Roumiantsev, S. A., Borisov, N. M., & Buzdin, A. A. (2019). Pathway instability is an effective new mutation-based type of cancer biomarkers. *Frontiers in Oncology*, *9*(JAN). https://doi.org/10.3389/fonc.2018.00658.

Zolotovskaia, M. A., Sorokin, M. I., Emelianova, A. A., Borisov, N. M., Kuzmin, D. V., Borger, P., et al. (2019). Pathway based analysis of mutation data is efficient for scoring target cancer drugs. *Frontiers in Pharmacology*, *9*(JAN). https://doi.org/10.3389/fphar.2019.00001.

Zolotovskaia, M. A., Sorokin, M. I., Petrov, I. V., Poddubskaya, E. V., Moiseev, A. A., Sekacheva, M. I., et al. (2020). Disparity between inter-patient molecular heterogeneity and repertoires of target drugs used for different types of cancer in clinical oncology. *International Journal of Molecular Sciences*, *21*(5), 1–18. https://doi.org/10.3390/IJMS21051580.

Zolotovskaia, M., Sorokin, M., Garazha, A., Borisov, N., & Buzdin, A. (2020). Molecular pathway analysis of mutation data for biomarkers discovery and scoring of target cancer drugs. *Methods in Molecular Biology (Clifton, N.J.)*, *2063*, 207–234. https://doi.org/10.1007/978-1-0716-0138-9_16.

Zolotovskaia, M. A., Tkachev, V. S., Seryakov, A. P., Kuzmin, D. V., Kamashev, D. E., Sorokin, M. I., et al. (2020). Mutation enrichment and transcriptomic activation signatures of 419 molecular pathways in cancer. *Cancers*, *12*(2), 271. https://doi.org/10.3390/cancers12020271.

CHAPTER TWO

Proteomics in systems toxicology

Carolina Madeira and Pedro M. Costa*
UCIBIO—Applied Molecular Biosciences Unit, Departamento de Ciências da Vida, Faculdade de Ciências e Tecnologia da Universidade Nova de Lisboa, Caparica, Portugal
*Corresponding author: e-mail address: pmcosta@fct.unl.pt

Contents

Abstract

Proteins are the ultimate product of gene expression. As they hinge between gene transcription and phenotype, they offer a more realistic perspective of toxicopathic effects, responses and even susceptibility to insult than targeting genes and mRNAs while dodging some inter-individual variability that hinders measuring downstream endpoints like metabolites or enzyme activity. Toxicologists have long focused on proteins as biomarkers but the advent of proteomics shifted risk assessment from narrow single-endpoint analyses to whole-proteome screening, enabling deriving protein-centric adverse outcome pathways (AOPs), which are pivotal for the derivation of Systems Biology informally named Systems Toxicology. Especially if coupled pathology, the identification of molecular initiating events (MIEs) and AOPs allow predictive modeling of toxicological pathways, which now stands as the frontier for the next generation of toxicologists. Advances in mass spectrometry, bioinformatics, protein databases and top-down proteomics create new opportunities for mechanistic and effects-oriented research in all fields, from ecotoxicology to pharmacotoxicology.

Advances in Protein Chemistry and Structural Biology, Volume 127
ISSN 1876-1623
https://doi.org/10.1016/bs.apcsb.2021.03.001

Abbreviations

2DE	two-dimensional gel electrophoresis
AO	adverse outcome
AOP	adverse outcome pathway
CID	collision-induced decay
DIA	data-independent acquisition
ECP	electron capture dissociation
ELISA	enzyme-linked immunosorbent assay
ERA	environmental risk assessment
FISH	*in situ* fluorescent hybridization
GE	gene enrichment
GO	gene ontology
HCS	high content screening
HMRS	high resolution mass spectrometry
HPLC	high-performance liquid chromatography
IHC	immunohistochemistry
iTRAQ	isobaric tag for relative and absolute quantitation
KE	key event
KEGG	Kyoto Encyclopedia of Genes and Genomes
LC	liquid chromatography
MALDI-TOF-MS	matrix-assisted laser desorption/ionization time-of-flight mass spectrometry
MIE	molecular initiating event
MS	mass spectrometry
MS/MS	tandem mass spectrometry
NM	nanomaterial
qPCR	quantitative polymerase chain reaction
SILAC	stable isotope labeling with amino-acids in cell culture
SWATH-MS	sequential window acquisition of all theoretical mass spectra
TMT	tandem mass tag

1. The proteome as toxicological endpoint

1.1 Proteoform as unifying concept and the rise of proteomics in toxicology

It is nowadays clear that the number of genes cannot account for the complexity of subcellular machinery, which means that the variability of molecules that are the cogs of every biological pathway falls upon the proteome, which, as the suffix "ome" implies, refers to the *globality* of proteins in a given living system. This issue led Smith and Kelleher (2013) to propose the term "proteoform" to describe all the different protein forms that results from the expression of a single gene as an alternative to "isoform" and "variant," which often have ambiguous or inconsistent meanings. Despite the obvious challenges of characterizing all proteoforms arising from a single gene, the proteome offers a more realistic overview to changes in metabolic pathways caused by toxicological challenge than the genome or even the transcriptome, as it is positioned downstream in the gene expression cascade and therefore more closely associated to phenotype and pathology. It is then not surprising that toxicologists long targeted changes in protein expression, modification and activity following chemical insult. Another important advantage of focusing on proteins to study toxicological effects (or responses), with particular respect to proteomics, is that a vast number of diverse proteins is secreted into bodily fluids of animals in response to external and internal changes (Bandara & Kennedy, 2002), enabling simple, non-invasive and ethical sample harvesting, which has evident implications for human toxicopharmacology and occupational toxicology (Hu, Loo, & Wong, 2006).

The rise of "omics" methods paved the way for a radical change in understanding toxicity and risk. Instead of the narrow scope of single-endpoint or few-endpoint approaches that had until then essentially directed toward effects, with the real possibility of producing negative or inconclusive results simply by missing the main target; the ability to retrieve multiple endpoints from the same subject in a single run enabled scientist to focus on entire metabolic networks. With this rose the possibility of constructing realistic and predictive models for risk assessment, since metabolic pathways and all their consequences downstream involve an intricate interplay between multiple enzymes and the molecular mechanisms that regulate their expression, metabolites and other substances that constitute the exposome. Evidently, it includes also all of the molecular and biochemical machinery that enables the cell to be a viable entity, such as energy production or protein folding.

Toxicology was thus pushed into the realm of Systems Biology, a concept first introduced two decades ago, in the aftermath of the Human Genome Project, envisaging that mathematical models that explain and predict the functioning of complex living systems can be produced by integrating data resulting from the monitoring of these systems after manipulation at various levels of biological organization (Ideker, Galitski, & Hood, 2001). Toxicologists, who are by definition accustomed to chemically perturb biological systems, swiftly responded to the call for a new paradigm that appeared to be a perfect match with the aims and scope of their field of research, giving birth to a discipline informally referred to as Systems Toxicology; which can be simply defined as the quantitative analysis of wide molecular networks and their functions impacted by toxicological insult (the reader is here redirected to the review by Sturla et al., 2014).

Quite naturally, researchers that were accustomed to analyze the expression (or activity) of proteins as biomarkers, from metallothioneins to cytochrome P4501A (CYP1A), just to quote a few classic examples, rapidly acknowledged the benefits of screening the proteome for global changes in protein expression. These apply whether the purpose was the discovery of novel biomarkers (the sub-individual changes that indicate the existence of toxicological challenge); a more comprehensive analysis of effects (what happens following toxicological challenge), determine susceptibility to intoxication or mechanistic toxicology (how do toxicological agents exert their effects and how do biological systems cope with challenge). It was no surprise, then, that the combination between toxicopathology and proteomics in living organisms or cell cultures quickly gave rise to "toxicoproteomics," an applied domain of proteomics that has been conceptualized almost 20 years ago (see for instance Wetmore & Merrick, 2004). Toxicoproteomics is now put together with genomics, transcriptomics, epigenomics metabolomics to form the wider domain of "toxicogenomics," which can be defined as a scientific domain dedicated to the study of toxicant-induced changes to genes and all products of their expression, regardless if gene-, transcript-, metabolite- or protein-centric (see for instance Costa & Fadeel, 2016; Liu, Huang, Roberts, & Tong, 2019; Martins, Dreij, & Costa, 2019; Waters & Fostel, 2004).

The application of toxicogenomics in virtually every subdomain of toxicology, from ecotoxicology to nanotoxicology and toxicopharmacology, and in a wide range of *in vitro* and *in vivo* systems, has been a major player in the evolution of analytical and computational methods associated with omics approaches. Transcriptomics and proteomics, pushed forward by microarray technology and mass spectrometry, respectively, pioneered toxicogenomics and, by far, still form the core of Systems Toxicology. Microarrays, the first true transcriptomic approach, were undisputedly on the forefront on toxicogenomics at its genesis, as no other method could provide a measure of expression for up to thousands of DNA probes, each pertaining to a single gene for a single biological sample. In fact, the first reviews on the subject date back to the late 1990s and virtually considered microarray technology to have given birth to toxicogenomics (e.g., Nuwaysir, Bittner, Trent, Barrett, & Afshari, 1999). It is undeniable that this technology, which is still in widespread use with much success, offers extraordinary possibilities to researchers who handle human tissue or cell cultures or murines and other conventional models. However, microarray technology is targeted and highly dependent of the degree of genomic annotation. Despite the possibility of obtaining customized cDNA chips from commercial suppliers, the lack of genomic resources for non-conventional model species is indeed a major limitation for many toxicologists, especially ecotoxicologists. On the other hand, mass spectrometry (MS)-based proteomics enables protein identification by peptide homology-matching against databases. As such, *a priori* knowledge on proteins is desirable but not mandatory. Proteomics, albeit far from devoid of limitations related to poor genomic annotation, allowed expanding toxicogenomics toward wildlife, feral organisms and unconventional laboratory models, contributing directly or indirectly to improve the knowledge on the "omes" of these organisms. We may also refer to the work by Ankley et al. (2010) who first introduced the term adverse outcome pathways (AOPs) as a novel framework toward risk assessment in ecotoxicology to replace or at least expand the traditional biomarker concept. The term is nowadays deeply associated with omics and their role in Systems Biology and expanded to virtually every domain of toxicological sciences. It illustrates perfectly the major role played by ecotoxicology in the implementation and development of omics methods.

Even though the impressive advances in proteomics methods, not just in tandem mass spectrometry (MS/MS) but also in protein quantification, now allow the matching of thousands of proteins with more confidence than early endeavors, next-generation RNA sequencing (RNA-Seq), which can analyze the entire transcriptome of a sample in a single run, may again seem to be favoring transcriptomics. Without neglecting the more recent applications of metabolomics and epigenomics, the choice for toxicotranscriptomics or toxicoproteomics, which are still the backbone of toxicogenomics, may result from the balance between the need to obtain the highest number of pieces of the puzzle or a more realist overview of the mechanism behind toxicopathological events. There are, however, other reasons that, at the resent stage, may favor transcriptomics. In a review that focuses on the application of toxicoproteomics in multi-omics studies, an issue that deserves a section of its own later in this chapter, Liang, Martyniuk, and Simmons (2020) noted that issues such as the lack of reproducibility between MS methods or protein databases lagging behind those for genomes and transcriptomes has a significant impact on the toxicologist's choice for the most appropriate omics. This issue is particularly pertinent to risk assessment for regulatory and monitoring purposes, as it implies high consistency between laboratories. However, the same authors again advocate the enormous asset given by the closer proximity of the proteome to the phenotype and therefore to pathophysiology. In this chapter, we will illustrate how proteomics earned its rightful place in Systems Toxicology and how the rapid advances in analytical methods and bioinformatics open new prospects.

1.2 Proteins as ultimate products of gene expression

Regardless of organism, the molecular pathways underlying gene expression are complex. Even in prokaryotes, who represent the summit of molecular efficiency, the way how genes are expressed into fully functional proteins is intricate and evidently interlinked with practically every major metabolic process in the cell. In eukaryotes, which are the effective target of toxicoproteomics, the genomes are infinitely larger (which is linearly correlated to organism complexity, though) and most of it is actually non-protein coding even though non-coding RNAs are now known to be active players in the regulation of expression (see for instance Mattick, 2001; Taft, Pheasant, & Mattick, 2007; for details on this frontier subject). Adding to this, the effective number of protein-coding genes is far from matching the number of proteins that mediate metabolism and structure. In fact, just to consider the human example, there are just over 20,000 protein-coding genes but the number of proteoforms in the human proteome may rise up to billions (see Ponomarenko et al., 2016; Smith & Kelleher, 2013).

In eukaryotes, the existence of introns and exons in protein-encoding genes is one of the primary factors that are responsible for such variation, as pre-mRNAs will endure maturation while still in the nucleus through a process called splicing. Splicing consists of removing introns, which are transcribed just as exons, to produce the final protein-encoding mRNA that has an open reading frame, i.e., a sequence that is translatable by ribosomes into a polypeptide. However, splicing does not always occur in the same way, alternative splicing is a non-random process that leads to multiple versions of a mature mRNA, originating, then, various proteoforms that may or not play different roles in a cell. Post-translational modifications are also responsible for major discrepancies between the proteome, the transcriptome ad the genome, to which we must add all the processes that silence gene expression. These include pre-transcriptional mechanisms such as DNA methylation and histone modification (e.g., acetylation) and post-transcriptional processes involving non-coding RNAs involved in

RNA interference (RNAi). Finally, we must also recall that peptides and proteins can be destroyed, either due to comply with new metabolic demands or as response to events such as misfolding and carbonylation, leading then to ubiquitination to flag damaged proteins to be digested by the proteasome. Either case is likely to occur following exposure to noxious chemicals. Even though it is not the place of this chapter to provide an in-depth review the molecular mechanisms of gene expression (the readers may refer to works such as Kornblihtt et al., 2013, on alternative splicing; or Inobe & Matouschek, 2014, about the ubiquitin-proteasome system), it is critical to understand why only too often there is no match between measuring gene expression at the level of protein and transcript and why proteins mediate genes and phenotype. Gene expression in eukaryotes and its relation the endpoint of omics is illustrated in Fig. 1. In face of the complexity of all the steps leading to a fully functional protein, it is obvious that downstream endpoints are more prone to be affected by variability. One can expect, then that, comparatively to transcriptomics, proteomics is more sensitive to the imbalance of all upstream processes that have an impact on gene expression, from energy production to chromatin methylation (or de-methylation) and post-translational silencing by interference RNAs. Furthermore, the toxicologist must expect issues like DNA breaks and adducts between DNA and toxicants (or their metabolites) that are unresolved by transcription-coupled nucleotide excision repair (TC-NER), therefore stalling or arresting RNA polymerase II during transcriptional elongation, just as well as toxicant-induced protein damage and misfolding, the latter an effect attributed to metal ions that can impact the entire enzymatic machinery of cells (e.g., Oh, Xu, Chong, & Wang, 2021; Sharma, Goloubinoff, & Christen, 2008). Altogether, direct or indirect action of toxicants can have a significant impact on gene expression that is magnified from the level of genome to that of protein.

Although depending on methods, proteomics is directed to post-translation products, whether referring to polypeptide chains, folded proteins of functionally modified proteins. If coupled to methods that allow asserting toxicopathic effects, proteomics can provide a robust overview of the changes induced by insult, therefore validating AOPs and the molecular initiating events (MIEs) upstream. Histopathology and related disciplines such as immunohistochemistry and *in situ* fluorescent hybridization (FISH) targeting specific mRNAs and proteins (e.g., Costa, Caeiro, Vale, Delvalls, & Costa, 2012; Costa et al., 2018; Young, Jackson, & Wyeth, 2020) are acknowledged to provide solid phenotypic anchoring in toxicological studies involving omics, proteomics included, as they provide a

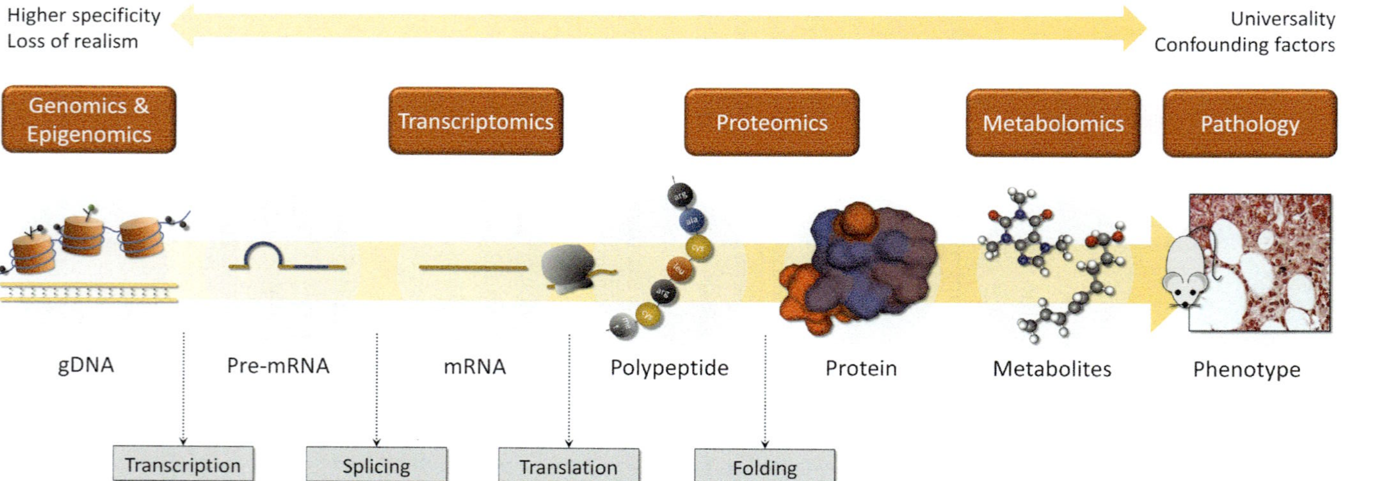

Fig. 1 A simple overview of gene expression in eukaryotes and its relation with omics methods and their analytical targets. Not all mechanisms intervening in the regulation of gene expression are represented (such as interference RNAs), as well as post-translational modifications (like phosphorylation and glycosylation) and eventual degradation of proteins by proteases and the proteasome. However, all these factors have important consequences for omics measurements.

realistic measure of damage. Advances in semi-quantitative and quantitative histopathological practices allow computational integration of data. However, tissue microarray technology, which enable placing hundreds of different tissue samples in a single block for sectioning brought histopathology into high-content screening (HCS) analyses such as automated variants of the Comet assay, oxidative stress biomarkers and other computer-assisted methods fitted for the analyses of a single-endpoint in many individuals or multiplexed biomarkers assays in a single biological sample (Li & Xia, 2019; Ryan, Mulrane, Rexhepaj, & Gallagher, 2011; Watson et al., 2014). These methods, coupled with machine learning and other state-of-the-art bioinformatics are regarded as powerful tools for integration with omics data (Sturla et al., 2014).

1.3 Aims and scope of proteomics in toxicology

Toxicogenomics implies the comparative quantification of amounts of a vast span of molecules between a control or reference and an exposure treatment, which commonly implies assessing statistical differences and imposing regulation thresholds. The *existence* of the molecule *per se* is often of scarce meaning as most proteins are constitutively expressed even if in trace amounts only. In the case of toxicoproteomics, it is this comparative assessment which allows inferring the most significantly up- or down-regulated proteins, i.e., with higher or lower amounts of the proteins in the aftermath of exposure, respectively. These shortlisted proteins have been fundamentally used for three different, but not necessarily independent, purposes: determination of effects and responses, biomarker discovery and mechanistic toxicology.

1.3.1 Proteomics in effects-oriented research

As defined by Paracelsus (1493–1541), the Swiss-German alchemist accredited for the foundation of toxicology, it is only a matter of dose which separates benefit from poison. Quite naturally, toxicologists spend a great deal of their work in trying to figure out what are the critical doses from which a substance becomes harmful and determine the deleterious effects caused by that substance or potentially even the responses to insult. By itself, purely effect-driven proteomics is the continuance of traditional approaches in the sense that focuses on *what* happens and not so much on *how* it happens. This strategy, which is mostly descriptive and not hypothesis-driven, is still applied to date involving both chronic and acute exposures. We can find a range of works between that of Bandara, Kelly, Lock, and Kennedy (2003) on the effects of nephrotoxicants in rats to that of Kwon et al. (2020) on the effects and responses of the fish *Oryzias latipes* (medaka) to a surfactant, for instance. The broad range of examples even include safety assessment of nanomaterials (NMs) for biomedical applications, for instance (e.g., Askri et al., 2019; Conde et al., 2014), resulting from the need to comprehensively screen for potentially harmful effects when reduced toxicity must be safeguarded.

Effects-driven research has been providing paramount information on major dysregulated pathways such as defense against oxidative stress, cell cycle, and cell death or energy production, therefore providing a measure of harm endured by target organisms in a far more complete manner than traditional studies with few endpoints. Still, quite often these applications of proteomics include the analyses of more traditional biomarker responses like histopathology, genotoxicity assessment and the activity of toxicologically relevant enzymes of phases I and II of detoxification that provide cross-leverage for the toxicopathological assessment.

1.3.2 Proteomics in the search for biomarkers

A toxicological biomarker can be simply defined as a sub-individual endpoint that provides a measure of exposure, susceptibility or harm of an organism subjected to toxic aggression (van Gestel & van Brummelen, 1996). Without prejudice of important factors such as dose-dependency, biomarkers should thus be expeditious, sensitive, biologically relevant targets ranging from molecules to behavior that hold reasonable specificity against a toxicological agent and be, as much as reasonably possible, universal. This means that biomarker analyses should be reproducible between laboratories and hold both sensitivity and specificity toward a given toxicant (see Amacher, 2010; Benninghoff, 2007). Whereas the advantage of swiftly and conclusively analyzing a single protein as biomarker for a given toxicant is self-explanatory, finding suitable biomarkers for the vast span of chemicals and nano-scaled materials potentially affecting organisms, humans included, is everything but simple. Proteomics directed to effects and responses swiftly led toxicologists to acknowledge that screening the proteome held high potential for biomarker discovery and both strategies rapidly became intertwined. In fact, quite often studies directed to effects provide clues on potential novel biomarkers in their concluding remarks. Nonetheless, the success of this enterprise is debatable not just due to different outcomes between the various subdisciplines among toxicological sciences but also due to the lack of follow-up to validate biomarkers for regulatory purposes (e.g., through inter-laboratory validation). The first reviews on toxicological biomarker discovery were published about to decades ago and tended to focus on pharmacotoxicology. Hale, Gelfanova, Ludwig, and Knierman (2003), for instance, debated that sample quality and protein quantification were paramount for the success of the strategy, which is highly reasonable in a day and age when methods were taking the first steps. In turn, Elrick, Walgren, Mitchell, and Thompson (2006) noted hindrances in analyzing the proteome of human blood plasma in search for biomarkers of disease, for being a particularly intricate matrix, making use of hitherto mainstream methods. In fact, Zhai et al. (2005) isolated five potential peptides or small proteins as potential biomarkers for occupational exposure to As and Pb in the blood plasma of workers but failed to identify them, which may not only reflect the aforesaid constraints but also the incipient stage of protein databases in early years. An interesting early example comes from Gao et al. (2004) who used proteomics on culture of normal human hepatocytes to isolated proteins as general biomarkers of drug-induced hepatotoxicity after testing multiple different approved drugs. The authors isolated two proteins, BMS-PTX-265 and BMS-PTX-837, whose overexpression was found suitable as biomarker of idiosyncratic drug-induced liver disease. An interesting aspect of the work is that it resulted in patents in Europe (EP1636338A2) and in the United States (US7452678B2). Still, it is unclear how the work was effectively turned into an effective commercial product. These highlight the initial focus on the discovery of biomarkers of human disease but they also illustrate the point at which proteomics (and likely other omics) yielded a seemingly disappointing number of effective, universal, toxicological biomarkers to the present day. In fact, despite the many technical advances, the need to give continuance to studies and promote inter-institutional calibration may have been overlooked.

The issues mentioned above pertain to virtually any subdomain of toxicology. Despite early promises about the potential of proteomics in biomarker discovery in ecotoxicology and environmental toxicology (e.g., López-Barea & Gómez-Ariza, 2006; Monsinjon & Knigge, 2007), the much literature on the subject has not yet been materialized in acknowledged endpoints for standard applications in (eco)toxicity testing and biomonitoring. A recent review by López-Pedrouso, Varela, Franco, Fernández, and Aboal (2020) focusing on the application of proteomics in the monitoring of aquatic ecosystems in the preceding decade, emphasized the purpose of this omics in biomarker discovery for environmental risk assessment, especially in the context of dealing with the increasing number of emerging pollutants and handling the wide span of species that best fit the extraordinary diversity of aquatic ecosystems. The authors noted that knowledge on the proteome of important groups of organisms, such as crustaceans, is lagging, just as the effort to validate novel proteins as biomarkers despite good indications resulting from proteomics. Even though the number of ecotoxicology papers reporting the use of proteomics and suggesting certain protein as biomarkers is too extensive to be comprehensively analyzed here, we may indicate a few recent examples to showcase the current trends and advances in the field, such as vitellogenin (Vg), male reproductive tract specific kazal type proteinase inhibitor (MRPINK); sperm gelatinase (MSG) and farnesoic acid *o*-methyltransferase (FAMeT) as potential biomarkers of exposure to an endocrine disrupting insecticide (chlordecone) in decapod crustaceans (Lafontaine et al., 2017), and globins as biomarkers of insecticide exposure in midge (Monteiro, Pestana, Soares, Devreese, & Lemos, 2020). Table 1 briefly illustrates the role of proteomics in biomarker discovery. Many recent works highlighted, nonetheless, that an issue that is certainly compromising biomarker discovery in toxicological sciences gaps in the knowledge on a toxicant's mode-of-action, molecular triggers and placement of the biomarker candidate in the cascade of events triggered by exposure. By simpler words: the lack of mechanistic insights.

1.3.3 Proteomics in mechanistic toxicology

Arguably the most important advantage of any omics is enabling the study of entire molecular pathways. This major leverage is allowed by the sheer number of molecules analyzed from a single sample, potentially providing an overview of full networks of genes and proteins that explain *how* does a substance become toxic or *how* cells respond to insult. This approach neither neglects determining basic effects and responses nor the search for expeditious biomarkers from empirical research. Instead, mechanistics expands them toward more realistic risk assessment strategies. Addressing biological networks means a bigger probability of finding the most specific biomarker or understanding how confounding factors can affect toxicopathological traits. Most importantly, it allows delivering robust models to predict outcomes and therefore estimate risk, which is simply defined as the *probability* of harm, therefore meeting the increasingly higher demands of risk assessment for regulatory purposes.

Rabilloud and Lescuyer (2015) rightfully highlighted that proteomics *per se* is not able to provide a definite measure of hierarchization between proteins, i.e., to separate essential responses to insult from fitness responses

Table 1 Potential protein biomarker proteins of toxicity inferred from proteomics (in single-omics studies).

Potential biomarker protein	Toxicant	Biological matrix	Assay	Organism	Reference
Unidentified peptides or proteins (≈2–5.6 kDa)	As, Pb	Blood serum	*In vivo*	Humans	Zhai et al. (2005)
BMS-PTX-265 BMS-PTX-837	Various pharmacological drugs	Hepatocytes	*In vitro*	Humans	Gao et al. (2004)
Adenylate kinase 2 (AK2)	Radiotherapy	T lymphocytes	*In vivo*	Humans	Lacombe et al. (2019)
Glutathione S-transferase Phosphatidylethanolamine-binding protein Testis-specific heat shock protein 70-2 Glyceraldehyde 3-phosphate dehydrogenase	Model reproductive toxicants	Testes	*In vivo*	Rat	Yamamoto, Fukushima, Kikkawa, Yamada, and Horii (2005)
Peroxiredoxins	Mixed pollutants	Liver	*In vivo*	Senegalese sole	Costa, Caeiro, et al. (2012)
Carbonyl reductase [NADPH] 1 Poly(RC)-binding protein 2 …	Non-mixed organic pollutants: Polychlorinated biphenyl 153 (PCB153) Perfluorononanoic acid (PFNA) Phenanthrene (Phe)	Primary cells (skin, liver, kidney, ovary and small intestine)	*In vitro*	Sea turtle	Chaousis, Leusch, Nouwens, Melvin, and van de Merwe (2021)
Vitellogenin Male reproductive tract specific kazal-type proteinase inhibitor Sperm gelatinase Farnesoic acid *o*-methyltransferase	Chlordecone (organochlorine insecticide)	Hepatopancreas	*In vivo*	Decapod crustacean	Lafontaine et al. (2017)
Globins	Spinosad (natural insecticide)	Whole-body	*In vivo*	Midge	Monteiro et al. (2020)

The list is not intended to be an exhaustive representation of case studies but rather a selection of different representative works according to toxicant and target organism.

(those that are not required for cell survival). The same authors also not that filtering proteins by their levels of expression is helpful but does not tell all about a toxicant's mode-of-action and downstream consequences. Mechanistic toxicology can address this question through solid biological knowledge and adequate bioinformatics. Without prejudice for adequate experimental planning (including the choice of omics and biological model) and phenotypical anchoring or validation, major advances in molecular databases, biostatistics and bioinformatics tools have been responsible for the revolution that renders them a pillar of Systems Toxicology, without which the labor-intensive analysis of massive amount of data commonly produced by omics and needed to derive molecular pathways and networks would be cumbersome if not utterly unfeasible. Essentially, mechanistic toxicology aims not only at finding the key molecules that intervene in a toxicant's mode-of-action but also their interactions. Considering that proteins arguably mediate every metabolic network in a cell, proteomics plays here a direct role to identify the cogs in the network and how they interact. Discovering toxicological mechanisms thus enables deriving toxicant AOPs (Fig. 2). Evidently, empirically derived measurements such as traditional biomarkers (activity of anti-oxidant enzymes, histopathology, etc.) provides important weight-of-evidence to validate and assist the interpretation of data.

The modern toxicologist can now benefit from a growing number of bioinformatics suites for analyzing "big data" in depth, from essential biostatistics to machine learning. For the purpose, the programming language R (Ihaka & Gentleman, 1996), which provides open-source dedicated packages for "omics" data is a tool of choice for bioinformaticians, at least for data processing, basic biostatistics and data mining. There are, however, more appropriate tools to derive molecular networks and pathways from the biological function of molecules, most of which apply to transcriptomics just as well as proteomics because in either case analysis is essentially protein-centric. The reader may consult the work by Wu, Hasan, and Chen (2014) on bioinformatics for analysis of proteomics data. Gene ontology (GO) associates biological functions to proteins but has essentially no topology, or by other words does not directly establish a network of molecular interactions. Online tools such as DAVID (Huang, Sherman, & Lempicki, 2009) became highly popular for the analysis of GO terms in large sets of genes, transcript or proteins for being expeditious, user-friendly and constantly updated. Gene enrichment (GE) goes a step further and allows matching sets of proteins against canonical pathways, making use of GO terms as flags, thus yielding some topology (Subramanian et al., 2005). The Kyoto Encyclopedia of

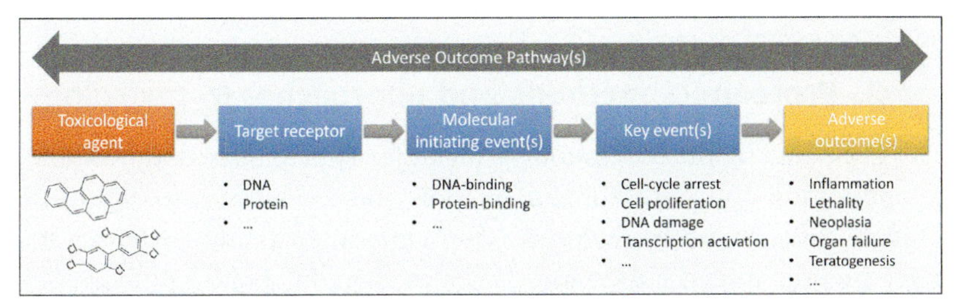

Fig. 2 Main elements that compose an adverse outcome pathway. Mechanistic toxicology identifies and associates molecular initiating events (MIEs) with the key events (KEs) that are responsible for adverse outcomes (AOs). Even if a toxicant holds specificity to a single target, such as a specific enzyme, and triggers a single MIE, e.g., by binding to the enzyme's active site, multiple sequential KEs can be activated, from inhibition of the enzyme's activity (upstream) to general metabolic imbalance (downstream), as examples. The cascade of resulting KEs is then responsible for one or more adverse outcomes that ultimately depend on factors such as susceptibility of the cell and dose- or time-responsiveness.

Genes Genomes (KEGG) databases for proteins and metabolites (Kanehisa et al., 2014) offers a very large array of biological pathways for proteins and metabolites and is also accessible through DAVID. However, tools that are entirely based on topology can derive dynamic protein networks, reveal signaling pathways and are, therefore, of very high value to mechanistic toxicogenomics, even if requiring good *a priori* knowledge on biological networks and a keen eye for statistics. It is the case of suites such as Cytoscape, STRING and Qiagen's Ingenuity Pathway Analysis (IPA), the two first being free, the latter benefitting from being a powerful tool associated to massive databases of molecular interactions, diseases, biological pathways (including toxicological) for a vast pan of *in vivo* and *in vitro* models (e.g., Costa & Fadeel, 2016; Szklarczyk et al., 2019, see also Qin & Zhao, 2014 for an overview of computational tools for molecular networks in human disease). Among bioinformatics tools, BLAST (Altschul, Gish, Miller, Myers, & Lipman, 1990), an old tool that is constantly evolving, must be mentioned for homology-based matching of proteins against databases such as UniProt, GenBank, RefSeq and the Human Reference Interactome Protein Database (see for instance Chen, Huang, & Wu, 2011), plus the resources integrated by the Human Proteome Project (Omenn et al., 2020). The list of available resources for proteomics in the "big data" era is too vast to be comprehensively discussed in this chapter and is continuously expanding.

2. Proteomics methods and approaches in toxicology

Pressured by the continuous demand for expediteness and robustness, the application and development of tools and inventories to address the immense diversity of proteoforms poses experimental and analytical challenges (see for instance Aksenov, Da Silva, Knight, Lopes, & Dorrestein, 2017). In this section we will emphasize the technological and conceptual advancements and requirements that allowed major breakthroughs in proteomics research in toxicology.

2.1 Experimental approaches to toxicoproteomics

Any experimental design must be delineated according to the goals and hypotheses set *a priori*. As an analytical procedure, toxicoproteomics is therefore a means to an end and can be categorized in four different approaches according to the aims of the study: (i) environmental; (ii) *in vivo*; (iii) *in vitro*, and (iv) *in silico*. Environmental approaches aim to establish a link between environmental exposure to pollutants and ecosystem or human health. Understanding how organisms (humans included) modulate protein biosynthesis upon exposure in their environment has major implications in predicting how organisms can cope with pollutants, with direct implications to biomonitoring, mitigation and remediation (López-Pedrouso et al., 2020; Martins et al., 2019; Tomanek, 2011, 2014). Even though the importance of environmental studies is beyond dispute, there are many pitfalls hindering cause-effect relationships due to the inherent complexity of any ecosystem.

Given that the effects of chemicals on biota are driven by their internal concentrations, which in turn is a function of their bioavailability, environmental or ecotoxicological studies are considered to require determining biological toxicant burden (internal dose) as well as the characterization of collection sites, e.g., levels of organic and inorganic toxicants, organic matter loading, temperature, pH, O_2, etc., as a necessary means to establish causation. In addition, Additionally, researchers are growing increasingly aware that endogenous variables like sex and reproductive stage or nutritional status (asides evident differences between target organs and tissues) are also important when subtracting background proteome variability from stress-induced proteome changes (Tomanek, 2011). Nonetheless, one of the major challenges set before environmental omics is that vast majority of the ecologically relevant species are non-model organisms, which results in high intraspecific variation, intricate populational phylogenetics plus incomplete or absent genomic annotation (Gouveia et al., 2019). On the other hand, any large-scale study of environmental proteomes requires high-quality spectral libraries and curated protein databases. Since such resources are virtually non-existing for wildlife, this issue may lead to biased homology-matching (Dowd, 2012) and poor representation of the high molecular divergence acquired by species during evolution, as pointed out by Gouveia et al. (2019). Nevertheless, species such as the mussel *Mytilus* spp., the tunicate *Ciona intestinalis*, micro- and macroalgae and many teleosts besides the acknowledged zebrafish model (e.g., *Takifugu rubripes*, *Pomatoschistus minutus*, salmonids, tilapia and medaka), to which is added the cladoceran *Daphnia*, are rising models in aquatic toxicology. Similarly, non-model plants, insects, annelids (such as the well-known *Eisenia fetida*) and wild amphibians or murines are conquering their space in the monitoring of terrestrial habitats, just to quote a few examples. Quite naturally, environmental toxicoproteomics holds a close association to ecotoxicology and environmental toxicology and does not necessarily imply field studies exclusively, as long as *environmental realism* is somehow safeguarded in the experimental design. With this respect, applying proteomics to determine human exposure to pollutants may also be allocated within environmental proteomics.

In vivo and *in vitro* toxicoproteomics aims at screening proteome-wide changes induced by exposure to toxicological agents (such as novel chemicals, drugs or toxins) to determine their relative potency, regulatory safety and mode-of-action based essentially on dose- and time-response controlled experiments. In many cases, these approaches follow the classic absorption (administration), distribution, metabolism and excretion (ADME) models to assess acute, sub-chronic and chronic toxicity. *In vivo* experiments imply testing toxicological agents in whole living systems, usually plants and non-human animals, including animals humanized by genetic manipulation or chimaeras that contain human cells (Kirchmair et al., 2015). In turn, *in vitro* testing implies using primary or immortalized cell-based or cultured tissue systems derived from prokaryote, mammals (human inclusively), fish, yeast and other models; normal or neoplasic, among which immortalized human cancer cell lines have been widely employed by toxicologists (Yahya, Hashim, Israf Ali, Chau Ling, & Cheema, 2021). Cell cultures can be the classical 2D (single or co-cultures grown in flat monolayers or plates); or 3D, in which cells are allowed to proliferate in any direction by mean of a biopolymer matrix, therefore better mimicking interaction with the extracellular milieu (Kapałczyńska et al., 2018). The latter include recent advances in spheroid and organoid cultures that are promising surrogates for toxicopharmacological testing. Multi-organ *in vitro* testing (i.e., "multi-organs-on-a-chip") is also becoming popular, evidencing the adaptable nature of *in vitro* systems. In these models, multiple organ-like tissues are cultured together on an array of microelectromechanical systems in a serum-free re-circulating medium in a way that resembles the metabolic organ circuitry (McAleer et al., 2019), providing a compromise between *in vivo* and *in vitro* testing. Whereas *In vivo* toxicological yields higher realism with respect to toxicant delivery and toxicopathological outcomes, *in vitro* systems eliminate inter-individual variability, are more expeditious and less subjected to ethical constraints, therefore highly compliant with the "3-R" principle (reduction, refinement and replacement) of laboratory animal use. In fact, *in vitro* testing is expected to become the gold standard in toxicology sciences on the long term. Although *in vitro* approaches cannot replicate the complex dynamics occurring in whole living systems, they may answer specific biological questions. With this respect, *in vitro* toxicoproteomics, due to the downstream positioning of the proteome in gene expression cascade can yield a more realist approach to mechanistic toxicology than transcriptomics or genomics.

In silico approaches are now at the epicenter of Systems Toxicology. These approaches are based on integrative computational models that use mathematical methods, including Boolean and Bayesian networks, ordinary differential equations, and stochastic Gillispie algorithms (Ramos, Martin, Ramos, & Rempala, 2018). Information collected *in vivo* and *in vitro* is mined and integrated with toxicopathological data using mathematics to construct knowledge-based predictive approaches (Hartung, 2009). *In silico* models can therefore be used to answer a variety of questions, including protein folding, interaction and signaling networks without conducting further bioassays (Kimber et al., 2011; Vishnoi, Matre, Garg, & Pandey, 2020). *In silico* modeling has also been useful to expand protein annotations, especially for protein families that exhibit defined and repeatable fragmentation patterns (Aksenov et al., 2017). Most importantly, software tools are available to build predictive models that simulate toxicological outcomes from single-cell level to whole-organism, focusing on both toxicodynamics (i.e., biological outcomes) and toxicokinetics (toxicant uptake, translocation, transformation and elimination). These software suites are now assisted by toxicology-direct databases such as PubChem, DrugBank, CHEMBL, ToxCast, COSMOS, RepDose, HSDB, ECHA or CTD, among many others (see Geenen et al., 2012; Schmidt, 2021 for reviews).

2.2 Analytical and technological approaches to proteome analysis

The standard proteomics workflow can be described as a five-step procedure: (i) protein harvesting; (ii) separation; (iii) mass spectrometry; (iv) computational analyses, and (v) validation. Pilot experiments are also important to determine the minimum number of samples and number of replicates necessary to detect an effect, through power analysis. Most importantly, experimental design, metadata assembly, storage and reporting should be made "MIAPE-compliant" (minimum information about a proteomics experiment) to safeguard reproducibility of results (Taylor et al., 2007).

Samples are collected from fresh or snap-frozen tissue, body fluids or cell cultures and proteins are extracted in appropriate buffers through physical disruption. It is important to bear in mind that cytoplasmatic (more water-soluble) and cell-membrane proteins (more hydrophobic) require different protocols for extraction. More hydrophobic proteins may be solubilized with chaotropic agents like urea or thiourea, surfactants and detergents such deoxycholate or triton (see Suder, Novák, Havlíček, & Bodzoń-Kułakowska, 2016). The addition of protease inhibitors can also be an important factor to preserve protein integrity.

Protein mixtures may be difficult to analyze, as more abundant proteoforms will overshadow rare proteins. For such reason, fractionation can reduce sample complexity. The choice of method largely depends of the aims of the study and evolved considerably over the last decade. One of the pioneering methods is 2D-dimensional gel electrophoresis (2DE), which is apt for untargeted proteomics and separates proteins based on their isoelectric point and molecular mass (Diniz, Madeira, & Araújo, 2017). After gel staining, image analysis can detect proteome-wide differentially regulated proteins between samples. The shortlisted "spots" are then excised from gels, trypsin-digested and subjected to MS. In targeted proteomics, a predefined set of identified proteins of interest are selected and monitored for accurate quantification. This requires the use of affinity-based or activity-based enrichment or isolation techniques, which employ chemically engineered probes (like antibodies) to capture proteins of interest (Borràs & Sabidó, 2017). Fractionation techniques, especially gel-based, are now being substituted by shotgun proteomics that directly analyses complex protein mixtures, which is made possible by advances in MS and protein databases. Most proteomics still employs *bottom-up* strategies, which means that proteins are broken down (digested) before ionization of fragments, e.g., using electrospray (ESI) or matrix-assisted laser desorption/ionization (MALDI), then referred to as precursor ions, which is required before MS. In contrast, in *top-down* approaches, proteins are not digested. Instead, they are ionized and fragmented, normally through collision-induced decay (CID) or electron capture dissociation (ECP), before MS. Top-down approaches retain the original polypeptidic sequence of the protein, rendering identification more accurate enable determining with precision the sies bearing post-translational modifications (Kar, Simonian, & Whitelegge, 2017). Fig. 3 illustrates the two approaches.

Qualitative proteomics aims at providing high-confidence high-throughput protein identification and assessment of post-translational modifications, whereas quantitative proteomics targets peptide or protein abundances. These approaches are complementary and rely on technical

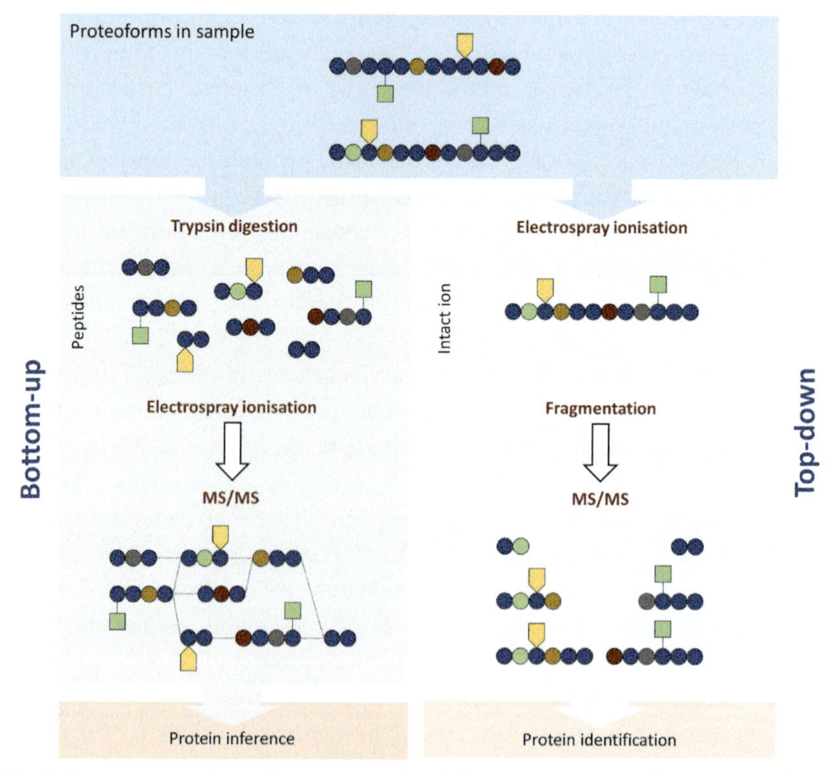

Fig. 3 Bottom-up and top-down proteomics. In bottom-up approaches, protein identification is made by matching peptidic sequences resulting from digestion, which does not immediately inform on their positioning in the protein, which may hinder the identification (and quantification) of proteoforms in the ample. Conversely, top-down retains structural detail, enabling more accurate matching and characterization of post-translational modifications.

advances in instrumentation and computation. Mass spectrometry is, however, the core of proteomics an is based on the measuring of mass-to-charge-ratios (*m/z*) of ions. Liquid chromatography coupled with tandem mass spectrometry (LC-MS/MS) is now considered the technique that provides greater dynamic range, higher versatility and efficiency (Liang et al., 2020). At this stage it is important to highlight the advantages that make tandem MS a required methodology over simple MS. Tandem MS, which consists of serial mass detectors, allow selecting precursor ions for peptides with a specific *m/z* to be fragmented and analyzed by a second mass analyzer. Consequently, precision and sensitivity are increased, with the additional advantage that samples do not have to be highly purified before analysis, therefore propelling shotgun proteomics. This represents a major innovation in comparison to older proteomics analyses based on simple MS, in which protein identification was based on peptide mass fingerprinting alone . Typically, in peptide mass fingerprinting, the mass of unknown proteins was determined using matching algorithms based on *in silico* protein digestion from a protein sequence database, yielding theoretical peptide masses that are compared and matched to real peptide masses from the samples. In contrast, in tandem MS, the fragmentation spectra and peptide masses are analyzed based on *in silico* fragmentation of known proteins and *de novo* sequencing, then matching amino-acid masses to fragment peaks in the chromatogram to reconstruct peptide sequences (e.g., Tran, Zhang, Xin, Shan, & Li, 2017). These *in silico* procedures use matching algorithms such as Mascot and protein databases (like UniProt) resorting to tools like Mascot Distiller and Rapid Denovo to perform protein identification (see Frank, Savitski, Nielsen, Zubarev, & Pevzner, 2007; Rajawat & Jhingan, 2019).

Determining differential protein expression (i.e., up- or-down regulation relatively to a control or reference) is a benchmark of toxicoproteomics. Protein quantification can be derived by MS from *m/z* ratios, as these are traced over time to build a quantifiable peak. For robust quantification, it is ideal to use a standard or calibrator to overcome uncontrolled sources of variation. Labeling-based methods are common alternatives for quantification, such as stable isotope labeling with amino-acids in cell culture (SILAC), which involves labeling with, e.g., ^{13}C and $^{1}5N$ and is especially suitable for *in vitro* experiments, as cell cultures can be grown in media enriched with radiolabeled amino-acids (Ong et al., 2002). Another common approach is chemical labeling using isobaric tags for relative and absolute quantification (iTRAQ), or tandem mass tags (TMT), which involve pre-labeling of protein extracts for quantification before MS (Lu et al., 2020). As alternative to label-free MS quantification, labeling is considered to offer higher reproducibility. However, modern label-free methods like sequential window acquisition of all theoretical mass spectra (SWATH-MS), a variation of data-independent acquisition (DIA), yield highly reliable and consistent quantification (Chen, Zhang, Fernie, Liu, & Zhu, 2020). Evidently, quantification of expression regardless if absolute or relative requires solid statistical analyses that require particular assumptions due to the large amount of proteoforms under scrutiny, such as the use of generalized linear models and methods for normalization and *p*-adjustment. It is worthwhile to check the paper by Gatto and Christoforou (2014) and references therein on R code, packages and statistics for proteomics. Finally, a word must be given to the importance of validating proteomics data (as for any omics). Immunolabeling methods such as Western Blot, enzyme-linked immunosorbent assay (ELISA) and even immunohistochemistry (IHC), due to their high specificity, are methods of choice to ascertain the quality of protein identification and quantification by cross-analyzing a few proteins of choice. Multi-omics studies may offer excellent cross-validation for omics and will be discussed in a separate section.

3. Applications

3.1 Ecotoxicology and environmental toxicology

Even though the distinction between ecotoxicology (more ecology-oriented) and environmental toxicology (more directed toward human health) is not entirely consensual, particularly under the current "One Health" perspective rightfully linking environmental and human health (see Mackenzie & Jeggo, 2019), it is acknowledged that they cojoin the hindrances and challenges of environmental science and toxicology. Perhaps surprisingly, ecotoxicologists and environmental toxicologists were not only among the pioneers for the application of omics approaches but also arguably contributed for major methodological advances. A practical reason for this was the need to retrieve multiple endpoints from organisms with reduced, if not absent, genomic resources. Until the rise of next-generation RNA sequencing, MS-based proteomics was more compelling for being non-targeted, even if yielding a far more reduced number of molecules. Another important reason for the success of "omics," proteomics in particular, within ecotoxicology and related areas relates to the natural complexity of the milieu in which humans and wildlife dwell. This issue has been addressed in the recent review by Gouveia et al. (2019), who coined the term "ecotoxicoproteomics" to ascribe the purpose and value of proteomics in ecotoxicology. As innumerable endogenous (biological) and exogenous (environmental) factors often leads to inconclusive or unexpected results from traditional biomarker analyses conducted in field (*in situ*) or laboratory (*ex situ*) studies, turning the determination of cause-effect relationships unachievable. The problem is made worse by the fact that environmental toxicants, natural or anthropogenic seldom occur alone. Addressing the effects of pollutants thus welcomes more integrative, holistic, approaches.

Most studies in the field involving proteomics (or other omics, in fact) have been more effects-oriented than mechanistic. In large part this strategy results from the need to determine when contamination (i.e., the levels of toxicants rise above background levels) turns into pollution (when the levels of toxicants begin inducing harm) when traditional endpoints fail to do so (see Chapman, 2007, for disambiguation of terms). Naturally, ecotoxicology should aim at ecologically relevant scenarios, from realistic doses of pollutants in laboratory assays to, most importantly, representative wildlife species for the testing of chemicals and for active (translocation) or passive (direct sampling) biomonitoring purposes. Among the wide span of biological models, we can find vertebrates (especially fish, amphibians and even wild murines), molluscs, crustaceans, annelids and plants. Due to the impossibility of covering all research in the field, we will illustrate the accomplishment with a few representative works.

In a study with juvenile Senegalese sole exposed *in situ* and in the laboratory to estuarine sediments moderately contaminated by organic and inorganic (metallic) pollutants, Costa, Caeiro, et al. (2012) and Costa, Chicano-Gálvez, et al. (2012), MS/MS-based proteomics following protein separation and regulation analysis by 2DE revealed an association between important anti-oxidant enzymes such as peroxiredoxins with genotoxicity and histopathological biomarkers, having found about 40 differentially regulated cytosolic proteins between the multiple experimental conditions. These same works also disclosed very significant changes between the metabolic profiles of fish exposed *in situ* and *ex situ*, the latter enduring down-regulation of enzymes associated to energy production and gene transcription, for instance, with serious implications for testing strategies in ERA. Proteomics has also been applied to terrestrial ecosystems for the purpose of ERA. It is the example of the work by Montes-Nieto et al. (2007) with wild *Mus spretus* collected from clean and polluted areas in Southern Spain, who used 2DE and MALDI-TOF-MS to disclose about 40 differentially expressed cytosolic proteins that revealed greater investment in anti-oxidant responses in animals collected from impacted sites. Albeit vastly outnumbered by research on animal models (*in vivo* and *in vitro*), plants (meaning microalgae to phanerogams) are excellent ecotoxicological models and bioindicator organisms, being resistant to many environmental toxicants (such as metals) and sharing well-conserved metabolic pathways for which there are excellent proteomic resources deriving from other areas of research (see for the reviews by Ceschin, Bellini, & Scalici, 2021; Kosakivska, Babenko, Romanenko, Korotka, & Potters, 2021; Singh et al., 2020; Wang, 2019). For example, Terzi and Yıldız (2021), endeavored a step beyond traditional effects-oriented research to study amelioration by cysteine in maize seedlings challenged in the laboratory with Cr using 2DE and MALDI-TOF-MS-based proteomics plus protein network analyses using bioinformatics complemented with a battery of single-endpoint biomarkers to provide phenotypic anchoring (e.g., oxidative stress). A less usual case study is that of Yu, Yin, Peng, Lu, and Dang (2020), who performed proteomics based on high resolution mass spectrometry (HRMS) and iTRAQ on a prokaryote (*Microbacterium*) common in contaminated aquatic sediments to verify its ability to degrade a brominated flame retardant.

Pushed forward by advances in MS methods, protein databases and bioinformatics, recent ecotoxicological research is venturing through mechanistics. An example pertaining to the deployment of proteomics in the survey for specific AOPs in ecologically relevant organisms is the work by Liu et al. (2021), who tested cladocerans with polystyrene nanoplastics and analyzed the digested and high-performance liquid chromatography (HPLC)-separated peptides from whole-body lysates by LC-MS/MS. The authors disclosed that more than 300 were differentially regulated by exposure and used a suite of bioinformatics tools to explore molecular pathways, e.g., by addressing protein-protein interactions. The results indicated oxidative stress as MIE, affecting several metabolic pathways downstream, some crustacean-specific (like molting), others more related to basal metabolic aspects, altogether yielding hindered growth and reproduction appear as ultimate AOPs. An interesting aspect of this work is not only the declared approach to AOPs but also the incorporation of modern concepts in toxicological studies involving proteomics, from validation by qPCR to modern bioinformatics and phenotypic anchoring *via* analysis of the activity to toxicity-relevant enzymes. This interesting example illustrates how much toxicoproteomics evolved in the last decade and that ecotoxicologists still occupy the frontline.

Among endogenous variables modulating the effects and responses to insult, Liang, Feswick, Simmons, and Martyniuk (2018) pertinently alerted that sex tends to be a neglected variable in ecotoxicological studies involve proteome profiling, whereas it is an acknowledged variable in human pharmacotoxicology even in non-gonad tissue and primary cell cultures. In fact, gender and maturation stage are nowadays required variables in biomonitoring and toxicity testing endeavors, which means that there is an important gap that must be filled within toxicoproteomics. There are however, a few indications that a substantial percentage of proteins of non-gonad tissue are differentially expressed between male and female ecotoxicologically relevant organisms. It is the case of the work by Viitaniemi and Leder (2011), who performed label-free proteomics based on LC-ESI-MS/MS on the livers of three-spine stickleback (*Gasterosteus aculeatus*) in a non-toxicological context, revealing then that ≈6% of the hepatic proteome was differentially expressed between males and females, mostly pertaining to the metabolism of aminoacids and proteins. Also of notice is the recent review by Soler and Oswald (2018) that alerts for a similar issue regarding the proteomics signature of organisms exposed to mycotoxins, which are ubiquitous natural toxins of fungal origin highly toxic to humans and other animals often found in food products. Altogether, these examples illustrate the need to isolate the effects of endogenous variables when interpreting omics data, with emphasis on proteomics as the expression of genes into full proteins is certainly more biased by confounding factors than transcripts.

3.2 Pharmacology and human toxicology

Major advances in the analysis of protein and peptides in fluids like blood plasma (see Geyer et al., 2016), expanded toxicoproteomics to human occupational exposure, disease and pharmacology, even though the distinction to environmental toxicology is often feeble. However, much research makes use of *in vitro* and *in vivo* surrogates research with acknowledged model organisms (from murines to zebrafish) for safety, risk assessment and biomarker discovery. Here we will highlight a few representative works.

The application of proteomics for the risk assessment of nanomaterials has been delivering many interesting case studies since the nanotoxicology boom in the last decade. As a recent subdomain within toxicology bearing its particular set of challenges for essentially focusing on a class of toxicological agent hinging between chemical and material and due to the relevance of many NMs for biomedical applications, nanotoxicology and nanosafety swiftly embraced the Systems Biology approach. Still, as in other realms of toxicology, it is arguable how much research has been able to effectively construct quantitative (or even qualitative) models that explain and predict the perturbation of biological systems by noxious agents (see Costa & Fadeel, 2016, 2018; Fadeel, 2015, and references therein). We may refer to recent works such as that of Juling et al. (2018), who combined 2DE-based proteomics with hepatic histopathology to address the toxicity of Ag nanoparticles to rats exposed *via* gavage for 28 days. An important aspect of this study is the use of IPA software suite mentioned in a previous section to identify several key events triggered by exposure involving, for instance, oxidative stress and metal homeostasis. Billing et al. (2020) used LC-MS/MS-based proteomics to study in the mouse model the mechanistics underlying toxicity of magnetic cobalt ferrite nanoparticles (NPs), due to the high interest in magnetic NPs as drug delivery agents in novel therapeutics (especially anti-cancer). This through work involved proteomics on bronchioalveolar lavage fluid (BALF), yielding almost 700 differentially regulated proteins. The authors used several state-of-the art bioinformatics suites to derive comprehensive information on toxicological pathways, protein networks and on the molecular mechanisms of neutrophil extracellular trap (NET) formation. These examples illustrate quite well some of the most fundamental principles of Systems Biology and the high relevance of proteomics in risk assessment and mechanistic studies.

Despite the growing awareness for environmentally induced neoplasic disease, associating cancer and exposure to pollutants has always been challenging and prone to controversy. In the past few decades, toxicoproteomics has often been deployed to address this issue either in a mechanistic or effects-oriented perspective. The reader may refer, for instance, to the review by George and Shukla (2011) on the application of proteomics in studies associating human cancer and exposure to pesticides. As an example, using shotgun proteomics based on LC-MS/MS Pizzatti et al. (2020) studied changes to the proteome in almost 250 breast cancer patients and its potential relation to chronic (occupational) exposure to pesticides, determined through questionnaires. The authors disclosed a relation between the de-regulation of proteins (more than 500 in total) related to several pathways, with emphasis on immune response and estrogen receptors, in patients that had been exposed to pesticides, then discussing a link between the down-regulation of such pathways with poor prognostics as consequence of exposure. Still within studies with more direct implication to human disease, Everson and Marsit (2018) noted that even though epigenomics has been the approach of choice to investigate pre-natal exposure to hazardous chemicals, other omics, proteomics included, isolated or in combination, have been producing solid outcomes to investigate the effects and mechanisms of exposure during critical developmental stages. Among other recent examples we may include the study by Antoniassi et al. (2020) on the effect of smoking on varicocele (a form of abnormal testicular vasodilation) based on proteomics performed on semen.

3.3 Integration of proteomics in multi-omics toxicology

The advantages of multi-omics studies have been amply debated within and outside toxicological sciences and mostly relate to the ability of scaffolding data from one omics with data from another. This feature yields not just cross-validation but also the possibility to fill-in gaps while providing a more complete outlook of molecular and metabolic pathways, which may be particularly relevant in complex scenarios such as those involving mixtures of toxicants (see for instance the reviews by Costa & Fadeel, 2016; Martins et al., 2019). Once more, ecotoxicologists have taken their place in the front row from the start, such as Dondero et al. (2010), which combined microarray and 2DE-based proteomics to study the effects of mixed insecticides in mussels. Computational approaches for the integration of multiple omics data have greatly evolved in the last decade, though.

Jiang et al. (2021) combined proteomics (using TMT and LC-MS/MS) and untargeted metabolomics in the serum of mice to study the toxicity of the synthetic opioid painkiller tramadol. The authors used bioinformatics to KEGG databases for proteins and metabolites to contrast common canonical pathways derived from the two omics, yielding then protein metabolism as the major pathway affected by the drug. Another interesting recent example that indicates future applications of proteomics is the work by Xu et al. (2021), who combined proteomics and metabolomics to discover the molecular (protein) targets of a phthalate (xenobiotics included in the rank of substances of emerging concern) in HepG2 cells. The authors, who classify their approach as "interactomics" used a suite of modern bioinformatics tools to integrate both omics with cell cycle regulation (as the substance promotes cell cycle arrest at G1 stage) and isolated three proteins associated with cell division and chromosomes (CPEB4, ANAPC5 and SPOUT1) that should be directly targeted by the pollutant. We can also add the work by Bannuscher et al. (2020), who integrated TMT-based proteomics and targeted HPLC-based metabolomics to study the specific effects and modes-of-action of seven different NMs onto a rat alveolar macrophage culture, highlighting, among other aspects, the consistency between the two omics. These three examples highlight a successful trend to integrate proteomics and metabolomics as the two omics that more realistically associate with toxicopathological effects. Titz et al. (2020) went further and combined iTRAQ-based proteomics with metabolomics, lipidomics and microarray-based transcriptomics to compare the toxicity of tobacco smoke vs heated tobacco in the mouse model. The authors used a wide array of bioinformatics tools, among which is included an R package for multi-omics factor analysis (MOFA), a novel statistical approach (represented in Fig. 4)

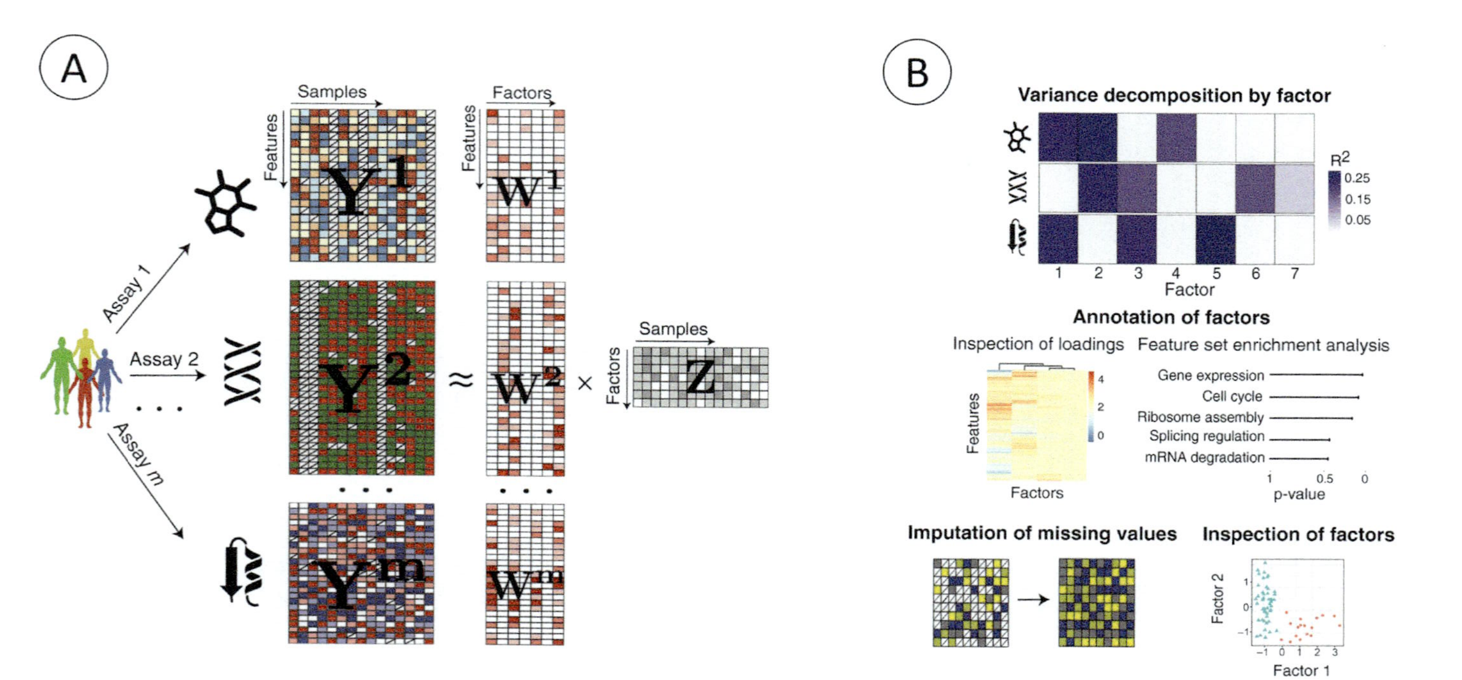

Fig. 4 Multi-omics factor analysis. (A) Model overview. The various matrices are expected to pertain do the same samples but may differ in length (number of "features"). (B) Illustration of possible downstream analyses. *Reprinted from Argelaguet, R., Velten, B., Arnol, D., Dietrich, S., Zenz, T., Marioni, J.C., et al. (2018). Multi-omics factor analysis—A framework for unsupervised integration of multi-omics data sets,* Molecular Systems Biology, 14, *e8124. http://doi.org/10.15252/msb.20178124 under a Creative Commons CC BY 4.0 license.*

based on Bayesian methods specifically conceived to integrate and analyze multi-omics data (Argelaguet et al., 2018).

4. Take-home lessons

Despite the technological advances in other omics, especially transcriptomics, which leaped from the targeted microarray analyses to the astonishing potential of RNA-Seq, which is untargeted and able to quantify the expression of many tens of thousands of transcripts in single runs, proteomics has an immense value in toxicogenomics. As proteins are positioned downstream in the gene expression cascade, they provide a more realistic proximity to true toxicopathic effects than mRNAs while yielding less interindividual variability than surveying metabolites or enzyme activities. Associated to the ability of analyzing proteins in complex biological matrices, extraordinary advances in protein detection and quantification from early 2DE-based approaches to modern label-free bottom-up or top-down methods are clearing new grounds toward a far more in-depth understanding and cataloging of functional proteoforms than could be expected two decades ago, when Systems Toxicology was in its début. With a substantial push from ecotoxicologists and environmental toxicologists, who endure greater challenges for working with unconventional biological models, toxicoproteomics very significantly contributed to the development of analytical tools, bioinformatics and databases that now form the core of Systems Biology. It must be noted, however, that most research claiming to be under the scope of Systems Biology falls short of producing quantitative (and predictive) models of changes induced by toxicants. Putting it simple, *omics alone to not turn any toxicological assessment into an application of Systems Toxicology*. Systems toxicology requires full knowledge on AOPs, establishing dose- and time-responsiveness and the ability to produce valid models that can estimate probabilities of occurring disturbance to biological systems. Altogether, when opting in or out for proteomics researcher should consider the following aspects:

Assets

- Closer relation to pathological traits, offering a good compromise between specificity and susceptibility to confounding factors.
- Proteins can be easily harvested from a variety of biological matrices, peripheral fluids inclusively, enabling non-invasive sampling.
- Provides excellent cross-validation and scaffolding with other omics and high-throughput methods, regardless of position in the gene expression cascade.

- Benefits from a growing investment in databases specifically designed for proteins.

Pitfalls

- Protein extraction and preservation is cumbersome, especially from intricate biological samples.
- Reduced output compared to next-gen DNA and RNA sequencing.
- Protein modifications are not identifiable by common bottom-up approaches.
- The high number of proteoforms makes protein identification and interpretation of results highly challenging.

From biomarker discovery to mechanistic toxicology, proteomics has gone a long way in risk assessment of pollutants, toxins, drugs and nanomaterials and it is far from having said its final word in what is now called the "post-genomic era." Recent toxicological applications of multi-omics have been demonstrating the value of proteomics in cross-validation and scaffolding with various other omics, leading to a much better fine-tuning of dysregulated molecular networks and far more sensitive computational models. As in our rapidly advancing society the number of new potentially harmful chemicals is rising at an alarming rate, toxicoproteomics will play in the future a major role in environmental and human health, now indisputably joined under the One Health paradigm.

Acknowledgments

The authors acknowledge the Portuguese Foundation for Science and Technology (FCT) for the grant CEECIND/01526/2018 to C.M. and for supporting the Applied Molecular Biosciences Unit-UCIBIO (UIDP/04378/2020 and UIDB/04378/2020).

References

Aksenov, A. A., Da Silva, R., Knight, R., Lopes, N. P., & Dorrestein, P. C. (2017). Global chemical analysis of biology by mass spectrometry. *Nature Reviews Chemistry*, *1*, 1–20. https://doi.org/10.1038/s41570-017-0054.

Altschul, S. F., Gish, W., Miller, W., Myers, E. W., & Lipman, D. J. (1990). Basic local alignment search tool. *Journal of Molecular Biology*, *215*, 403–410. https://doi.org/10.1016/S0022-2836(05)80360-2.

Amacher, D. E. (2010). The discovery and development of proteomic safety biomarkers for the detection of drug-induced liver toxicity. *Toxicology and Applied Pharmacology*, *245*, 134–142. https://doi.org/10.1016/j.taap.2010.02.011.

Ankley, G. T., Bennett, R. S., Erickson, R. J., Hoff, D. J., Hornung, M. W., Johnson, R. D., et al. (2010). Adverse outcome pathways: A conceptual framework to support ecotoxicology research and risk assessment. *Environmental Toxicology and Chemistry*, *29*, 730–741. https://doi.org/10.1002/etc.34.

Antoniassi, M. P., Belardin, L. B., Camargo, M., Intasqui, P., Carvalho, V. M., Cardozo, K. H. M., et al. (2020). Seminal plasma protein networks and enriched functions in varicocele: Effect of smoking. *Andrologia*, *52*, e13562. https://doi.org/10.1111/and.13562.

Argelaguet, R., Velten, B., Arnol, D., Dietrich, S., Zenz, T., Marioni, J. C., et al. (2018). Multi-omics factor analysis—A framework for unsupervised integration of multi-omics data sets. *Molecular Systems Biology*, *14*, e8124. https://doi.org/10.15252/msb.20178124.

Askri, D., Cunin, V., Ouni, S., Béal, D., Rachidi, W., Sakly, M., et al. (2019). Effects of iron oxide nanoparticles (γ-Fe2O3) on liver, lung and brain proteomes following sub-acute intranasal exposure: A new toxicological assessment in rat model using iTRAQ-based quantitative proteomics. *International Journal of Molecular Sciences*, *20*, 5186. https://doi.org/10.3390/ijms20205186.

Bandara, L. R., Kelly, M. D., Lock, E. A., & Kennedy, S. (2003). A correlation between a proteomic evaluation and conventional measurements in the assessment of renal proximal tubular toxicity. *Toxicological Sciences*, *73*, 195–206. https://doi.org/10.1093/toxsci/kfg068.

Bandara, L. R., & Kennedy, S. (2002). Toxicoproteomics—A new preclinical tool. *Drug Discovery Today*, *7*, 411–418. https://doi.org/10.1016/S1359-6446(02)02211-0.

Bannuscher, A., Karkossa, I., Buhs, S., Nollau, P., Kettler, K., Balas, M., et al. (2020). A multi-omics approach reveals mechanisms of nanomaterial toxicity and structure-activity relationships in alveolar macrophages. *Nanotoxicology*, *14*, 181–195. https://doi.org/10.1080/17435390.2019.1684592.

Benninghoff, A. D. (2007). Toxicoproteomics—The next step in the evolution of environmental biomarkers? *Toxicological Sciences*, *95*, 1–4. https://doi.org/10.1093/toxsci/kfl157.

Billing, A. M., Knudsen, K. B., Chetwynd, A. J., Ellis, L. A., Tang, S. V. Y., Berthing, T., et al. (2020). Fast and robust proteome screening platform identifies neutrophil extracellular trap formation in the lung in response to cobalt ferrite nanoparticles. *ACS Nano*, *14*, 4096–4110. https://doi.org/10.1021/acsnano.9b08818.

Borràs, E., & Sabidó, E. (2017). What is targeted proteomics? A concise revision of targeted acquisition and targeted data analysis in mass spectrometry. *Proteomics*, *17*, 1–13. https://doi.org/10.1002/pmic.201700180.

Ceschin, S., Bellini, A., & Scalici, M. (2021). Aquatic plants and ecotoxicological assessment in freshwater ecosystems: A review. *Environmental Science and Pollution Research*, *28*, 4975–4988. https://doi.org/10.1007/s11356-020-11496-3.

Chaousis, S., Leusch, F. D. L., Nouwens, A., Melvin, S. D., & van de Merwe, J. P. (2021). Changes in global protein expression in sea turtle cells exposed to common contaminants indicates new biomarkers of chemical exposure. *The Science of the Total Environment*, *751*, 141680. https://doi.org/10.1016/j.scitotenv.2020.141680.

Chapman, P. M. (2007). Determining when contamination is pollution—Weight of evidence determinations for sediments and effluents. *Environment International*, *33*, 492–501. https://doi.org/10.1016/j.envint.2006.09.001.

Chen, C., Huang, H., & Wu, C. H. (2011). Protein bioinformatics databases and resources. *Methods in Molecular Biology*, *694*, 3–24. https://doi.org/10.1007/978-1-60761-977-2_1.

Chen, M. X., Zhang, Y., Fernie, A. R., Liu, Y. G., & Zhu, F. Y. (2020). SWATH-MS-based proteomics: Strategies and applications in plants. *Trends in Biotechnology*. https://doi.org/10.1016/j.tibtech.2020.09.002. in press.

Conde, J., Larguinho, M., Cordeiro, A., Raposo, L. R., Costa, P. M., Santo, S., et al. (2014). Gold-nanobeacons for gene therapy: Evaluation of genotoxicity, cell toxicity and proteome profiling analysis. *Nanotoxicology*, *8*, 521–532. https://doi.org/10.3109/17435390.2013.802821.

Costa, P. M., Caeiro, S., Vale, C., Delvalls, T.À., & Costa, M. H. (2012). Can the integration of multiple biomarkers and sediment geochemistry aid solving the complexity of sediment risk assessment? A case study with a benthic fish. *Environmental Pollution, 2012*(161), 107–120. https://doi.org/10.1016/j.envpol.2011.10.010.

Costa, P. M., Chicano-Gálvez, E., Caeiro, S., Lobo, J., Martins, M., Ferreira, A. M., et al. (2012). Hepatic proteome changes in *Solea senegalensis* exposed to contaminated estuarine sediments: A laboratory and *in situ* survey. *Ecotoxicology, 21*, 1194–1207. https://doi.org/10.1007/s10646-012-0874-7.

Costa, P. M., & Fadeel, B. (2016). Emerging systems biology approaches in nanotoxicology: Towards a mechanism-based understanding of nanomaterial hazard and risk. *Toxicology and Applied Pharmacology, 299*, 101–111. https://doi.org/10.1016/j.taap.2015.12.014.

Costa, P. M., & Fadeel, B. (2018). Emerging systems toxicology approaches in nanosafety assessment. In A. Dhawan, D. Anderson, & R. Shanker (Eds.), *Nanotoxicology: Experimental and computational perspectives* (pp. 174–202). Cambridge, UK: Royal Society of Chemistry. https://doi.org/10.1039/9781782623922-00174.

Costa, P. M., Gosens, I., Williams, A., Farcal, L., Pantano, D., Brown, D. M., et al. (2018). Transcriptional profiling reveals gene expression changes associated with inflammation and cell proliferation following short-term inhalation exposure to copper oxide nanoparticles. *Journal of Applied Toxicology, 38*, 385–397. https://doi.org/10.1002/jat.3548.

Diniz, M. S., Madeira, D., & Araújo, J. E. (2017). Effects of climate change in marine organisms: A proteomic approach. In T. G. Barrera, & J. L. G. Ariza (Eds.), *Environmental problems in marine biology: Methodological aspects and applications* (pp. 190–212). Boca Raton, FL, USA: CRC Press.

Dondero, F., Negri, A., Boatti, L., Marsano, F., Mignone, F., & Viarengo, A. (2010). Transcriptomic and proteomic effects of a neonicotinoid insecticide mixture in the marine mussel (*Mytilus galloprovincialis*, Lam.). *The Science of the Total Environment, 408*, 3775–3786. https://doi.org/10.1016/j.scitotenv.2010.03.040.

Dowd, W. W. (2012). Challenges for biological interpretation of environmental proteomics data in non-model organisms. *Integrative and Comparative Biology, 52*, 705–720. https://doi.org/10.1093/icb/ics093.

Elrick, M. M., Walgren, J. L., Mitchell, M. D., & Thompson, D. C. (2006). Proteomics: Recent applications and new technologies. *Basic & Clinical Pharmacology & Toxicology, 98*, 432–441. https://doi.org/10.1111/j.1742-7843.2006.pto_391.x.

Everson, T. M., & Marsit, C. J. (2018). Integrating-omics approaches into human population-based studies of prenatal and early-life exposures. *Current Environmental Health Reports, 5*, 328–337. https://doi.org/10.1007/s40572-018-0204-1.

Fadeel, B. (2015). Systems biology in nanosafety research. *Nanomedicine, 10*, 1039–1041. https://doi.org/10.2217/nnm.15.17.

Frank, A. M., Savitski, M. M., Nielsen, M. N., Zubarev, R. A., & Pevzner, P. A. (2007). De novo peptide sequencing and identification with precision mass spectrometry. *Journal of Proteome Research, 6*, 1–23. https://doi.org/10.1021/pr060271u.De.

Gao, J., Garulacan, L. A., Storm, S. M., Hefta, S. A., Opiteck, G. J., Lin, J. H., et al. (2004). Identification of *in vitro* protein biomarkers of idiosyncratic liver toxicity. *Toxicology in Vitro, 18*, 533–541. https://doi.org/10.1016/j.tiv.2004.01.012.

Gatto, L., & Christoforou, A. (2014). Using R and bioconductor for proteomics data analysis. *Biochimica et Biophysica Acta, 1844*(1A), 42–51. https://doi.org/10.1016/j.bbapap.2013.04.032.

Geenen, S., Taylor, P. N., Snoep, J. L., Wilson, I. D., Kenna, J. G., & Westerhoff, H. V. (2012). Systems biology tools for toxicology. *Archives of Toxicology, 86*, 1251–1271. https://doi.org/10.1007/s00204-012-0857-8.

George, J., & Shukla, Y. (2011). Pesticides and cancer: Insights into toxicoproteomic-based findings. *Journal of Proteomics, 74*, 2713–2722. https://doi.org/10.1016/j.jprot.2011.09.024.

Geyer, P. E., Kulak, N. A., Pichler, G., Holdt, L. M., Teupser, D., & Mann, M. (2016). Plasma proteome profiling to assess human health and disease. *Cell Systems, 2*, 185–195. https://doi.org/10.1016/j.cels.2016.02.015.

Gouveia, D., Almunia, C., Cogne, Y., Pible, O., Degli-Esposti, D., Salvador, A., et al. (2019). Ecotoxicoproteomics: A decade of progress in our understanding of anthropogenic impact on the environment. *Journal of Proteomics, 198*, 66–77. https://doi.org/10.1016/j.jprot.2018.12.001.

Hale, J. E., Gelfanova, V., Ludwig, J. R., & Knierman, M. D. (2003). Application of proteomics for discovery of protein biomarkers. *Briefings in Functional Genomics & Proteomics, 2*, 185–193. https://doi.org/10.1093/bfgp/2.3.185.

Hartung, T. (2009). Toxicology for the twenty-first century. *Nature, 460*, 208–212. https://doi.org/10.1038/460208a.

Hu, S., Loo, J. A., & Wong, D. T. (2006). Human body fluid proteome analysis. *Proteomics, 6*, 6326–6365. https://doi.org/10.1002/pmic.200600284.Human.

Huang, D. W., Sherman, B. T., & Lempicki, R. A. (2009). Systematic and integrative analysis of large gene lists using DAVID bioinformatics resources. *Nature Protocols, 4*, 44–57. https://doi.org/10.1038/nprot.2008.211.

Ideker, T., Galitski, T., & Hood, L. (2001). A new approach to decoding life: Systems biology. *Annual Review of Genomics and Human Genetics, 2*, 343–372. https://doi.org/10.1146/annurev.genom.2.1.343.

Ihaka, R., & Gentleman, R. (1996). R: A language for data analysis and graphics. *Journal of Computational and Graphical Statistics, 5*, 299–314. https://doi.org/10.1080/10618600.1996.10474713.

Inobe, T., & Matouschek, A. (2014). Paradigms of protein degradation by the proteasome. *Current Opinion in Structural Biology, 24*, 156–164. https://doi.org/10.1016/j.sbi.2014.02.002.

Jiang, S., Liu, G., Yuan, H., Xu, E., Xia, W., Zhang, X., et al. (2021). Changes on proteomic and metabolomic profile in serum of mice induced by chronic exposure to tramadol. *Scientific Reports, 14*, 1454. https://doi.org/10.1038/s41598-021-81109-7.

Juling, S., Böhmert, L., Lichtenstein, D., Oberemm, A., Creutzenberg, O., Thünemann, A. F., et al. (2018). Comparative proteomic analysis of hepatic effects induced by nanosilver, silver ions and nanoparticle coating in rats. *Food and Chemical Toxicology, 113*, 255–266. https://doi.org/10.1016/j.fct.2018.01.056.

Kanehisa, M., Goto, S., Sato, Y., Kawashima, M., Furumichi, M., & Tanabe, M. (2014). Data, information, knowledge and principle: Back to metabolism in KEGG. *Nucleic Acids Research, 42*, D199–D205. https://doi.org/10.1093/nar/gkt1076.

Kapałczyńska, M., Kolenda, T., Przybyła, W., Zajączkowska, M., Teresiak, A., Filas, V., et al. (2018). 2D and 3D cell cultures—A comparison of different types of cancer cell cultures. *Archives of Medical Science, 14*, 910–919. https://doi.org/10.5114/aoms.2016.63743.

Kar, U. K., Simonian, M., & Whitelegge, J. P. (2017). Integral membrane proteins: Bottom-up, top-down and structural proteomics. *Expert Review of Proteomics, 14*, 715–723. https://doi.org/10.1080/14789450.2017.1359545.

Kimber, I., Humphris, C., Westmoreland, C., Alepee, N., Negro, G. D., & Manou, I. (2011). Computational chemistry, systems biology and toxicology. Harnessing the chemistry of life: Revolutionizing toxicology. A commentary. *Journal of Applied Toxicology, 31*, 206–209. https://doi.org/10.1002/jat.1666.

Kirchmair, J., Göller, A. H., Lang, D., Kunze, J., Testa, B., Wilson, I. D., et al. (2015). Predicting drug metabolism: Experiment and/or computation? *Nature Reviews. Drug Discovery*, *14*, 387–404. https://doi.org/10.1038/nrd4581.

Kornblihtt, A. R., Schor, I. E., Alló, M., Dujardin, G., Petrillo, E., & Muñoz, M. J. (2013). Alternative splicing: A pivotal step between eukaryotic transcription and translation. *Nature Reviews. Molecular Cell Biology*, *14*, 153–165. https://doi.org/10.1038/nrm3525.

Kosakivska, I. V., Babenko, L. M., Romanenko, K. O., Korotka, I. Y., & Potters, G. (2021). Molecular mechanisms of plant adaptive responses to heavy metals stress. *Cell Biology International*, *45*, 258–272. https://doi.org/10.1002/cbin.11503.

Kwon, Y. S., Jung, J. W., Kim, Y. J., Park, C. B., Shon, J. C., Kim, J. H., et al. (2020). Proteomic analysis of whole-body responses in medaka (*Oryzias latipes*) exposed to benzalkonium chloride. *Journal of Environmental Science and Health. Part A, Toxic/Hazardous Substances & Environmental Engineering*, *55*, 1387–1397. https://doi.org/10.1080/10934529.2020.1796117.

Lacombe, J., Brengues, M., Mangé, A., Bourgier, C., Gourgou, S., Pèlegrin, A., et al. (2019). Quantitative proteomic analysis reveals AK2 as potential biomarker for late normal tissue radiotoxicity. *Radiation Oncology*, *14*, 142. https://doi.org/10.1186/s13014-019-1351-8.

Lafontaine, A., Baiwir, D., Joaquim-Justo, C., De Pauw, E., Lemoine, S., Boulangé-Lecomte, C., et al. (2017). Proteomic response of *Macrobrachium rosenbergii* hepatopancreas exposed to chlordecone: Identification of endocrine disruption biomarkers? *Ecotoxicology and Environmental Safety*, *141*, 306–314. https://doi.org/10.1016/j.ecoenv.2017.03.043.

Li, S., & Xia, M. (2019). Review of high-content screening applications in toxicology. *Archives of Toxicology*, *93*, 3387–3396. https://doi.org/10.1007/s00204-019-02593-5.

Liang, X., Feswick, A., Simmons, D., & Martyniuk, C. J. (2018). Environmental toxicology and omics: A question of sex. *Journal of Proteomics*, *172*, 152–164. https://doi.org/10.1016/j.jprot.2017.09.010.

Liang, X., Martyniuk, C. J., & Simmons, D. B. D. (2020). Are we forgetting the "proteomics" in multi-omics ecotoxicology? *Comparative Biochemistry and Physiology Part D: Genomics and Proteomics*, *36*, 100751. https://doi.org/10.1016/j.cbd.2020.100751.

Liu, Z., Huang, R., Roberts, R., & Tong, W. (2019). Toxicogenomics: A 2020 vision. *Trends in Pharmacological Sciences*, *40*, 92–103. https://doi.org/10.1016/j.tips.2018.12.001.

Liu, Z., Li, Y., Sepúlveda, M. S., Jiang, Q., Jiao, Y., Chen, Q., et al. (2021). Development of an adverse outcome pathway for nanoplastic toxicity in *Daphnia pulex* using proteomics. *The Science of the Total Environment*, *766*, 144249. https://doi.org/10.1016/j.scitotenv.2020.144249.

López-Barea, J., & Gómez-Ariza, J. L. (2006). Environmental proteomics and metallomics. *Proteomics*, *6*, S51–S62. https://doi.org/10.1002/pmic.200500374.

López-Pedrouso, M., Varela, Z., Franco, D., Fernández, J. A., & Aboal, J. R. (2020). Can proteomics contribute to biomonitoring of aquatic pollution? A critical review. *Environmental Pollution*, *267*, 115473. https://doi.org/10.1016/j.envpol.2020.115473.

Lu, Z., Wang, S., Ji, C., Li, F., Cong, M., Shan, X., et al. (2020). iTRAQ-based proteomic analysis on the mitochondrial responses in gill tissues of juvenile olive flounder *Paralichthys olivaceus* exposed to cadmium. *Environmental Pollution*, *257*, 113591. https://doi.org/10.1016/j.envpol.2019.113591.

Mackenzie, J. S., & Jeggo, M. (2019). The one health approach—Why is it so important? *Tropical Medicine and Infectious Disease*, *4*, 88. https://doi.org/10.3390/tropicalmed4020088.

Martins, C., Dreij, K., & Costa, P. M. (2019). The state-of-the art of environmental toxicogenomics: Challenges and perspectives of "omics" approaches directed to toxicant mixtures. *International Journal of Environmental Research and Public Health*, *16*, 4718. https://doi.org/10.3390/ijerph16234718.

Mattick, J. S. (2001). Non-coding RNAs: The architects of eukaryotic complexity. *EMBO Reports*, *2*, 986–991. https://doi.org/10.1093/embo-reports/kve230.

McAleer, C. W., Long, C. J., Elbrecht, D., Sasserath, T., Bridges, L. R., Rumsey, J. W., et al. (2019). Multi-organ system for the evaluation of efficacy and off-target toxicity of anticancer therapeutics. *Science Translational Medicine*, *11*, eaav1386. https://doi.org/10.1126/scitranslmed.aav1386.

Monsinjon, T., & Knigge, T. (2007). Proteomic applications in ecotoxicology. *Proteomics*, *7*, 2997–3009. https://doi.org/10.1002/pmic.200700101.

Monteiro, H. R., Pestana, J. L. T., Soares, A. M. V. M., Devreese, B., & Lemos, M. F. L. (2020). *Chironomus riparius* proteome responses to spinosad exposure. *Toxics*, *8*, 117. https://doi.org/10.3390/toxics8040117.

Montes-Nieto, R., Fuentes-Almagro, C. A., Bonilla-Valverde, D., Prieto-Alamo, M. J., Jurado, J., Carrascal, M., et al. (2007). Proteomics in free-living *Mus spretus* to monitor terrestrial ecosystems. *Proteomics*, *7*, 4376–4387. https://doi.org/10.1002/pmic.200700409.

Nuwaysir, E. F., Bittner, M., Trent, J., Barrett, J. C., & Afshari, C. A. (1999). Microarrays and toxicology: The advent of toxicogenomics. *Molecular Carcinogenesis*, *23*, 153–159. https://doi.org/10.1002/(sici)1098-2744(199903)24:3<153::aid-mc1>3.0.co;2-p.

Oh, J., Xu, J., Chong, J., & Wang, D. (2021). Molecular basis of transcriptional pausing, stalling, and transcription-coupled repair initiation. *Biochimica et Biophysica Acta, Gene Regulatory Mechanisms*, *1864*, 194659. https://doi.org/10.1016/j.bbagrm.2020.194659.

Omenn, G. S., Lane, L., Overall, C. M., Cristea, I. M., Corrales, F. J., Lindskog, C., et al. (2020). Research on the human proteome reaches a major milestone: >90% of predicted human proteins now credibly detected, according to the HUPO Human Proteome Project. *Journal of Proteome Research*, *19*, 4735–4746. https://doi.org/10.1021/acs.jproteome.0c00485.

Ong, S. E., Blagoev, B., Kratchmarova, I., Kristensen, D. B., Steen, H., Pandey, A., et al. (2002). Stable isotope labelling by amino acids in cell culture, SILAC, as a simple and accurate approach to expression proteomics. *Molecular & Cellular Proteomics*, *1*, 376–386. https://doi.org/10.1074/mcp.M200025-MCP200.

Pizzatti, L., Kawassaki, A. C. B., Fadel, B., Nogueira, F. C. S., Evaristo, J. A. M., Woldmar, N., et al. (2020). Toxicoproteomics disclose pesticides as downregulators of TNF-α, IL-1β and estrogen receptor pathways in breast cancer women chronically exposed. *Frontiers in Oncology*, *10*, 1698. https://doi.org/10.3389/fonc.2020.01698.

Ponomarenko, E. A., Poverennaya, E. V., Ilgisonis, E. V., Pyatnitskiy, M. A., Kopylov, A. T., Zgoda, V. G., et al. (2016). The size of the human proteome: The width and depth. *International Journal of Analytical Chemistry*, *2016*, 7436849. https://doi.org/10.1155/2016/7436849.

Qin, G., & Zhao, X. M. (2014). A survey on computational approaches to identifying disease biomarkers based on molecular networks. *Journal of Theoretical Biology*, 7(362), 9–16. https://doi.org/10.1016/j.jtbi.2014.06.007. Epub 2014 Jun 12 24931674.

Rabilloud, T., & Lescuyer, P. (2015). Proteomics in mechanistic toxicology: History, concepts, achievements, caveats, and potential. *Proteomics*, *15*, 1051–1074. https://doi.org/10.1002/pmic.201400288.

Rajawat, J., & Jhingan, G. (2019). Mass spectroscopy. In G. Misra (Ed.), *Data processing handbook for complex biological data sources* (pp. 1–20). Cambridge, MA, USA: Academic Press. https://doi.org/10.1016/B978-0-12-816548-5.00001-0.

Ramos, K. S., Martin, M., Ramos, I. N., & Rempala, G. A. (2018). Bioinformatics and computational biology in toxicology: Gateways for precision medicine. In C. A. McQueen (Ed.), *Comprehensive toxicology* (2nd ed., pp. 720–728). Amsterdam, Netherlands: Elsevier Science. https://doi.org/10.1016/b978-0-12-801238-3.99176-1.

Ryan, D., Mulrane, L., Rexhepaj, E., & Gallagher, W. M. (2011). Tissue microarrays and digital image analysis. *Methods in Molecular Biology*, *691*, 97–112. https://doi.org/10.1007/978-1-60761-849-2_6.

Schmidt, F. (2021). Computational toxicology. In O. Wolkenhauer (Ed.), *Systems medicine: Integrative qualitative and computational approaches* (pp. 283–300). Cambridge, MA, USA: Academic Press. https://doi.org/10.1016/b978-0-12-801238-3.11534-x.

Sharma, S. K., Goloubinoff, P., & Christen, P. (2008). Heavy metal ions are potent inhibitors of protein folding. *Biochemical and Biophysical Research Communications*, *372*, 341–345. https://doi.org/10.1016/j.bbrc.2008.05.052.

Singh, S., Kumar, V., Datta, S., Dhanjal, D. S., Singh, S., Kumar, S., et al. (2020). Physiological responses, tolerance, and remediation strategies in plants exposed to metalloids. *Environmental Science and Pollution Research*. https://doi.org/10.1007/s11356-020-10293-2. in press.

Smith, L. M., & Kelleher, N. L. (2013). Proteoform: A single term describing protein complexity. *Nature Methods*, *10*, 186–187. https://doi.org/10.1038/nmeth.2369.

Soler, L., & Oswald, I. P. (2018). The importance of accounting for sex in the search of proteomic signatures of mycotoxin exposure. *Journal of Proteomics*, *178*, 114–122. https://doi.org/10.1016/j.jprot.2017.12.017.

Sturla, S. J., Boobis, A. R., Fitzgerald, R. E., Hoeng, J., Kavlock, R. J., Schirmer, K., et al. (2014). Systems toxicology: From basic research to risk assessment. *Chemical Research in Toxicology*, *27*, 314–329. https://doi.org/10.1021/tx400410s.

Subramanian, A., Tamayo, P., Mootha, V. K., Mukherjee, S., Ebert, B. L., Gillette, M. A., et al. (2005). Gene set enrichment analysis: A knowledge-based approach for interpreting genome-wide expression profiles. *Proceedings. National Academy of Sciences of the United States of America*, *102*, 15545–15550. https://doi.org/10.1073/pnas.0506580102.

Suder, P., Novák, P., Havlíček, V., & Bodzoń-Kułakowska, A. (2016). General strategies for proteomic sample preparation. In P. Ciborowski, & J. Silberring (Eds.), *Proteomic profiling and analytical chemistry: The crossroads* (2nd ed., pp. 25–49). Cambridge, MA, USA: Academic Press. https://doi.org/10.1016/B978-0-444-63688-1.00003-3.

Szklarczyk, D., Gable, A. L., Lyon, D., Junge, A., Wyder, S., Huerta-Cepas, J., et al. (2019). STRING v11: Protein-protein association networks with increased coverage, supporting functional discovery in genome-wide experimental datasets. *Nucleic Acids Research*, *47*(D1), D607–D613. https://doi.org/10.1093/nar/gky1131.

Taft, R. J., Pheasant, M., & Mattick, J. S. (2007). The relationship between non-protein-coding DNA and eukaryotic complexity. *BioEssays*, *29*, 288–299. https://doi.org/10.1002/bies.20544.

Taylor, C. F., Paton, N. W., Lilley, K. S., Binz, P. A., Julian, R. K., Jr., Jones, A. R., et al. (2007). The minimum information about a proteomics experiment (MIAPE). *Nature Biotechnology*, *25*, 887–893. https://doi.org/10.1038/nbt1329.

Terzi, H., & Yıldız, M. (2021). Proteomic analysis reveals the role of exogenous cysteine in alleviating chromium stress in maize seedlings. *Ecotoxicology and Environmental Safety*, *209*, 111784. https://doi.org/10.1016/j.ecoenv.2020.111784.

Titz, B., Szostak, J., Sewer, A., Phillips, B., Nury, C., Schneider, T., et al. (2020). Multi-omics systems toxicology study of mouse lung assessing the effects of aerosols from two heat-not-burn tobacco products and cigarette smoke. *Computational and Structural Biotechnology Journal*, *18*, 1056–1073. https://doi.org/10.1016/j.csbj.2020.04.011.

Tomanek, L. (2011). Environmental proteomics: Changes in the proteome of marine organisms in response to environmental stress, pollutants, infection, symbiosis, and development. *Annual Review of Marine Science*, *3*, 373–399. https://doi.org/10.1146/annurev-marine-120709-142729.

Tomanek, L. (2014). Proteomics to study adaptations in marine organisms to environmental stress. *Journal of Proteomics*, *105*, 92–106. https://doi.org/10.1016/j.jprot.2014.04.009.
Tran, N. H., Zhang, X., Xin, L., Shan, B., & Li, M. (2017). De novo peptide sequencing by deep learning. *Proceedings. National Academy of Sciences of the United States of America*, *114*, 8247–8252. https://doi.org/10.1073/pnas.1705691114.
van Gestel, C. A., & van Brummelen, T. C. (1996). Incorporation of the biomarker concept in ecotoxicology calls for a redefinition of terms. *Ecotoxicology*, *5*, 217–225. https://doi.org/10.1007/BF00118992.
Viitaniemi, H. M., & Leder, E. H. (2011). Sex-biased protein expression in threespine stickleback, *Gasterosteus aculeatus*. *Journal of Proteome Research*, *10*, 4033–4040. https://doi.org/10.1021/pr200234a.
Vishnoi, S., Matre, H., Garg, P., & Pandey, S. K. (2020). Artificial intelligence and machine learning for protein toxicity prediction using proteomics data. *Chemical Biology & Drug Design*, *96*, 902–920. https://doi.org/10.1111/cbdd.13701.
Wang, X. (2019). Protein and proteome atlas for plants under stresses: New highlights and ways for integrated omics in post-genomics era. *International Journal of Molecular Sciences*, *20*, 5222. https://doi.org/10.3390/ijms20205222.
Waters, M., & Fostel, J. (2004). Toxicogenomics and systems toxicology: Aims and prospects. *Nature Reviews. Genetics*, *5*, 936–948. https://doi.org/10.1038/nrg1493.
Watson, C., Ge, J., Cohen, J., Pyrgiotakis, G., Engelward, B. P., & Demokritou, P. (2014). High-throughput screening platform for engineered nanoparticle-mediated genotoxicity using CometChip technology. *ACS Nano*, *8*, 2118–2133. https://doi.org/10.1021/nn404871p.
Wetmore, B. A., & Merrick, B. A. (2004). Toxicoproteomics: Proteomics applied to toxicology and pathology. *Toxicologic Pathology*, *32*(6), 619–642. https://doi.org/10.1080/01926230490518244. 15580702.
Wu, X., Hasan, M. A., & Chen, J. Y. (2014). Pathway and network analysis in proteomics. *Journal of Theoretical Biology*, *362*, 44–52. https://doi.org/10.1016/j.jtbi.2014.05.031.
Xu, T., Chen, L., Lim, Y. T., Zhao, H., Chen, H., Chen, M. W., et al. (2021). System biology-guided chemical proteomics to discover protein targets of monoethylhexyl phthalate in regulating cell cycle. *Environmental Science & Technology*, *55*, 1842–1851. https://doi.org/10.1021/acs.est.0c05832.
Yahya, F. A., Hashim, N. F. M., Israf Ali, D. A., Chau Ling, T., & Cheema, M. S. (2021). A brief overview to systems biology in toxicology: The journey from in to vivo, in-vitro and –omics. *Journal of King Saud University-Science*, *33*, 101254. https://doi.org/10.1016/j.jksus.2020.101254.
Yamamoto, T., Fukushima, T., Kikkawa, R., Yamada, H., & Horii, I. (2005). Protein expression analysis of rat testes induced testicular toxicity with several reproductive toxicants. *The Journal of Toxicological Sciences*, *30*, 111–126. https://doi.org/10.2131/jts.30.111.
Young, A. P., Jackson, D. J., & Wyeth, R. C. (2020). A technical review and guide to RNA fluorescence in situ hybridization. *PeerJ*, *8*, e8806. https://doi.org/10.7717/peerj.8806.
Yu, Y., Yin, H., Peng, H., Lu, G., & Dang, Z. (2020). Proteomic mechanism of decabromodiphenyl ether (BDE-209) biodegradation by *Microbacterium* Y2 and its potential in remediation of BDE-209 contaminated water-sediment system. *Journal of Hazardous Materials*, *387*, 121708. https://doi.org/10.1016/j.jhazmat.2019.121708.
Zhai, R., Su, S., Lu, X., Liao, R., Ge, X., He, M., et al. (2005). Proteomic profiling in the sera of workers occupationally exposed to arsenic and lead: Identification of potential biomarkers. *Biometals*, *18*, 603–613. https://doi.org/10.1007/s10534-005-3001-x.

CHAPTER THREE

Posttranslational modifications in systems biology

Suruchi Aggarwal[a,b], Priya Tolani[a], Srishti Gupta[a,c], and Amit Kumar Yadav[a,*]

[a]Translational Health Science and Technology Institute, NCR Biotech Science Cluster, Faridabad, Haryana, India
[b]Department of Molecular Biology and Biotechnology, Cotton University, Guwahati, Assam, India
[c]School of Biosciences and Technology, Vellore Institute of Technology, Vellore, India
*Corresponding author: e-mail address: amit.yadav@thsti.res.in

Contents

Advances in Protein Chemistry and Structural Biology, Volume 127
ISSN 1876-1623
https://doi.org/10.1016/bs.apcsb.2021.03.005

Abstract

The biological complexity cannot be captured by genes or proteins alone. The protein posttranslational modifications (PTMs) impart functional diversity to the proteome and regulate protein structure, activity, localization and interactions. Their dynamics drive cellular signaling, growth and development while their dysregulation causes many diseases. Mass spectrometry based quantitative profiling of PTMs and bioinformatics analysis tools allow systems level insights into their network architecture. High-resolution profiling of PTM networks will advance disease understanding and precision medicine. It can accelerate the discovery of biomarkers and drug targets. This requires better tools for unbiased, high-throughput and accurate PTM identification, site localization and automated annotation on a systems level.

1. Introduction

Connecting genotype to phenotype has been the holy grail of molecular biology. The phenotype is a result of interaction between molecular players (DNA, RNA, proteins and metabolites), which can be accurately identified and quantified by high-throughput omics, like next-generation sequencing (genomics and transcriptomics), ribosome profiling (Ribo-Seq) and mass spectrometry (proteomics, metabolomics and lipidomics).

Cells constantly receive, process and respond to environmental stimuli. This response to stimuli requires fast and reversible switches, a functionality provided by covalent protein posttranslational modifications (Prabakaran, Lippens, Steen, & Gunawardena, 2012). The genome and the transcriptome provide limited functional information. Mass spectrometry allows the capture of the context-specific proteome diversity from protein posttranslational modifications (PTMs) that can rapidly respond to environmental cues (Legrain et al., 2011). The functional diversity of an organism is thus defined by the diversity of their proteoforms (variants and modifications) (Bludau & Aebersold, 2020).

Modifications like phosphorylation, ubiquitination, acetylation, etc., alter protein functions through their concerted effects. The PTMs demonstrate crosstalks among themselves to regulate a plethora of biological functions (Kandpal, Aggarwal, Jamwal, & Yadav, 2017; Schwammle, Aspalter, Sidoli, & Jensen, 2014). The PTMs and their interactions need to be measured in a systems biology paradigm, with mathematical approaches for inferring their functional modules.

In shotgun proteomics, the proteins from a sample are digested with trypsin. Quantitation is performed with metabolic labels (SILAC, NeuCode, etc.), chemical labels (iTRAQ, TMT, etc.) or label-free approaches. The peptides are analyzed by high-performance liquid-chromatography (HPLC), followed by tandem mass spectrometry (MS^2 or MS/MS) sequencing. The fragmentation breaks the peptide backbone to generate b and y ions (in collision-induced dissociation, CID; and high energy C-trap dissociation, HCD) or c- and z-type ions (in electron transfer dissociation, ETD). The tandem mass spectra are searched against an *in silico* peptide database to select and score candidate peptides, based on their theoretical MS/MS matches, using computational search algorithms. For searching modifications, the tools can adjust the candidate peptide mass accordingly. There are downstream processing tools for statistical validation of identified peptides and their quantitation.

We first describe the most important and widely studied PTM types, followed by their mass spectrometry detection, computational algorithms and tools for data analysis. We will also discuss the advances boosting the systems level PTM studies.

2. Major PTM types

The PTMs are chemical moieties added to a protein after translation and regulate protein folding, degradation, signaling, localization, stability, enzymatic activity and protein-protein interactions (PPIs) (Olsen & Mann, 2013). There are hundreds of known PTM types (>400) found in humans with the help of biophysical and biochemical methods (Aebersold et al., 2018).

Phosphorylation is the most widely studied PTM, which can regulate signaling pathways, cell cycle, metabolism, immune response, protein synthesis, degradation, cellular growth and differentiation (Bodenmiller et al., 2010). Acetylation regulates protein stability, degradation, localization, apoptosis, mediating protein-protein interactions, transcription control in cancers and immune system dysregulation (Choudhary & Mann, 2010;

Choudhary, Weinert, Nishida, Verdin, & Mann, 2014; Weinert, Moustafa, Iesmantavicius, Zechner, & Choudhary, 2015). Acetylation is known to crosstalk with several PTMs and their implication in cardiovascular diseases have been recently highlighted (Aggarwal, Banerjee, Talukdar, & Yadav, 2020). Ubiquitin (Ub) is a small, highly conserved protein (76 amino acids) that attaches to a substrate lysine through an isopeptide bond (Walsh, Garneau-Tsodikova, & Gatto Jr., 2005), involving a complex conjugation process that requires ubiquitin-activating enzyme (E1), ubiquitin-conjugating enzyme (E2) and ubiquitin ligase (E3). It regulates the protein life cycle by degradation of its substrates, their subcellular localization and formation of protein-protein complexes (Wagner et al., 2012). Small ubiquitin-like modifier (SUMO) can also form an isopeptide linkage with substrate lysine residues like ubiquitin (Hammond, Cai, & Verhey, 2008). It regulates protein-protein interactions, promotes intracellular protein and mRNA trafficking, subcellular localization, prevents protein degradation, regulates endocytosis and apoptosis, etc. (Abdel-Hafiz, Dudevoir, Perez, Abdel-Hafiz, & Horwitz, 2018). Another conjugating protein, NEDD8 (Kumar, Yoshida, & Noda, 1993) is reported to be important for neurodegenerative disorders as well as cancers. Methylation of proteins by methyl transferases (Paik, Paik, & Kim, 2007) was usually studied with the histone code (Jenuwein & Allis, 2001). Glycosylation is the process of covalently attaching a carbohydrate sugar (donor) to a hydroxyl group or other functional group of lipids, proteins or any other organic molecule (acceptor) through an enzymatic reaction (by glycosyl-transferases). It can be N-linked, O-linked, C-mannosylation, or glycosylphosphatidylinositol (GPI) anchor linked. The GPI anchors act as receptors, adhesion molecules, transporters and participate in cell-cell and molecular interactions as well as signal transduction (Hart & Copeland, 2010).

3. PTM identification and quantification

Shotgun proteomics (bottom-up proteomics) deals with large-scale identification and quantification of proteins which are digested into peptides, analyzed by liquid-chromatography tandem mass spectrometry (LC-MS/MS) and later inferred from data.

3.1 Shotgun proteomics for PTM analysis

For analysis of PTMs, sample preparation is followed by PTM enrichment and trypsin digestion. The peptides are separated by 2D gel-electrophoresis or liquid-chromatography (LC) and are ionized using electrospray ionization (ESI). The mass-to-charge ratios (*m/z*) of the peptides is measured in the first mass analyzer (MS^1).

In the data-dependent acquisition (DDA) mode, the *top n* highest intensity peaks (peptide ions) are selected for fragmentation by the second mass analyzer (MS/MS or MS^2) and the precursors already analyzed are dynamically excluded for ~60s. The selected ions are fragmented using collision-induced dissociation (CID), electron transfer dissociation (ETD) or high energy C-trap dissociation (HCD), to generate specific N- and C-term ladders of fragmented peptides. These fragmentation techniques break the peptide at the backbone to generate b and y ions (CID/HCD) or c and z ions (ETD). The acquired MS/MS spectra are analyzed using computational tools for peptide identification, quantitation and protein assembly. These tools identify and localize the site of modification as well. Using stable isotope labels, either in cell culture (SILAC) or chemical labels (iTRAQ/TMT), different biological states can be quantitatively compared.

In data-independent acquisition (DIA) method, there is no dynamic exclusion. In a wide *m/z* window of approximately 25 Da, all precursors are selected for MS/MS fragmentation together. The machine performs multiplexed sequencing and the analysis can reveal each peptide that generates a unique fragment ion in the high resolution MS/MS. The DIA method called SWATH, can be used to quantify low-abundance proteins and their modified forms in label-free manner.

3.2 Fragmentation modes for PTM analysis

The peptide ions are sequenced using b and y ions from CID spectra (Steen & Mann, 2004). The modified peptides may be unsuitable for fragmentation due to side chains (Zhou, Dong, & Vachet, 2011), or higher charges (>+2) or labile group (phosphorylation gets scrambled in CID making site localization challenging). Alternate modes of fragmentation like-electron capture dissociation (ECD), ETD and HCD, are gaining popularity for modified peptide analysis for higher charge states, preserving the localization information. The alternate enzymes Lys-C and Asp-N are good alternatives to trypsin for ETD analysis of modified peptides (Riley & Coon, 2018).

HCD generates b, y and immonium ions with PTM-specific diagnostic ions, such as oxonium ions for glycopeptides (Toghi Eshghi et al., 2016). EThcD, a technique combining CID and ETD fragmentation, generates b, y, c and z ions, for phosphopeptide and sialylated peptide analysis

(Frese et al., 2013). A new method utilizing high energy by ultraviolet photodissociation (UVPD) can fragment peptide backbone while keeping the modification site unaffected for better localization of O-GlcNAc (Escobar et al., 2020).

Diagnostic-ion scanning in CID or HCD can enrich modified spectra during mass spectrometry without biochemical enrichment. A dominant neutral loss peak of 98 Da is a characteristic of Ser and Thr phosphopeptides (Mann & Jensen, 2003). A diagnostic immonium ion peak at 126.0931 is used to validate acetyl-lysine peptides. The SUMmon tool can scan SUMOylated peptides by their fragmentation patterns (Pedrioli et al., 2006) while PTM MarkerFinder (Nanni et al., 2013) can specifically search for diagnostic ions of certain PTMs from CID, HCD or ETD spectra.

3.3 System-level PTM discovery

A system-level study of PTMs comprises of sample preparation, enrichment, separation using LC and analysis using MS. The modified peptides coexist with unmodified ones with low relative abundance. The modified spectra contain the mass-shifted ion-ladders that can identify the modification site.

The analysis relies on a database search algorithm to match the experimental MS/MS spectra to the candidate theoretical spectra in an appropriate database (Kumar, Yadav, & Dash, 2017). Shotgun PTM profiling requires analysis of MS/MS data to infer proteins, modifications and their interconnected networks. The millions of spectra thus generated, are searched using a database search tool like X!Tandem (Craig & Beavis, 2004), MassWiz (Yadav, Kumar, & Dash, 2011), MaxQuant (Cox & Mann, 2008), etc. The results are then filtered using a target-decoy approach with false discovery rate (FDR) control (Aggarwal & Yadav, 2016b; Yadav, Kadimi, Kumar, & Dash, 2013), after the use of post-processing tools that improve the search results (Kall, Canterbury, Weston, Noble, & MacCoss, 2007; Yadav, Kumar, & Dash, 2012). The identified peptides are used to infer the proteins using the rules of parsimony (Nesvizhskii & Aebersold, 2005; Yadav et al., 2011).

4. Data analysis strategies

For DIA data, the bioinformatics tools utilize a library of peptides for identification, which is first built using the DDA techniques. Therefore, the accuracy of DDA data analysis is of paramount importance. There are three major ways to analyze PTMs through DDA, i.e. *de novo* sequencing, database search and a spectral library search. The DIA data is searched against a spectral library as reference.

4.1 *De novo* sequencing

The manual *de novo* sequencing of peptides (including PTMs) was a viable method to identify peptides in early days of MS until data avalanche rendered this method untenable. The modified spectra contain mass shifts equal to the modification mass. In low resolution, several residues and their combinations appear isobaric. Modern instruments generate high resolution spectra that help resolve sequencing ambiguities. Several algorithms for automated *de novo* sequencing that can also pinpoint modifications are available, utilizing graph theoretical approaches.

Lutefisk was the first tool for automated de novo sequencing (Taylor & Johnson, 1997), soon followed by SHERENGA (Dancik, Addona, Clauser, Vath, & Pevzner, 1999). There have been seminal advances in computational *de novo* sequencing, that include the use of dynamic programming, hidden Markov models, spectrum graph and deep learning implemented in tools like PEAKS (Ma et al., 2003), pNovo (Yang, Chi, Zeng, Zhou, & He, 2019), DeepNovo (Tran, Zhang, Xin, Shan, & Li, 2017), etc.

4.2 Variable modification search

In the database search, each MS/MS data spectrum is searched against an *in silico* digested proteome database, to find candidate peptides within instrument mass error. For every candidate peptide, a theoretical MS/MS is generated to match against the experimentally acquired spectrum. The search tool can specify the modification mass shift during search as a variable modification, which can shift each corresponding ion ladder by the PTM mass to generate the modified theoretical spectrum for matching the experimental spectrum.

The PTMs in a biological sample can be present in sub-stoichiometric amount making its direct detection challenging. PTM enrichment techniques enable their detection, otherwise there is a masking effect of abundant, non-modified peptide ions. The anticipated PTM (with or without enrichment) can be searched as a *variable modification* in a database search (Yates 3rd, Eng, McCormack, & Schieltz, 1995). The modified theoretical fragment ions are help in correct identification and localization. This search creates a compute overhead since every potentially modified peptide will be searched for every possible mod-position, causing exponential increase in search space and thus more false positives.

4.3 Open modification search

In an open modification search (blind search), unanticipated modifications can be discovered by their mass shifts. The error-tolerant search strategy uses the partial "sequence tags" for search. The pre and post-fix masses along with the partial sequence tag help in identifying peptides, while unaccounted masses can be ascribed to modifications (Mann & Wilm, 1994). The method limits the search space explosion and is used in tools like InsPecT (Tanner et al., 2005).

In another approach, a primary search (without modifications) was employed to quickly identify the proteins of interest. These proteins were then searched again to identify modifications and variants from the unidentified spectra. X!Tandem, Paragon, InsPect, etc., have a built-in refinement option to carry out "two-pass" searches. It is a common practice in proteogenomics tools to use transcriptome-proteome integration for identifying variants and modifications (Kumar, Mondal, Yadav, & Dash, 2014; Kumar, Yadav, Jia, Mulvenna, & Dash, 2016). Shrinking the database in iterative steps, to identify the previously unidentified spectra, can also be used in the so-called "multi-pass" search, employed in MetaMorpheus (Solntsev, Shortreed, Frey, & Smith, 2018) and PeaksPTM (Han, He, Xin, Shan, & Ma, 2011).

In the open search, there is no restriction of search space to hunt for unknown mass shifts. These tools search in a wide precursor mass tolerance window, but the fragment mass tolerance is kept narrow to prevent excessive false matching. Due to its heavy requirement of compute-power, the open searches are coupled to inventive methods to check the false positives using tag-based or multi-pass searches. MODa (Na, Bandeira, & Paek, 2012), SpecOMS (David, Fertin, Rogniaux, & Tessier, 2017), MSFragger (Kong, Leprevost, Avtonomov, Mellacheruvu, & Nesvizhskii, 2017), TagGraph (Devabhaktuni et al., 2019), ANN-Solo (Bittremieux, Laukens, & Noble, 2019), Open-pFind (Sun et al., 2019), etc., are tools that can perform an open modification search without any assumptions on the number or type of modifications. Due to higher false positives, it is necessary to have accurate PTM site localization scores (Na et al., 2012).

4.4 Open spectral library search

Instead of a FASTA database, the experimental spectra can be directly queried against a library of previously identified spectra. This prevents the identification of unobserved or modified peptides, but is a much faster approach for commonly observed peptide species because of limited library size. The spectral libraries were constructed from database search results for non-modified peptides. With variable modification searches, the database search enabled the construction of spectral libraries for modified peptides as well.

The cosine similarity of spectra was shown to be useful for peptide identifications (Stein & Scott, 1994). The proteomics spectral libraries could be constructed at greater scale owing to the comparative ease of identifying peptides through database search. Attempts to create modification-specific libraries have been made, most notably for phosphopeptides. The spectral libraries generally contain common modifications, particularly N-term acetylation, oxidation and phosphorylation. The initial spectral libraries were limited in size and scope but efforts by NIST, GPMDB and PeptideAtlas have created a comprehensive array of species-specific and modification-specific spectral libraries. Many tools have been developed that allow spectral library based open searches, like QuickMod (Ahrne, Nikitin, Lisacek, & Muller, 2011), SpectraST (Ma & Lam, 2014), MzMod (Apache Spark version of QuickMod) (Horlacher, Lisacek, & Muller, 2016), etc. Some recent tools like ANN-SoLo use GPU computing (Bittremieux et al., 2019) for faster searches. A fragment-ion indexing method matches the *m/z* of ions and mass losses to find the PTMs along with insertions or deletions (Burke et al., 2017). The method is gaining popularity and implemented in MSFragger and MetaMorpheus as well.

4.5 Global PTM discovery searches

The global PTM identification (G-PTM) is a recent approach for exhaustive search of curated modifications from the UniProt database (Shortreed et al., 2015). Instead of finding a novel PTM site, this method aims for faster identifications of the known sites (as per UniProt). Evidently, the search speed is better than blind PTM, due to the limited query database. A caveat is the inability to identify previously unknown PTM sites and types. The approach was advanced into *global posttranslational modification (PTM) discovery strategy (G-PTM-D)* by coupling it with a first-stage blind-mode search. In second stage, the G-PTM method was used to search UniProt along with previously discovered PTM sites. This enhanced database allowed the PTM discovery in the samples (Li et al., 2017). The MetaMorpheus tool was further enhanced by calibrating the input and implementing the *multi-notch search* in the first stage in which, the Unimod PTM masses are searched exclusively in specific notches of observable PTM mass difference. This is an ingenious approach that prevents the compute overhead of a full-fledged open search (Solntsev et al., 2018).

4.6 DIA-based PTM analysis

The DIA methods are gaining popularity due to their depth of coverage, reproducibility and low cost. It does not suffer from stochastic sampling issue of DDA. A highly multiplexed MS/MS spectrum is acquired, fragmenting all precursor masses in a window of 25 Da, called as Sequential Window Acquisition of all Theoretical Mass Spectra (SWATH). DIA-SWATH analysis requires a DDA-based library, to match the unique fragment ions from the precursors for their identification and quantification. SWATH has been instrumental in finding and quantifying the low abundant histone variants and modifications from human and mouse stem cells (Sidoli et al., 2015). The SWATHProphet^{PTM} is a tool for automated PTM analysis using SWATH-DIA data, incorporated into the SWATHProphet software (Keller et al., 2016). A new tool, Thesaurus, can identify and quantify the isomers of phosphopeptides from parallel reaction monitoring acquisition in the DIA mode (Searle, Lawrence, MacCoss, & Villen, 2019). PIQED is another pipeline for processing the DIA data for identifying and quantifying the unknown modified peptides, which are then combined into proteins (Meyer et al., 2017). The PTM site localization, scoring and filtering of data is carried out in an integrated pipeline. Subsequently, the data compilation step also incorporates the correction of modification-specific changes from the changes observed in modified peptides and the global proteins levels.

4.7 Statistical validation

Identification and validation of peptide spectra is a critical part of proteomics data analysis. The FASTA sequences of the organism under study are called *target* sequences, while their reversed counterparts are called *decoys*, used for *multiple testing correction*. Each candidate peptide match to a spectrum is called a peptide spectrum match (PSM). The decoy PSM scores are used as threshold to assess the false discovery rate of target PSMs. However, the target-decoy (TD) strategy has shortcomings in proteogenomics or open-PTM searches. The equality assumptions of TD competition break down as the database size increases. It undermines the search sensitivity resulting in inaccurate FDR estimates (Hart-Smith, Yagoub, Tay, Pickford, & Wilkins, 2016; Kumar, Yadav, et al., 2016). A transferred subgroup false discovery rate (tsFDR) method was proposed to calculate accurate FDR in blind-mode PTM search (Fu & Qian, 2014). Accurate estimation of FDR for rare PTMs is difficult due to the low number of spectra and does not match global FDR. The authors demonstrated an empirical relationship between subgroup and global FDRs which could be leveraged to calculate subgroup PTM FDR estimates from global FDR.

The correct inference of PTM sites is crucial using the false localization rate (FLR) for statistical confidence. Open searches usually sacrifice the accuracy for speed, leading to incorrect site localization, which is usually corrected post search. Probabilistic localization score was first used by Ascore algorithm for large-scale phosphoproteomics data (Beausoleil, Villen, Gerber, Rush, & Gygi, 2006). Tools like phosphoRS (later ptmRS) (Taus et al., 2011), LuciPhor (Fermin, Walmsley, Gingras, Choi, & Nesvizhskii, 2013), PTMProphet (Shteynberg et al., 2019), etc., use CID data. The SLoMo tool performs site localization from ETD/ECD data (Bailey et al., 2009).

5. PTM code

The PTMs impart a distinct function to their target proteins or facilitate protein-protein interactions that cohesively bring about a biological function. This coordinated PTM-mediated regulation of transcription by controlling histones, transcription factors and RNA polymerase has been labeled as the *PTM Code*. The tails of these histone proteins (H2A, H2B, H3 and H4) contain various PTMs like acetylation, methylation (mono-, di- or tri-methyl), phosphorylation, ubiquitination, SUMOylation, NEDDylation and ADP-ribosylation, etc. These PTM marks represent a "code" which can be read by chromatin modifying enzymes and transcriptional regulators that can switch gene expression "on" or "off" based on the specificity encoded into these PTM marks (Strahl & Allis, 2000). The PTMs control transcriptional regulation through combinatorial interplay on histone tails (histone code), but also on RNA polymerase II C-terminal domain (CTD code), dynamic assembly and regulation of tubulins (tubulin code) to drive several important biological processes.

In this so-called histone code, the PTM marks form the input signal, the reader domains act as their corresponding adaptors and the chromatin state (on or off) acts like the output (Strahl & Allis, 2000).

The carboxy-terminus of the large subunit (RPB1) of RNA polymerase II (called CTD) is highly conserved across life forms, transcribes RNA from DNA (Eick & Geyer, 2013). The CTD and its PTMs coordinate the transcription process by a platform for the reading writing and erasing of the PTMs, called the "CTD code" (Zaborowska, Egloff, & Murphy, 2016).

The microtubules maintain the cytoskeleton of the cell which demands that microtubules have a conserved structure, made of repeating units of tubulin. The functions of the tubulin are controlled by the isotypic forms and the various posttranslational modifications, generally referred to as the *tubulin code* (Janke & Magiera, 2020).

PTMs orchestrate biological functions through *positive*, *negative* or *processive (step-wise cooperative)* interactions (Hunter, 2007) called as crosstalk. Broadly, a positive crosstalk enables the addition of another modification near a PTM site. A negative/competitive crosstalk inhibits the occurrence of another modification on the same or adjacent site. In the processive mechanism, a PTM addition may form a platform for further combinatorial PTMs to form a scaffold in a step-wise manner.

6. PTM and PPIs networks

Protein-protein interactions extend the functional repertoire of a cell. The PTMs regulate binding affinities in PPIs. Elucidation of the interconnected network of proteins and PTMs, can reveal functional insights into disease mechanisms. The PPIs may be *direct* or *indirect* based on physical contact, or may be *transient* or *permanent* based on their stability. In a PPI network model, the participating proteins are represented as *nodes* while their interaction is shown as an *edge*. The PPI networks are generally measured as snapshots of biological processes in space and time. The ability to measure PPIs is greatly limited by changing abundance, posttranslational modification (PTM) status of proteins and their transient nature (Peng, Wang, Peng, Wu, & Pan, 2017). Since PTM-mediated PPIs play several important biological roles, it is a sought-after goal to connect these two regulatory layers in a coherent manner.

6.1 Integration with biological networks

Affinity-purification mass spectrometry and yeast-2 hybrid systems (Y2H) are the most widely used techniques for profiling of PPIs. PTM networks involve the readers, writers and erasers of the modifications, all forming the nodes of the PTM interactome connecting each node to the other through a well-defined PPI edge. The perturbation studies on this PTM-PPI interactome can map the disease dynamics and propagation of cellular signals. This approach can enable the finding of PTM-based drug targets and pharmacological interventions in clinical trials (Cohen & Tcherpakov, 2010). The interactome can also reveal the obscure enzyme-substrate relationships. The progress remains poor and limited to kinases and their substrates (Miller & Turk, 2018) (like BCR-ABL and LRKK2), even though more PTMs are now being systematically targeted (Shi et al., 2021).

The phosphorylation substrates are easier to discover due to their direct enzyme-substrate relationship, but ubiquitin and related proteins are challenging. From the E1 conjugating enzyme to E2 and E3, the permutations keep increasing that diminishes the chances of facile discovery of their specific interactor/substrate attachment. Several E2 conjugating enzymes were discovered to be workhorses for a wide variety of E3 ligases (RINGs) (Woodsmith, Jenn, & Sanderson, 2012) while others were found to be specific to few (e.g. ARIH1-UBE2L3/6, MARCH10-UBE2N/K).

6.2 Network-based function prediction

The interactome can be mined to predict the protein functions based on their network context. The PTM co-expression networks can also be used to predict functions, based on the assumption that the quantitative correlation between nodes is functional. The quantitative PTM interactome provides a natural way to analyze and visualize functional linkages between proteins. Many types of interactomes can be integrated using Bayesian and kernel techniques to construct highly stringent network linkages for predicting function at genome-wide scale. This has enabled the development of rich resources for data mining—like HPRD, STRING, MINT, BioGRID, etc. (Janga, Diaz-Mejia, & Moreno-Hagelsieb, 2011).

6.3 Kinase substrate-networks

A kinase-substrate interactome regulates cellular functions like growth, development and response to stimuli. The kinases form a large family of enzymes (>500 proteins in the humans) which transfer phosphate group from ATP to S/T/Y residues in the substrate. Phosphorylation is a transient molecular switch that controls protein activity, structure, subcellular localization and turnover (Damle & Mohanty, 2014). Mass spectrometry phosphoproteomics data analysis permits the inference of the kinase activity

(Savage & Zhang, 2020). The dysregulation of kinases in complex diseases like cancers, neurodegenerative diseases, fatty liver disease, diabetes, etc., can be measured by phosphoproteomics (Yilmaz et al., 2021). Several computational methods have been developed to identify kinases-substrate interactions. For example, NetworKIN uses a combination of consensus sequence motifs and protein-association networks for predicting the kinase for the experimentally identified sites (Linding et al., 2008). Another algorithm, GPS, uses substitution matrix and Markov cluster algorithm to predict phosphorylation sites. PhosNetConstruct can identify the kinase-substrate relationship by analysis of the domain-specific phosphorylation network (Damle & Mohanty, 2014).

7. Advances in PTM studies

There have been many large-scale systematic studies on PTMs, few of which are demonstrated here.

7.1 PTM functional prioritization

To ascribe function to the plethora of modified sites is a massive challenge. A compilation of ~200,000 sites of phosphorylation, acetylation and ubiquitination from 11 eukaryotes were used (Beltrao et al., 2012) to prioritize their functions based on their cross-regulatory events, domain activity regulation, mediation of PPIs and conservation. Regulatory hotspots were discovered in functionally important regions and predictions were validated in HSP70 domain family. The analysis found that only a fraction of sites is functional, and transient PTM networks may be an evolutionary mechanism to explore functional diversity. The PTMFunc (http://ptmfunc.com) resource was developed to disseminate how mutations may cause diseases.

7.2 Serial-enrichment for PTM crosstalk

Studying PTM crosstalk is experimentally challenging due to technical limitations. The serial enrichment method was demonstrated to circumvent this limitation, using enrichment followed by shotgun proteomics. In the SEPTM strategy, serial enrichment of phosphorylation, ubiquitination and acetylation was carried out from human leukemia cells treated with bortezomib (Mertins et al., 2013). Nearly 8000 proteins, with 20,000 phosphorylation sites, 15,000 ubiquitination and 3000 acetylation sites were quantitatively profiled, using flow-through of first enrichment step for second PTM, and the subsequent one for third PTM. This process enriched phosphopeptides first, followed by ubiquitinated peptides and eventually, acetylated peptides. The coverage was similar to individual PTM enrichment. Another group developed similar strategy to use double enrichment to study co-modified phosphorylated and ubiquitinated peptides to study protein degradation and ubiquitylation-phosphorylation crosstalk (Swaney et al., 2013). Frequently co-occurring sites carrying phosphorylation and ubiquitination were found, that also led to the discovery of several phosphodegrons.

7.3 Global network of PTM crosstalk

A network approach was applied to define the global association of PTMs in a functional context. The interplay of 13 frequent PTMs in 8 eukaryotes was used to compare their co-evolution as a measure of functional association (Minguez et al., 2012). The PTMs were highly interconnected, and formed a global network of ~6000 proteins carrying 450,000 PTM sites in humans. There were 35 PTM types found to display extensive crosstalk-network. These predictions covered well-known crosstalks, but also discovered new ones in secretory and membrane proteins. The PTM-domain information and linear motifs were connected to unambiguous pairs of PTM interplay in a functional context, which brought spatio-temporal specificity of crosstalk. More PTM data can bring information on the missing links between PTM types for combinatorial regulation.

7.4 Regulatory human phosphoproteome

Enriched phosphoproteomics studies containing 112 datasets from 104 different cell types or tissues were compiled into the functional human phosphoproteome landscape (Ochoa et al., 2020). The reanalysis of 6801 proteomics experiments was performed to create a reference human phosphoproteome containing 119,809 phosphosites. To discern the functional sites, machine learning was applied to a 59-feature panel to combine into a *functional score*, from the features related to conservation, structure, regulation and evolutionary information. Diverse types of phosphosites were identified as functional regulatory sites which also revealed genetic predisposition to diseases for the whole genome. Many sites were experimentally validated, like SMARCC2 phosphosites. This protein is a member of the chromatin-remodeling complex SWI/SNF.

7.5 Multiplexing in system-level PTM dynamics

Quantitative shotgun proteomics techniques have been used for PTM dynamic measurements (Aggarwal, Kandpal, Asthana, & Yadav, 2017; Aggarwal & Yadav, 2016a; Gajadhar & White, 2014; Yadav, 2017). The multiplexing capacity of quantitative techniques is a limitation for studying PTM dynamics in a robust manner, which can be overcome by the recent higher-order multiplexing technique that combines MS1 and MS2 labels in one experiment (Aggarwal et al., 2020; Aggarwal, Talukdar, & Yadav, 2019; Jamval, Aggarwal, Yadav, Subba Rao, & Kumar, 2016; Kumar, Jamwal, et al., 2016; Welle et al., 2016). The cost of experiment is a great bargain with high quality data, higher reproducibility and lesser run-to-run variation. Even the tools for facile computational analysis from such complex experiments are beginning to emerge (Aggarwal, Kumar, et al., 2020).

8. Resources for systems analysis

The research communities around the world have been collecting, collating and warehousing data in repositories, accessible to anyone for analysis. It is now easier to find emergent properties from large-scale data analysis. This allows researchers to ask questions from data without worrying about data cleaning, curation and organizing efforts on raw data. Here, we compile some resources and tools for large-scale PTM data.

8.1 PTM prediction tools

The computational PTM analysis has been an alternate research approach since a complete catalog of PTMs is near impossible to achieve. The technological advances ensure that training models keeps getting better input data (primary sequence features or 3D structural information), which also increases the accuracy of predictions. Several algorithms are used for PTM prediction, such as position-specific scoring matrix (PSSM), hidden Markov models (HMM), support vector machines (SVMs), artificial neural network (ANN) and Bayesian decision theory, etc. (He, Wei, & Zou, 2018).

With enormous information present on the precision of domains and their short linear motifs to code for a PTM site, machine learning is a popular tool to predict PTM sites. Prediction tools like NetworKIN (Linding et al., 2008) can utilize the interaction domains of the kinases from their substrate proteins, to predict probable phosphorylation sites. There are several reviews that discuss in detail the various PTM prediction approaches and tools (Ivanisenko et al., 2019; Xu et al., 2014). Many tools can predict modification sites, e.g., ProAcePred (acetylation) (Chen et al., 2018), DeepUbi (ubiquitination) (Fu, Yang, Wang, Wang, & Xu, 2019), C-iSUMO (SUMOylation) (Lopez, Dehzangi, Reddy, & Sharma, 2020), Met-predictor (methylation) (Zheng, Wuyun, Cheng, Hu, & Zhang, 2020), GlyStruct (glycosylation) (Reddy et al., 2019). Some predictors for NEDDylation, palmitoylation, malonylation, etc. (http://www.biocuckoo.org) are also available.

8.2 PTM databases

With an increase in high-throughput mass spectrometry studies, public availability of raw data for data reanalysis and community-wide sharing, have been developed over the years and are described briefly.

8.2.1 PhosphoSitePlus (PSP)

PhosphoSitePlus is a comprehensive database and web resource for exploring the structure and function of experimentally determined human and mouse PTM sites (Hornbeck et al., 2012). Initially launched as PhosphoSite database (Hornbeck, Chabra, Kornhauser, Skrzypek, & Zhang, 2004), it aggregated structure and regulatory interaction data for phosphorylation sites from 1200 reference articles. The updated version, PhosphoSitePlus (PSP), curated >12,000 references for PTM interactions, biological roles, disease association, kinase-substrate interactions and structural information, for acetylation, ubiquitination and O-GlcNAcylation. It permitted querying of proteins for disease, treatments or cellular compartments, custom data downloads, monthly releases, visual sequence logo analysis, structural annotation of PTM sites and pathway visualization of kinase- substrate interactions.

Updated with >330,000 PTM sites, including phosphorylation, acetylation, ubiquitination and methylation with >95% of the sites from mass spectrometry experiments (Hornbeck et al., 2015), it later allowed the additional download of"Regulatory sites" that encompassed PTM sites involved in cellular processes, molecular functions and PPIs. Another download for "PTMVar" data integrated missense mutations (from UniPROTKB, TCGA and other sources) with PTM sites data and comprised of >25,000 PTMVars which can rewire cellular signaling pathways. The unique PTM sites in PSP rose to >450,000 from >22,000 articles and thousands of mass spectrometry datasets. The disease and isoform information increased depth and coverage, with additions on somatic cancer mutations, viewed along with quantitative PTM information as lollipop plots (Hornbeck et al., 2015).

8.2.2 dbPTM

This database compiled information on PTM sites such as catalytic sites, solvent accessibility, structure (secondary and tertiary), protein domains and also the variants (Lee et al., 2006). The database compiled information from Swiss-Prot, PhosphoELM and O-GLYCBASE. The Swiss-Prot proteins were annotated with PTM information for a very small number of proteins at that time. The database contained structural and functional information on three major PTMs, i.e., phosphorylation, glycosylation and sulfation, with computational prediction of solvent accessibility and secondary structure to map and overlay with the PTM sites.

The updated dbPTM2.0 was released with protein domains, disordered regions, and variation features (Lee et al., 2009). It also benchmarked several PTMs (phosphorylation, methylation, glycosylation and acetylation) against an evaluation dataset. The version 3.0 (Lu et al., 2013) integrated experimental PTM sites from public resources and manual curation of MS/MS spectra from literature. The exploration of PTM substrate specificity and their functional association with interacting proteins using PPI as well as domain-domain interactions was possible. The structural topologies of transmembrane proteins and TopPTM module for identification of functional PTM sites on transmembrane proteins (Su et al., 2014) was added. The 2016 update of dbPTM, provided integrated access to many PTM resources (14 databases) and manually curated 12,000 modified peptides containing many PTMs that included S-nitrosylation, S-glutathionylation and succinylation, from text mining of ~500 articles. The PDB mapped PTM sites allowed the users to analyze spatially close amino acids, solvent accessible surface area, side-chain orientation with their substrate sites and drug-binding. The metabolic and PPI interaction network analysis was also integrated in this release (Huang et al., 2016).

In the 2019 release, dbPTM (http://dbPTM.mbc.nctu.edu.tw) increased the number of experimentally validated PTM sites (Huang et al., 2019) and added PTM-disease association information for non-synonymous single nucleotide polymorphisms (nsSNPs) from the dbSNP database. The authors also explored PTM crosstalks and their functional relevance by motif search and enrichment analysis.

8.2.3 SysPTM

SysPTM is an integrated systematic data resource for PTM research consisting of a manually curated knowledgebase of nearly 50 PTM types and data mining tools (Li et al., 2009). It contains information on >100,000 sites from >33,000 proteins with protein annotations combined from Pfam, KEGG, GO based functional classification and orthology information. It provides PTMBlast for comparing user input with SysPTM data, PTMPathway for mapping these to KEGG pathways, PTMPhylog for discovery of conserved sites and PTMCluster to locate multi-site PTM clusters. The roles of PTMs can be inspected fully in its biological context through automatic data collection from five databases, i.e., Swiss-Prot 56.2, Phospho.ELM 8.0, HPRD 7, O-GLYCBASE 6.0, and Ubiprot version 1.0. Four webservers were also used, namely SUMOsp 1.0, Memo 2.0, NetAcet 1.0, and LysAcet 1.1. These were combined into the database SysPTM-A, while SysPTM-B was constructed by pubmed literature mining through Perl program with the keywords-mass spectrometry, proteomics, and seven modifications (acetylation, methylation, glycosylation, phosphorylation, S-nitrosylation, SUMOylation and ubiquitination). After data quality checks and manual curation, the sites were extracted and mapped to protein sequences.

In 2014, SysPTM 2.0 was published with major updates to its data that nearly doubled in 4 years (http://lifecenter.sgst.cn/SysPTM/) and enhanced the analysis tools PTMBlast, PTMPathway, PTMPhylog and PTMCluster. A new tool SysPTM-H represented the combinatorial histone modifications and their regulation. Another new tool PTM-GO was integrated for functional and enrichment analysis of PTM sites. The focus expanded from being a data resource to allowing the users to perform a systematic functional investigation (Li et al., 2014).

8.2.4 PTMCode

The PTMCode database is an integrative compilation of results on known as well as novel functional associations on PTM sites from several large-scale studies and predictions. It assembled results on 13 diverse PTM types across 8 eukaryotes to assist studies on evolutionary, network and emergent properties of PTMs in various model organisms (Minguez, Letunic, Parca, & Bork, 2013). Known and predicted PTM sites with their functional association from co-evolving sites across many eukaryotes, their proximity in protein structure, competing PTM sites, literature mining of crosstalk information and sites within PTM hotspots (high density PTM regions) were integrated. This database provided a searchable protein-based interactive web-interface (http://ptmcode.embl.de) with information on co-regulation of ~75,000 PTM containing residues in >10,000 proteins. In PTMCode v2, a new orthology-based strategy for propagating PTMs from validated PTM sites was incorporated. The species count increased from 8 to 19 eukaryotes and collected 300,000 verified PTM sites (with >1.300,000 propagated PTM sites). It housed 69 types of PTMs with regulatory information on >100,000 proteins and >100,000 interactions, totally accounting for ~8 million single-protein PTM associations and >9.4 million PPIs being regulated by the PTM sites (Minguez et al., 2015).

8.2.5 PTMD

PTMD is a database (http://ptmd.biocuckoo.org) that collects and integrates human PTMs associated with diseases (Xu et al., 2018). The authors compiled 1950 known PTM-disease associations (PDAs) through literature curation, in 749 proteins for wide range of PTM types (23 types) and diseases (275 disease types). The analysis of database discovered phosphorylation to carry the highest number of known disease associations, while neurological disorders carry the highest frequency of PTM associations. The database also contained detailed annotations for many aspects of the PTMs for an in-depth analysis, by PTM or by disease. Aberrant signaling events can be analyzed in a disease-gene network with 1437 interaction edges. The extensive PTM crosstalk in complex diseases like prostate cancers and bladder cancers were discovered. The top 10 genes with highest number of disease associations controlled important signaling pathways. The p53 protein expectedly, connected with 21 types of cancers. The analysis demonstrated a strong association of PTMs and diseases.

8.2.6 iPTMnet

The iPTMnet database (https://research.bioinformatics.udel.edu/iptmnet) is an integrated resource for PTMs based on literature and other knowledge resources, proteoforms level representation of PTM proteins and PTM enzymes, quality score for the PTM integration data, network visualization of PPIs in the PTM-enzyme-substrate relationships and sequence alignment of (single, multiple or overlapping) PTM sites across species (Huang et al., 2018). The fragmented PTM information in literature is all connected into a biologically semantic context for systematic analysis and knowledge discovery. The workflow collates information from text mining of literature, ontologies and biological databases and organizes into a well-curated iPTMnet web knowledgebase to cater to many important modifications for humans and other model organisms. The latest version (iPTMnet v5.1) was updated with 737,803 PTM sites spread across 63,490 proteins and 12,167 proteoforms. There are 12,111 enzyme-substrate pairs, 23,315 enzyme-substrate sites and 1444 PTM-dependent PPIs in this release.

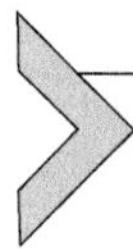

9. PTMs and systems biology

9.1 Flux-balance analysis

The rate of flow of metabolites through a metabolic network is analyzed through a mathematical approach known as flux balance analysis (FBA). The flow of metabolites in a network is calculated and is used to predict the growth rate of organism or the production of important metabolites. The approach is bound by mathematical constraints depicted as equations that have to balance the input for the reactions with the possible outputs, within the bounds of the system. It is also being used for PTMs since their effect on pathway output can be more sustained than a simple regulatory effect of metabolites. Only about half a dozen metabolites in *E. coli* are known to be regulated by PTMs to control the flux. The complex PTM-mediated interactions are a big challenge for biomedical sciences. Using a multiscale-workflow with genome-scale metabolic modeling, and molecular dynamics studies, the effect of PTMs on metabolites and microbial fitness in *E. coli* metabolic network were investigated (Brunk et al., 2018) to check how PTMs impacted cellular fitness caused by nutrient change.

9.2 PTM-metabolite interactions

Biological systems display a robust character to easily adapt to perturbation through their crosstalks with the metabolic and signaling networks. The protein-metabolite interactions (PMIs) can also regulate protein activity and function through a change in their conformation. The metabolites when bound to proteins can alter PTMs to regulate or fine-tune their metabolic activity. These PMIs can be covalent PTMs or non-covalent PMIs (Piazza et al., 2018). The covalent PTMs are specifically non-enzymatic in nature as they are induced by the reactive metabolites directly. The nutrient perception can stimulate the signal transduction by the receptors at plasma membrane, a hotbed for crosstalk between metabolism and signaling pathways. This can often result in gene expression and modification of several enzymes that are involved in metabolism, like glucose perception in *Saccharomyces cerevisiae* by Snf3 and Rgt2, and the discernment of amino acids, sulfate, ammonium and phosphate by trans-receptors. The Ras/PKA pathway, calcium release, TOR activity in yeast, mitochondrial retrograde response, etc., are controlled by PMIs (Milanesi, Coccetti, & Tripodi, 2020).

9.3 Mathematical modeling

Biological systems are highly complex and compartmentalized. The dedicated cells or organelles perform specific biological functions cohesively, which can be mathematically modeled to understand and predict the complex behavior (Motta & Pappalardo, 2013). A mathematical model has some basic features like-state variables or model parameters, steady-state or transient behavior, linearity or nonlinearity, and are either deterministic or stochastic.

The mathematical structures are used in a formal model which consists of these steps—model implementation (formal description in mathematical form or computer code), use of model to forecast the behavior of the system, and assess how closely the model follows the reality by evaluating the predictions with real data (Motta & Pappalardo, 2013). Mathematical modeling has been applied to metabolic processes, signaling and regulatory pathways among many others.

Due to poor capture techniques, low abundance and stoichiometry of PTMs, understanding their functions has been particularly challenging. Measuring the network dynamics through the cue-signal-response type experiments, can reveal the network nodes and their quantitative read-outs that can be modeled to understand the system more comprehensively. For example, the response of tyrosine phosphorylation network on HER2 expression in EGF stimulated cells was monitored through the quantitative study of cellular migration and proliferation (Wolf-Yadlin et al., 2006). The temporal measurement of hundreds of pY sites was measured across 4 time-points, and the responses were modeled through partial least square regression (PLSR). This identified important functional sites by correlating the response to a loss of given phosphosite. The perturbation responses can be understood at the network level using the PTM dynamics studies.

9.4 PTMs in proteoforms diversity

The PTMs fine-tune protein structure for altering the stability, activity, cellular localization and interactions. The proteoform is a unique molecular entity which differs from the canonical gene product in length, sequence, variants (nsSNPs) or modifications. The genomic blueprint does not fully explain the functional diversity. Alternate splicing and non-synonymous variations increase diversity of the proteome. The PTMs induce mind-boggling combinatorial possibilities (Aebersold et al., 2018) to increase proteome diversity. Characterizing the proteoforms can provide a greater insight into proteome function. These proteoform profiling studies are being pursued to delineate which of the proteoforms is the dominant regulator in a given biological context (Bludau & Aebersold, 2020). The proteoform-specific functions of the disease-associated proteins would be a challenge worth pursuing for therapeutics. Proteoforms with different combinations of PTMs are being targeted by blind PTM approaches regularly.

The PTMs can rapidly expand the diversity of proteome by scaling up the proteoforms. In the human histone H4, a combinatorial explosion occurs if all the forms of the 58 Swiss-Prot-annotated PTMs at 17 known amino acid sites are allowed, to give rise to $>10^{10}$ proteoforms in theory. The 13 most common PTM sites from the literature, when used with a single variant E64Q (with a minor allele frequency of ~0.001%), can still create 98,304 proteoforms (Aebersold et al., 2018).

Shotgun proteomics is cumbersome for proteoform characterization and top-down approach is most appropriate (Compton, Kelleher, & Gunawardena, 2018). The proteoforms identification will be highly useful for early disease detection, staging, prevention and cure. Proteoform characterization for apolipoproteins, transferrin, hemoglobin, cystatin C-truncated proteoforms, vitamin D-binding protein, C-reactive protein and immunoglobulin G (NISTmAb), etc., is required in many diseases.

10. Challenges and outlook for PTM systems biology

The PTM studies ought to reach patients in the form of useful biomarkers or drug targets. This requires a deeper understanding of PTM biology with respect to the proteoform distribution. Connecting the molecular profiles to discernable phenotypes is important. The systems biology perspective guides that PTM as a regulatory layer can be better understood by integrative omics (den Ridder, Daran-Lapujade, & Pabst, 2020). Proteogenomics has already been very useful for mapping cancer mutations (Mnatsakanyan et al., 2018) and novel proteoforms with alternate start sites, novel translation products, modifications and variants in diverse organisms (Jaffe, Berg, & Church, 2004; Kelkar et al., 2011; Kumar et al., 2013).

The recent surge in AI and ML techniques assists drug target prediction and finding kinase networks from PTM data. Using the functionally important sites, disease progression can be monitored with the help of kinome. It should extend to other PTM types in near future for accurate disease classification, biomarker panels, monitoring drug responses and testing candidate drugs and drug repurposing (Pagel, Loroch, Sickmann, & Zahedi, 2015; Thygesen, Boll, Finsen, Modzel, & Larsen, 2018). Cell culture models of drug sensitivity testing and biomarker panels that profile the specific signaling responses (for example in cancers) can quickly guide the diagnostic and therapeutic clinical decisions (Ramroop, Stein, & Drake, 2018). Even mechanism of action of novel drugs can be elucidated in a facile manner. This requires deeper PTM profiling and understanding the mechanistic details before these discoveries can be leveraged in a clinical setting. The plethora of data thus generated can be used for deep learning algorithms to guide precision medicine and therapy (Shilo, Rossman, & Segal, 2020; Wen et al., 2020).

11. Conclusion

While the identification and quantitation of PTMs is challenging, their interrelatedness and crosstalks drive cellular signaling, which makes it imperative for quantitative studies to focus on discerning their network architecture. Multiplexing techniques will be useful in enhancing the throughput and accuracy of such an endeavor providing both coverage and depth requisite for gaining biological insights. Algorithms for accurate PTM identification, localization and annotation in an unanticipated approach (blind search) is beginning to provide the insights into this hidden biological layer of regulation. The identified sites are compiled into rich databases but their biological annotations lag behind due to size and scale. Here, the advances in deep learning and other machine learning algorithms can be useful to predict functional sites and cluster PTMs in unsupervised manner as per their emergent properties, hitherto unseen in the mountains of data. Tool development in these areas will also benefit biological interpretations of data. These algorithms can also be used to employ PPIs and network properties for functional interpretations from large-scale PTM sites. The fundamental discoveries need to be translated into effective biomarkers and therapeutic targets. Such goals depend heavily on the data integration (single as well as multi-omics) to decode the biological processes that are perturbed in a complex disease and how the PPI and PTM crosstalk networks are rewired in disease progression. Measuring the dynamics and interactions of these PTMs as a reflection of biological state can uniquely contribute to the disease understanding and therapeutic development.

Acknowledgments

A.K.Y. is supported by DBT-Big Data Initiative grant (BT/PR16456/BID/7/624/2016). This grant also supports P.T. Translational Research Program (TRP) at THSTI funded by DBT also supports A.K.Y. and S.A.

Conflict of interest

The authors declare no competing interests.

References

Abdel-Hafiz, H. A., Dudevoir, M. L., Perez, D., Abdel-Hafiz, M., & Horwitz, K. B. (2018). SUMOylation regulates transcription by the progesterone receptor a isoform in a target gene selective manner. *Diseases*, *6*(1), 5. https://doi.org/10.3390/diseases6010005.

Aebersold, R., Agar, J. N., Amster, I. J., Baker, M. S., Bertozzi, C. R., Boja, E. S., et al. (2018). How many human proteoforms are there? *Nature Chemical Biology*, *14*(3), 206–214. https://doi.org/10.1038/nchembio.2576.

Aggarwal, S., Banerjee, S. K., Talukdar, N. C., & Yadav, A. K. (2020). Post-translational modification crosstalk and hotspots in sirtuin interactors implicated in cardiovascular diseases. *Frontiers in Genetics*, *11*, 356. https://doi.org/10.3389/fgene.2020.00356.

Aggarwal, S., Kandpal, M., Asthana, S., & Yadav, A. K. (2017). Perturbed signaling and role of posttranslational modifications in cancer drug resistance. In *Drug resistance in bacteria, fungi, malaria, and cancer* (pp. 483–510). Springer.

Aggarwal, S., Kumar, A., Jamwal, S., Midha, M. K., Talukdar, N. C., & Yadav, A. K. (2020). HyperQuant-A computational pipeline for higher order multiplexed quantitative proteomics. *ACS Omega*, *5*(19), 10857–10867. https://doi.org/10.1021/acsomega.0c00515.

Aggarwal, S., Talukdar, N. C., & Yadav, A. K. (2019). Advances in higher order multiplexing techniques in proteomics. *Journal of Proteome Research*, *18*(6), 2360–2369. https://doi.org/10.1021/acs.jproteome.9b00228.

Aggarwal, S., & Yadav, A. K. (2016a). Dissecting the iTRAQ data analysis. *Methods in Molecular Biology*, *1362*, 277–291. https://doi.org/10.1007/978-1-4939-3106-4_18.

Aggarwal, S., & Yadav, A. K. (2016b). False discovery rate estimation in proteomics. *Methods in Molecular Biology*, *1362*, 119–128. https://doi.org/10.1007/978-1-4939-3106-4_7.

Ahrne, E., Nikitin, F., Lisacek, F., & Muller, M. (2011). QuickMod: A tool for open modification spectrum library searches. *Journal of Proteome Research*, *10*(7), 2913–2921. https://doi.org/10.1021/pr200152g.

Bailey, C. M., Sweet, S. M., Cunningham, D. L., Zeller, M., Heath, J. K., & Cooper, H. J. (2009). SLoMo: Automated site localization of modifications from ETD/ECD mass spectra. *Journal of Proteome Research*, *8*(4), 1965–1971. https://doi.org/10.1021/pr800917p.

Beausoleil, S. A., Villen, J., Gerber, S. A., Rush, J., & Gygi, S. P. (2006). A probability-based approach for high-throughput protein phosphorylation analysis and site localization. *Nature Biotechnology*, *24*(10), 1285–1292. https://doi.org/10.1038/nbt1240.

Beltrao, P., Albanese, V., Kenner, L. R., Swaney, D. L., Burlingame, A., Villen, J., et al. (2012). Systematic functional prioritization of protein posttranslational modifications. *Cell*, *150*(2), 413–425. https://doi.org/10.1016/j.cell.2012.05.036.

Bittremieux, W., Laukens, K., & Noble, W. S. (2019). Extremely fast and accurate open modification spectral library searching of high-resolution mass spectra using feature hashing and graphics processing units. *Journal of Proteome Research*, *18*(10), 3792–3799. https://doi.org/10.1021/acs.jproteome.9b00291.

Bludau, I., & Aebersold, R. (2020). Proteomic and interactomic insights into the molecular basis of cell functional diversity. *Nature Reviews. Molecular Cell Biology*, *21*(6), 327–340. https://doi.org/10.1038/s41580-020-0231-2.

Bodenmiller, B., Wanka, S., Kraft, C., Urban, J., Campbell, D., Pedrioli, P. G., et al. (2010). Phosphoproteomic analysis reveals interconnected system-wide responses to perturbations of kinases and phosphatases in yeast. *Science Signaling*, *3*(153), rs4. https://doi.org/10.1126/scisignal.2001182.

Brunk, E., Chang, R. L., Xia, J., Hefzi, H., Yurkovich, J. T., Kim, D., et al. (2018). Characterizing posttranslational modifications in prokaryotic metabolism using a multiscale workflow. *Proceedings of the National Academy of Sciences of the United States of America*, *115*(43), 11096–11101. https://doi.org/10.1073/pnas.1811971115.

Burke, M. C., Mirokhin, Y. A., Tchekhovskoi, D. V., Markey, S. P., Heidbrink Thompson, J., Larkin, C., et al. (2017). The hybrid search: A mass spectral library search method for discovery of modifications in proteomics. *Journal of Proteome Research*, *16*(5), 1924–1935. https://doi.org/10.1021/acs.jproteome.6b00988.

Chen, G., Cao, M., Luo, K., Wang, L., Wen, P., & Shi, S. (2018). ProAcePred: Prokaryote lysine acetylation sites prediction based on elastic net feature optimization. *Bioinformatics*, *34*(23), 3999–4006. https://doi.org/10.1093/bioinformatics/bty444.

Choudhary, C., & Mann, M. (2010). Decoding signalling networks by mass spectrometry-based proteomics. *Nature Reviews. Molecular Cell Biology*, *11*(6), 427–439. https://doi.org/10.1038/nrm2900.

Choudhary, C., Weinert, B. T., Nishida, Y., Verdin, E., & Mann, M. (2014). The growing landscape of lysine acetylation links metabolism and cell signalling. *Nature Reviews. Molecular Cell Biology*, *15*(8), 536–550. https://doi.org/10.1038/nrm3841.

Cohen, P., & Tcherpakov, M. (2010). Will the ubiquitin system furnish as many drug targets as protein kinases? *Cell*, *143*(5), 686–693. https://doi.org/10.1016/j.cell.2010.11.016.

Compton, P. D., Kelleher, N. L., & Gunawardena, J. (2018). Estimating the distribution of protein post-translational modification states by mass spectrometry. *Journal of Proteome Research*, *17*(8), 2727–2734. https://doi.org/10.1021/acs.jproteome.8b00150.

Cox, J., & Mann, M. (2008). MaxQuant enables high peptide identification rates, individualized p.p.b.-range mass accuracies and proteome-wide protein quantification. *Nature Biotechnology*, *26*(12), 1367–1372. https://doi.org/10.1038/nbt.1511.

Craig, R., & Beavis, R. C. (2004). TANDEM: Matching proteins with tandem mass spectra. *Bioinformatics*, *20*(9), 1466–1467. https://doi.org/10.1093/bioinformatics/bth092.

Damle, N. P., & Mohanty, D. (2014). Deciphering kinase-substrate relationships by analysis of domain-specific phosphorylation network. *Bioinformatics*, *30*(12), 1730–1738. https://doi.org/10.1093/bioinformatics/btu112.

Dancik, V., Addona, T. A., Clauser, K. R., Vath, J. E., & Pevzner, P. A. (1999). De novo peptide sequencing via tandem mass spectrometry. *Journal of Computational Biology*, *6*(3–4), 327–342. https://doi.org/10.1089/106652799318300.

David, M., Fertin, G., Rogniaux, H., & Tessier, D. (2017). SpecOMS: A full open modification search method performing all-to-all spectra comparisons within minutes. *Journal of Proteome Research*, *16*(8), 3030–3038. https://doi.org/10.1021/acs.jproteome.7b00308.

den Ridder, M., Daran-Lapujade, P., & Pabst, M. (2020). Shot-gun proteomics: Why thousands of unidentified signals matter. *FEMS Yeast Research*, *20*(1), 1–9. https://doi.org/10.1093/femsyr/foz088.

Devabhaktuni, A., Lin, S., Zhang, L., Swaminathan, K., Gonzalez, C. G., Olsson, N., et al. (2019). TagGraph reveals vast protein modification landscapes from large tandem mass spectrometry datasets. *Nature Biotechnology*, *37*(4), 469–479. https://doi.org/10.1038/s41587-019-0067-5.

Eick, D., & Geyer, M. (2013). The RNA polymerase II carboxy-terminal domain (CTD) code. *Chemical Reviews*, *113*(11), 8456–8490. https://doi.org/10.1021/cr400071f.

Escobar, E. E., King, D. T., Serrano-Negron, J. E., Alteen, M. G., Vocadlo, D. J., & Brodbelt, J. S. (2020). Precision mapping of O-linked N-acetylglucosamine sites in proteins using ultraviolet photodissociation mass spectrometry. *Journal of the American Chemical Society*, *142*(26), 11569–11577. https://doi.org/10.1021/jacs.0c04710.

Fermin, D., Walmsley, S. J., Gingras, A. C., Choi, H., & Nesvizhskii, A. I. (2013). LuciPHOr: Algorithm for phosphorylation site localization with false localization rate estimation using modified target-decoy approach. *Molecular & Cellular Proteomics*, *12*(11), 3409–3419. https://doi.org/10.1074/mcp.M113.028928.

Frese, C. K., Zhou, H., Taus, T., Altelaar, A. F., Mechtler, K., Heck, A. J., et al. (2013). Unambiguous phosphosite localization using electron-transfer/higher-energy collision dissociation (EThcD). *Journal of Proteome Research*, *12*(3), 1520–1525. https://doi.org/10.1021/pr301130k.

Fu, Y., & Qian, X. (2014). Transferred subgroup false discovery rate for rare post-translational modifications detected by mass spectrometry. *Molecular & Cellular Proteomics*, *13*(5), 1359–1368. https://doi.org/10.1074/mcp.O113.030189.

Fu, H., Yang, Y., Wang, X., Wang, H., & Xu, Y. (2019). DeepUbi: A deep learning framework for prediction of ubiquitination sites in proteins. *BMC Bioinformatics*, *20*(1), 86. https://doi.org/10.1186/s12859-019-2677-9.

Gajadhar, A. S., & White, F. M. (2014). System level dynamics of post-translational modifications. *Current Opinion in Biotechnology*, *28*, 83–87. https://doi.org/10.1016/j.copbio.2013.12.009.

Hammond, J. W., Cai, D., & Verhey, K. J. (2008). Tubulin modifications and their cellular functions. *Current Opinion in Cell Biology*, *20*(1), 71–76. https://doi.org/10.1016/j.ceb.2007.11.010.

Han, X., He, L., Xin, L., Shan, B., & Ma, B. (2011). PeaksPTM: Mass spectrometry-based identification of peptides with unspecified modifications. *Journal of Proteome Research*, *10*(7), 2930–2936. https://doi.org/10.1021/pr200153k.

Hart, G. W., & Copeland, R. J. (2010). Glycomics hits the big time. *Cell*, *143*(5), 672–676. https://doi.org/10.1016/j.cell.2010.11.008.

Hart-Smith, G., Yagoub, D., Tay, A. P., Pickford, R., & Wilkins, M. R. (2016). Large scale mass spectrometry-based identifications of enzyme-mediated protein methylation are subject to high false discovery rates. *Molecular & Cellular Proteomics*, *15*(3), 989–1006. https://doi.org/10.1074/mcp.M115.055384.

He, W., Wei, L., & Zou, Q. (2018). Research progress in protein posttranslational modification site prediction. *Briefings in Functional Genomics*, *18*(4), 220–229. https://doi.org/10.1093/bfgp/ely039.

Horlacher, O., Lisacek, F., & Muller, M. (2016). Mining large scale tandem mass spectrometry data for protein modifications using spectral libraries. *Journal of Proteome Research*, *15*(3), 721–731. https://doi.org/10.1021/acs.jproteome.5b00877.

Hornbeck, P. V., Chabra, I., Kornhauser, J. M., Skrzypek, E., & Zhang, B. (2004). PhosphoSite: A bioinformatics resource dedicated to physiological protein phosphorylation. *Proteomics*, *4*(6), 1551–1561. https://doi.org/10.1002/pmic.200300772.

Hornbeck, P. V., Kornhauser, J. M., Tkachev, S., Zhang, B., Skrzypek, E., Murray, B., et al. (2012). PhosphoSitePlus: A comprehensive resource for investigating the structure and function of experimentally determined post-translational modifications in man and mouse. *Nucleic Acids Research*, *40*(Database issue), D261–D270. https://doi.org/10.1093/nar/gkr1122.

Hornbeck, P. V., Zhang, B., Murray, B., Kornhauser, J. M., Latham, V., & Skrzypek, E. (2015). PhosphoSitePlus, 2014: Mutations, PTMs and recalibrations. *Nucleic Acids Research*, *43*(Database issue), D512–D520. https://doi.org/10.1093/nar/gku1267.

Huang, H., Arighi, C. N., Ross, K. E., Ren, J., Li, G., Chen, S. C., et al. (2018). iPTMnet: An integrated resource for protein post-translational modification network discovery. *Nucleic Acids Research*, *46*(D1), D542–D550. https://doi.org/10.1093/nar/gkx1104.

Huang, K. Y., Lee, T. Y., Kao, H. J., Ma, C. T., Lee, C. C., Lin, T. H., et al. (2019). dbPTM in 2019: Exploring disease association and cross-talk of post-translational modifications. *Nucleic Acids Research*, *47*(D1), D298–D308. https://doi.org/10.1093/nar/gky1074.

Huang, K. Y., Su, M. G., Kao, H. J., Hsieh, Y. C., Jhong, J. H., Cheng, K. H., et al. (2016). dbPTM 2016: 10-year anniversary of a resource for post-translational modification of proteins. *Nucleic Acids Research*, *44*(D1), D435–D446. https://doi.org/10.1093/nar/gkv1240.

Hunter, T. (2007). The age of crosstalk: Phosphorylation, ubiquitination, and beyond. *Molecular Cell*, *28*(5), 730–738. https://doi.org/10.1016/j.molcel.2007.11.019.

Ivanisenko, V. A., Ivanisenko, T. V., Saik, O. V., Demenkov, P. S., Afonnikov, D. A., & Kolchanov, N. A. (2019). Web-based computational tools for the prediction and analysis of posttranslational modifications of proteins. *Methods in Molecular Biology*, *1934*, 1–20. https://doi.org/10.1007/978-1-4939-9055-9_1.

Jaffe, J. D., Berg, H. C., & Church, G. M. (2004). Proteogenomic mapping as a complementary method to perform genome annotation. *Proteomics*, *4*(1), 59–77. https://doi.org/10.1002/pmic.200300511.

Jamval, S., Aggarwal, S., Yadav, A., Subba Rao, K., & Kumar, A. (2016). *Method of hyperplexing in mass spectrometry to elucidate temporal dynamics of proteome*. India Patent IN 201611029904.

Janga, S. C., Diaz-Mejia, J. J., & Moreno-Hagelsieb, G. (2011). Network-based function prediction and interactomics: The case for metabolic enzymes. *Metabolic Engineering*, *13*(1), 1–10. https://doi.org/10.1016/j.ymben.2010.07.001.

Janke, C., & Magiera, M. M. (2020). The tubulin code and its role in controlling microtubule properties and functions. *Nature Reviews. Molecular Cell Biology*, *21*(6), 307–326. https://doi.org/10.1038/s41580-020-0214-3.

Jenuwein, T., & Allis, C. D. (2001). Translating the histone code. *Science*, *293*(5532), 1074–1080. https://doi.org/10.1126/science.1063127.

Kall, L., Canterbury, J. D., Weston, J., Noble, W. S., & MacCoss, M. J. (2007). Semi-supervised learning for peptide identification from shotgun proteomics datasets. *Nature Methods*, *4*(11), 923–925. https://doi.org/10.1038/nmeth1113.

Kandpal, M., Aggarwal, S., Jamwal, S., & Yadav, A. K. (2017). Emergence of drug resistance in mycobacterium and other bacterial pathogens: The posttranslational modification perspective. In *Drug resistance in bacteria, fungi, malaria, and cancer* (pp. 209–231). Springer.

Kelkar, D. S., Kumar, D., Kumar, P., Balakrishnan, L., Muthusamy, B., Yadav, A. K., et al. (2011). Proteogenomic analysis of mycobacterium tuberculosis by high resolution mass spectrometry. *Molecular & Cellular Proteomics*, *10*(12), M111.011627. https://doi.org/10.1074/mcp.M111.011445.

Keller, A., Bader, S. L., Kusebauch, U., Shteynberg, D., Hood, L., & Moritz, R. L. (2016). Opening a SWATH window on posttranslational modifications: Automated pursuit of modified peptides. *Molecular & Cellular Proteomics*, *15*(3), 1151–1163. https://doi.org/10.1074/mcp.M115.054478.

Kong, A. T., Leprevost, F. V., Avtonomov, D. M., Mellacheruvu, D., & Nesvizhskii, A. I. (2017). MSFragger: Ultrafast and comprehensive peptide identification in mass spectrometry-based proteomics. *Nature Methods*, *14*(5), 513–520. https://doi.org/10.1038/nmeth.4256.

Kumar, A., Jamwal, S., Midha, M. K., Hamza, B., Aggarwal, S., Yadav, A. K., et al. (2016). Dataset generated using hyperplexing and click chemistry to monitor temporal dynamics of newly synthesized macrophage secretome post infection by mycobacterial strains. *Data in Brief*, *9*, 349–354. https://doi.org/10.1016/j.dib.2016.08.055.

Kumar, D., Mondal, A. K., Yadav, A. K., & Dash, D. (2014). Discovery of rare protein-coding genes in model methylotroph Methylobacterium extorquens AM1. *Proteomics*, *14*(23–24), 2790–2794. https://doi.org/10.1002/pmic.201400153.

Kumar, D., Yadav, A. K., & Dash, D. (2017). Choosing an optimal database for protein identification from tandem mass spectrometry data. *Methods in Molecular Biology*, *1549*, 17–29. https://doi.org/10.1007/978-1-4939-6740-7_3.

Kumar, D., Yadav, A. K., Jia, X., Mulvenna, J., & Dash, D. (2016). Integrated transcriptomic-proteomic analysis using a proteogenomic workflow refines rat genome annotation. *Molecular & Cellular Proteomics*, *15*(1), 329–339. https://doi.org/10.1074/mcp.M114.047126.

Kumar, D., Yadav, A. K., Kadimi, P. K., Nagaraj, S. H., Grimmond, S. M., & Dash, D. (2013). Proteogenomic analysis of Bradyrhizobium japonicum USDA110 using GenoSuite, an automated multi-algorithmic pipeline. *Molecular & Cellular Proteomics*, *12*(11), 3388–3397. https://doi.org/10.1074/mcp.M112.027169.

Kumar, S., Yoshida, Y., & Noda, M. (1993). Cloning of a cDNA which encodes a novel ubiquitin-like protein. *Biochemical and Biophysical Research Communications*, *195*(1), 393–399. https://doi.org/10.1006/bbrc.1993.2056.

Lee, T. Y., Hsu, J. B., Chang, W. C., Wang, T. Y., Hsu, P. C., & Huang, H. D. (2009). A comprehensive resource for integrating and displaying protein post-translational modifications. *BMC Research Notes*, *2*, 111. https://doi.org/10.1186/1756-0500-2-111.

Lee, T. Y., Huang, H. D., Hung, J. H., Huang, H. Y., Yang, Y. S., & Wang, T. H. (2006). dbPTM: An information repository of protein post-translational modification. *Nucleic Acids Research*, *34*(Database issue), D622–D627. https://doi.org/10.1093/nar/gkj083.

Legrain, P., Aebersold, R., Archakov, A., Bairoch, A., Bala, K., Beretta, L., et al. (2011). The human proteome project: Current state and future direction. *Molecular & Cellular Proteomics*, *10*(7), M111.009993. https://doi.org/10.1074/mcp.M111.009993.

Li, J., Jia, J., Li, H., Yu, J., Sun, H., He, Y., et al. (2014). SysPTM 2.0: An updated systematic resource for post-translational modification. *Database: The Journal of Biological Databases and Curation*, *2014*, bau025. https://doi.org/10.1093/database/bau025.

Li, Q., Shortreed, M. R., Wenger, C. D., Frey, B. L., Schaffer, L. V., Scalf, M., et al. (2017). Global post-translational modification discovery. *Journal of Proteome Research*, *16*(4), 1383–1390. https://doi.org/10.1021/acs.jproteome.6b00034.

Li, H., Xing, X., Ding, G., Li, Q., Wang, C., Xie, L., et al. (2009). SysPTM: A systematic resource for proteomic research on post-translational modifications. *Molecular & Cellular Proteomics*, *8*(8), 1839–1849. https://doi.org/10.1074/mcp.M900030-MCP200.

Linding, R., Jensen, L. J., Pasculescu, A., Olhovsky, M., Colwill, K., Bork, P., et al. (2008). NetworKIN: A resource for exploring cellular phosphorylation networks. *Nucleic Acids Research*, *36*(Database issue), D695–D699. https://doi.org/10.1093/nar/gkm902.

Lopez, Y., Dehzangi, A., Reddy, H. M., & Sharma, A. (2020). C-iSUMO: A sumoylation site predictor that incorporates intrinsic characteristics of amino acid sequences. *Computational Biology and Chemistry*, *87*, 107235. https://doi.org/10.1016/j.compbiolchem.2020.107235.

Lu, C. T., Huang, K. Y., Su, M. G., Lee, T. Y., Bretana, N. A., Chang, W. C., et al. (2013). DbPTM 3.0: An informative resource for investigating substrate site specificity and functional association of protein post-translational modifications. *Nucleic Acids Research*, *41*(Database issue), D295–D305. https://doi.org/10.1093/nar/gks1229.

Ma, C. W., & Lam, H. (2014). Hunting for unexpected post-translational modifications by spectral library searching with tier-wise scoring. *Journal of Proteome Research*, *13*(5), 2262–2271. https://doi.org/10.1021/pr401006g.

Ma, B., Zhang, K., Hendrie, C., Liang, C., Li, M., Doherty-Kirby, A., et al. (2003). PEAKS: Powerful software for peptide de novo sequencing by tandem mass spectrometry. *Rapid Communications in Mass Spectrometry*, *17*(20), 2337–2342. https://doi.org/10.1002/rcm.1196.

Mann, M., & Jensen, O. N. (2003). Proteomic analysis of post-translational modifications. *Nature Biotechnology*, *21*(3), 255–261. https://doi.org/10.1038/nbt0303-255.

Mann, M., & Wilm, M. (1994). Error-tolerant identification of peptides in sequence databases by peptide sequence tags. *Analytical Chemistry*, *66*(24), 4390–4399. https://doi.org/10.1021/ac00096a002.

Mertins, P., Qiao, J. W., Patel, J., Udeshi, N. D., Clauser, K. R., Mani, D. R., et al. (2013). Integrated proteomic analysis of post-translational modifications by serial enrichment. *Nature Methods*, *10*(7), 634–637. https://doi.org/10.1038/nmeth.2518.

Meyer, J. G., Mukkamalla, S., Steen, H., Nesvizhskii, A. I., Gibson, B. W., & Schilling, B. (2017). PIQED: Automated identification and quantification of protein modifications from DIA-MS data. *Nature Methods*, *14*(7), 646–647. https://doi.org/10.1038/nmeth.4334.

Milanesi, R., Coccetti, P., & Tripodi, F. (2020). The regulatory role of key metabolites in the control of cell signaling. *Biomolecules*, *10*(6), 862. https://doi.org/10.3390/biom10060862.

Miller, C. J., & Turk, B. E. (2018). Homing in: Mechanisms of substrate targeting by protein kinases. *Trends in Biochemical Sciences*, *43*(5), 380–394. https://doi.org/10.1016/j.tibs.2018.02.009.

Minguez, P., Letunic, I., Parca, L., & Bork, P. (2013). PTMcode: A database of known and predicted functional associations between post-translational modifications in proteins. *Nucleic Acids Research*, *41*(Database issue), D306–D311. https://doi.org/10.1093/nar/gks1230.

Minguez, P., Letunic, I., Parca, L., Garcia-Alonso, L., Dopazo, J., Huerta-Cepas, J., et al. (2015). PTMcode v2: A resource for functional associations of post-translational modifications within and between proteins. *Nucleic Acids Research*, *43*(Database issue), D494–D502. https://doi.org/10.1093/nar/gku1081.

Minguez, P., Parca, L., Diella, F., Mende, D. R., Kumar, R., Helmer-Citterich, M., et al. (2012). Deciphering a global network of functionally associated post-translational modifications. *Molecular Systems Biology*, *8*, 599. https://doi.org/10.1038/msb.2012.31.

Mnatsakanyan, R., Shema, G., Basik, M., Batist, G., Borchers, C. H., Sickmann, A., et al. (2018). Detecting post-translational modification signatures as potential biomarkers in clinical mass spectrometry. *Expert Review of Proteomics*, *15*(6), 515–535. https://doi.org/10.1080/14789450.2018.1483340.

Motta, S., & Pappalardo, F. (2013). Mathematical modeling of biological systems. *Briefings in Bioinformatics*, *14*(4), 411–422. https://doi.org/10.1093/bib/bbs061.

Na, S., Bandeira, N., & Paek, E. (2012). Fast multi-blind modification search through tandem mass spectrometry. *Molecular & Cellular Proteomics*, *11*(4), M111.010199. https://doi.org/10.1074/mcp.M111.010199.

Nanni, P., Panse, C., Gehrig, P., Mueller, S., Grossmann, J., & Schlapbach, R. (2013). PTM MarkerFinder, a software tool to detect and validate spectra from peptides carrying post-translational modifications. *Proteomics*, *13*(15), 2251–2255. https://doi.org/10.1002/pmic.201300036.

Nesvizhskii, A. I., & Aebersold, R. (2005). Interpretation of shotgun proteomic data: The protein inference problem. *Molecular & Cellular Proteomics*, *4*(10), 1419–1440. https://doi.org/10.1074/mcp.R500012-MCP200.

Ochoa, D., Jarnuczak, A. F., Vieitez, C., Gehre, M., Soucheray, M., Mateus, A., et al. (2020). The functional landscape of the human phosphoproteome. *Nature Biotechnology*, *38*(3), 365–373. https://doi.org/10.1038/s41587-019-0344-3.

Olsen, J. V., & Mann, M. (2013). Status of large-scale analysis of post-translational modifications by mass spectrometry. *Molecular & Cellular Proteomics*, *12*(12), 3444–3452. https://doi.org/10.1074/mcp.O113.034181.

Pagel, O., Loroch, S., Sickmann, A., & Zahedi, R. P. (2015). Current strategies and findings in clinically relevant post-translational modification-specific proteomics. *Expert Review of Proteomics*, *12*(3), 235–253. https://doi.org/10.1586/14789450.2015.1042867.

Paik, W. K., Paik, D. C., & Kim, S. (2007). Historical review: The field of protein methylation. *Trends in Biochemical Sciences*, *32*(3), 146–152. https://doi.org/10.1016/j.tibs.2007.01.006.

Pedrioli, P. G., Raught, B., Zhang, X. D., Rogers, R., Aitchison, J., Matunis, M., et al. (2006). Automated identification of SUMOylation sites using mass spectrometry and SUMmOn pattern recognition software. *Nature Methods*, *3*(7), 533–539. https://doi.org/10.1038/nmeth891.

Peng, X., Wang, J., Peng, W., Wu, F. X., & Pan, Y. (2017). Protein-protein interactions: Detection, reliability assessment and applications. *Briefings in Bioinformatics*, *18*(5), 798–819. https://doi.org/10.1093/bib/bbw066.

Piazza, I., Kochanowski, K., Cappelletti, V., Fuhrer, T., Noor, E., Sauer, U., et al. (2018). A map of protein-metabolite interactions reveals principles of chemical communication. *Cell*, *172*(1–2), 358–372.e323. https://doi.org/10.1016/j.cell.2017.12.006.

Prabakaran, S., Lippens, G., Steen, H., & Gunawardena, J. (2012). Post-translational modification: Nature's escape from genetic imprisonment and the basis for dynamic information encoding. *Wiley Interdisciplinary Reviews. Systems Biology and Medicine*, *4*(6), 565–583. https://doi.org/10.1002/wsbm.1185.

Ramroop, J. R., Stein, M. N., & Drake, J. M. (2018). Impact of phosphoproteomics in the era of precision medicine for prostate cancer. *Frontiers in Oncology*, *8*, 28. https://doi.org/10.3389/fonc.2018.00028.

Reddy, H. M., Sharma, A., Dehzangi, A., Shigemizu, D., Chandra, A. A., & Tsunoda, T. (2019). GlyStruct: Glycation prediction using structural properties of amino acid residues. *BMC Bioinformatics*, *19*(Suppl. 13), 547. https://doi.org/10.1186/s12859-018-2547-x.

Riley, N. M., & Coon, J. J. (2018). The role of electron transfer dissociation in modern proteomics. *Analytical Chemistry*, *90*(1), 40–64. https://doi.org/10.1021/acs.analchem.7b04810.

Savage, S. R., & Zhang, B. (2020). Using phosphoproteomics data to understand cellular signaling: A comprehensive guide to bioinformatics resources. *Clinical Proteomics*, *17*, 27. https://doi.org/10.1186/s12014-020-09290-x.

Schwammle, V., Aspalter, C. M., Sidoli, S., & Jensen, O. N. (2014). Large scale analysis of co-existing post-translational modifications in histone tails reveals global fine structure of cross-talk. *Molecular & Cellular Proteomics*, *13*(7), 1855–1865. https://doi.org/10.1074/mcp.O113.036335.

Searle, B. C., Lawrence, R. T., MacCoss, M. J., & Villen, J. (2019). Thesaurus: Quantifying phosphopeptide positional isomers. *Nature Methods*, *16*(8), 703–706. https://doi.org/10.1038/s41592-019-0498-4.

Shi, Y., Zhu, J., Xu, Y., Tang, X., Yang, Z., & Huang, A. (2021). Malonyl-proteome profiles of Staphylococcus aureus reveal lysine malonylation modification in enzymes involved in energy metabolism. *Proteome Science*, *19*(1), 1. https://doi.org/10.1186/s12953-020-00169-1.

Shilo, S., Rossman, H., & Segal, E. (2020). Axes of a revolution: Challenges and promises of big data in healthcare. *Nature Medicine*, *26*(1), 29–38. https://doi.org/10.1038/s41591-019-0727-5.

Shortreed, M. R., Wenger, C. D., Frey, B. L., Sheynkman, G. M., Scalf, M., Keller, M. P., et al. (2015). Global identification of protein post-translational modifications in a single-pass database search. *Journal of Proteome Research*, *14*(11), 4714–4720. https://doi.org/10.1021/acs.jproteome.5b00599.

Shteynberg, D. D., Deutsch, E. W., Campbell, D. S., Hoopmann, M. R., Kusebauch, U., Lee, D., et al. (2019). PTMProphet: Fast and accurate mass modification localization for the trans-proteomic pipeline. *Journal of Proteome Research*, *18*(12), 4262–4272. https://doi.org/10.1021/acs.jproteome.9b00205.

Sidoli, S., Lin, S., Xiong, L., Bhanu, N. V., Karch, K. R., Johansen, E., et al. (2015). Sequential window acquisition of all theoretical mass spectra (SWATH) analysis for characterization and quantification of histone post-translational modifications. *Molecular & Cellular Proteomics*, *14*(9), 2420–2428. https://doi.org/10.1074/mcp.O114.046102.

Solntsev, S. K., Shortreed, M. R., Frey, B. L., & Smith, L. M. (2018). Enhanced global post-translational modification discovery with MetaMorpheus. *Journal of Proteome Research*, *17*(5), 1844–1851. https://doi.org/10.1021/acs.jproteome.7b00873.

Steen, H., & Mann, M. (2004). The ABC's (and XYZ's) of peptide sequencing. *Nature Reviews. Molecular Cell Biology*, *5*(9), 699–711. https://doi.org/10.1038/nrm1468.

Stein, S. E., & Scott, D. R. (1994). Optimization and testing of mass spectral library search algorithms for compound identification. *Journal of the American Society for Mass Spectrometry*, *5*(9), 859–866. https://doi.org/10.1016/1044-0305(94)87009-8.

Strahl, B. D., & Allis, C. D. (2000). The language of covalent histone modifications. *Nature*, *403*(6765), 41–45. https://doi.org/10.1038/47412.

Su, M. G., Huang, K. Y., Lu, C. T., Kao, H. J., Chang, Y. H., & Lee, T. Y. (2014). topPTM: A new module of dbPTM for identifying functional post-translational modifications in transmembrane proteins. *Nucleic Acids Research*, *42*(Database issue), D537–D545. https://doi.org/10.1093/nar/gkt1221.

Sun, J., Shi, J., Wang, Y., Wu, S., Zhao, L., Li, Y., et al. (2019). Open-pFind enhances the identification of missing proteins from human testis tissue. *Journal of Proteome Research*, *18*(12), 4189–4196. https://doi.org/10.1021/acs.jproteome.9b00376.

Swaney, D. L., Beltrao, P., Starita, L., Guo, A., Rush, J., Fields, S., et al. (2013). Global analysis of phosphorylation and ubiquitylation cross-talk in protein degradation. *Nature Methods*, *10*(7), 676–682. https://doi.org/10.1038/nmeth.2519.

Tanner, S., Shu, H., Frank, A., Wang, L. C., Zandi, E., Mumby, M., et al. (2005). InsPecT: Identification of posttranslationally modified peptides from tandem mass spectra. *Analytical Chemistry*, 77(14), 4626–4639. https://doi.org/10.1021/ac050102d.

Taus, T., Kocher, T., Pichler, P., Paschke, C., Schmidt, A., Henrich, C., et al. (2011). Universal and confident phosphorylation site localization using phosphoRS. *Journal of Proteome Research*, *10*(12), 5354–5362. https://doi.org/10.1021/pr200611n.

Taylor, J. A., & Johnson, R. S. (1997). Sequence database searches via de novo peptide sequencing by tandem mass spectrometry. *Rapid Communications in Mass Spectrometry*, *11*(9), 1067–1075. https://doi.org/10.1002/(SICI)1097-0231(19970615)11:9<1067::AID-RCM953>3.0.CO;2-L.

Thygesen, C., Boll, I., Finsen, B., Modzel, M., & Larsen, M. R. (2018). Characterizing disease-associated changes in post-translational modifications by mass spectrometry. *Expert Review of Proteomics*, *15*(3), 245–258. https://doi.org/10.1080/14789450.2018.1433036.

Toghi Eshghi, S., Yang, W., Hu, Y., Shah, P., Sun, S., Li, X., et al. (2016). Classification of tandem mass spectra for identification of N- and O-linked glycopeptides. *Scientific Reports*, *6*, 37189. https://doi.org/10.1038/srep37189.

Tran, N. H., Zhang, X., Xin, L., Shan, B., & Li, M. (2017). De novo peptide sequencing by deep learning. *Proceedings of the National Academy of Sciences of the United States of America, 114*(31), 8247–8252. https://doi.org/10.1073/pnas.1705691114.

Wagner, S. A., Beli, P., Weinert, B. T., Scholz, C., Kelstrup, C. D., Young, C., et al. (2012). Proteomic analyses reveal divergent ubiquitylation site patterns in murine tissues. *Molecular & Cellular Proteomics, 11*(12), 1578–1585. https://doi.org/10.1074/mcp.M112.017905.

Walsh, C. T., Garneau-Tsodikova, S., & Gatto, G. J., Jr. (2005). Protein posttranslational modifications: The chemistry of proteome diversifications. *Angewandte Chemie (International Ed. in English), 44*(45), 7342–7372. https://doi.org/10.1002/anie.200501023.

Weinert, B. T., Moustafa, T., Iesmantavicius, V., Zechner, R., & Choudhary, C. (2015). Analysis of acetylation stoichiometry suggests that SIRT3 repairs nonenzymatic acetylation lesions. *The EMBO Journal, 34*(21), 2620–2632. https://doi.org/10.15252/embj.201591271.

Welle, K. A., Zhang, T., Hryhorenko, J. R., Shen, S., Qu, J., & Ghaemmaghami, S. (2016). Time-resolved analysis of proteome dynamics by tandem mass tags and stable isotope labeling in cell culture (TMT-SILAC) hyperplexing. *Molecular & Cellular Proteomics, 15*(12), 3551–3563. https://doi.org/10.1074/mcp.M116.063230.

Wen, B., Zeng, W. F., Liao, Y., Shi, Z., Savage, S. R., Jiang, W., et al. (2020). Deep learning in proteomics. *Proteomics, 20*(21 − 22), e1900335. https://doi.org/10.1002/pmic.201900335.

Wolf-Yadlin, A., Kumar, N., Zhang, Y., Hautaniemi, S., Zaman, M., Kim, H. D., et al. (2006). Effects of HER2 overexpression on cell signaling networks governing proliferation and migration. *Molecular Systems Biology, 2*, 54. https://doi.org/10.1038/msb4100094.

Woodsmith, J., Jenn, R. C., & Sanderson, C. M. (2012). Systematic analysis of dimeric E3-RING interactions reveals increased combinatorial complexity in human ubiquitination networks. *Molecular & Cellular Proteomics, 11*(7). https://doi.org/10.1074/mcp.M111.016162, M111.016162.

Xu, H., Wang, Y., Lin, S., Deng, W., Peng, D., Cui, Q., et al. (2018). PTMD: A database of human disease-associated post-translational modifications. *Genomics, Proteomics & Bioinformatics, 16*(4), 244–251. https://doi.org/10.1016/j.gpb.2018.06.004.

Xu, Y., Wang, X., Wang, Y., Tian, Y., Shao, X., Wu, L. Y., et al. (2014). Prediction of posttranslational modification sites from amino acid sequences with kernel methods. *Journal of Theoretical Biology, 344*, 78–87. https://doi.org/10.1016/j.jtbi.2013.11.012.

Yadav, A. K. (2017). Commentary: Deep phosphoproteomic measurements pinpointing drug induced protective mechanisms in neuronal cells. *Frontiers in Physiology, 8*, 174. https://doi.org/10.3389/fphys.2017.00174.

Yadav, A. K., Bhardwaj, G., Basak, T., Kumar, D., Ahmad, S., Priyadarshini, R., et al. (2011). A systematic analysis of eluted fraction of plasma post immunoaffinity depletion: Implications in biomarker discovery. *PLoS One, 6*(9), e24442. https://doi.org/10.1371/journal.pone.0024442.

Yadav, A. K., Kadimi, P. K., Kumar, D., & Dash, D. (2013). ProteoStats—A library for estimating false discovery rates in proteomics pipelines. *Bioinformatics, 29*(21), 2799–2800. https://doi.org/10.1093/bioinformatics/btt490.

Yadav, A. K., Kumar, D., & Dash, D. (2011). MassWiz: A novel scoring algorithm with target-decoy based analysis pipeline for tandem mass spectrometry. *Journal of Proteome Research, 10*(5), 2154–2160. https://doi.org/10.1021/pr200031z.

Yadav, A. K., Kumar, D., & Dash, D. (2012). Learning from decoys to improve the sensitivity and specificity of proteomics database search results. *PLoS One, 7*(11), e50651. https://doi.org/10.1371/journal.pone.0050651.

Yang, H., Chi, H., Zeng, W. F., Zhou, W. J., & He, S. M. (2019). pNovo 3: Precise de novo peptide sequencing using a learning-to-rank framework. *Bioinformatics*, *35*(14), i183–i190. https://doi.org/10.1093/bioinformatics/btz366.

Yates, J. R., 3rd, Eng, J. K., McCormack, A. L., & Schieltz, D. (1995). Method to correlate tandem mass spectra of modified peptides to amino acid sequences in the protein database. *Analytical Chemistry*, *67*(8), 1426–1436. https://doi.org/10.1021/ac00104a020.

Yilmaz, S., Ayati, M., Schlatzer, D., Cicek, A. E., Chance, M. R., & Koyuturk, M. (2021). Robust inference of kinase activity using functional networks. *Nature Communications*, *12*(1), 1177. https://doi.org/10.1038/s41467-021-21211-6.

Zaborowska, J., Egloff, S., & Murphy, S. (2016). The pol II CTD: New twists in the tail. *Nature Structural & Molecular Biology*, *23*(9), 771–777. https://doi.org/10.1038/nsmb.3285.

Zheng, W., Wuyun, Q., Cheng, M., Hu, G., & Zhang, Y. (2020). Two-level protein methylation prediction using structure model-based features. *Scientific Reports*, *10*(1), 6008. https://doi.org/10.1038/s41598-020-62883-2.

Zhou, Y., Dong, J., & Vachet, R. W. (2011). Electron transfer dissociation of modified peptides and proteins. *Current Pharmaceutical Biotechnology*, *12*(10), 1558–1567. https://doi.org/10.2174/138920111798357230.

Big data, integrative omics and network biology

Priya Tolani[a], Srishti Gupta[a,b], Kirti Yadav[a,c], Suruchi Aggarwal[a,d], and Amit Kumar Yadav[a,]*

[a]Translational Health Science and Technology Institute, NCR Biotech Science Cluster, Faridabad, Haryana, India
[b]School of Biosciences and Technology, Vellore Institute of Technology, Vellore, India
[c]Department of Pharmaceutical Biotechnology, Delhi Pharmaceutical Sciences and Research University, New Delhi, India
[d]Department of Molecular Biology and Biotechnology, Cotton University, Guwahati, Assam, India
*Corresponding author: e-mail address: amit.yadav@thsti.res.in

Contents

Advances in Protein Chemistry and Structural Biology, Volume 127
ISSN 1876-1623
https://doi.org/10.1016/bs.apcsb.2021.03.006

Abstract

A cell integrates various signals through a network of biomolecules that crosstalk to synergistically regulate the replication, transcription, translation and other metabolic activities of a cell. These networks regulate signal perception and processing that drives biological functions. The biological complexity cannot be fully captured by a single -omics discipline. The holistic study of an organism—in health, perturbation, exposure to environment and disease, is studied under systems biology. The bottom-up molecular approaches (genes, mRNA, protein, metabolite, etc.) have laid the foundation of current biological knowledge covering the horizon from viruses, bacteria, fungi, plants and animals. Yet, these techniques provide a rather myopic view of biology at the molecular level. To understand how the interconnected molecular components are formed and rewired in disease or exposure to environmental stimuli is the holy grail of modern biology. The omics era was heralded by the genomics revolution but advanced sequencing techniques are now also ubiquitous in transcriptomics, proteomics, metabolomics and lipidomics. Multi-omics data analysis and integration techniques are driving the quest for deeper insights into how the different layers of biomolecules talk to each other in diverse contexts.

1. Introduction

The sequencing revolution in genomics, enabled a high-throughput concomitant measurement of all genes in a single experiment. Many other biomolecules such as transcripts, proteins, and metabolites soon followed the omics path, driven by technological advances that permitted large-scale and cost-efficient analysis. The ability to survey global gene expression patterns quickly found application in disease biology. A combination of high-throughput genotyping, statistical tools, high quality reference map of the human genome, and large cohorts of thousands of patients, has enabled the mapping of genetic variants contributing to diseases (Hasin, Seldin, &

Lusis, 2017). Proteomics connects the genes with their functionally diverse protein products (Tyers & Mann, 2003). Proteogenomics integrates genomic knowledge with novel protein products (alternate start/end sites, modifications, splice-junction peptides and amino-acid variants) for deeper functional insights (Kelkar et al., 2011; Kumar et al., 2013; Kumar, Mondal, Yadav, & Dash, 2014; Kumar, Yadav, Jia, Mulvenna, & Dash, 2016). The reflection of these omics states is seen in metabolome and lipidome, which is driven by diverse proteoforms to regulate cellular metabolism in health and disease. Systems biology aims to model complex biological interactions by integrating information from interdisciplinary fields in a holistic manner. In order to thoroughly understand disease, conventional reductionist approach is not enough. The network biology of normal and abnormal phenotype allows predictive, preventive, personalized, and participatory (P4) medicine for the proactive conservation of well-being relevant to the individual. It is important to integrate multi-omics (genomics, transcriptomics, proteomics, metabolomics, lipidomics, etc.) data for holistic understanding to reach the goals of P4 medicine (Alyass, Turcotte, & Meyre, 2015; Flores, Glusman, Brogaard, Price, & Hood, 2013). Employment of multi-omics approach has resulted in the development of tools, methods, and platforms provisioning multi-omics data analysis, visualization, and interpretation (Huang, Chaudhary, & Garmire, 2017). Methods for integration of multi-omics datasets can resolve various challenges related to disease such as, disease subtyping and classification, prediction of biomarkers, etc. The approaches or methods can be largely classified into: network, Bayesian, fusion, similarity-based, correlation-based, and other multivariate methods (Subramanian, Verma, Kumar, Jere, & Anamika, 2020). The microbiome plays important role in disease biology, as the microbial small molecules and metabolites affect the physiology of an individual. The great diversity of microbial chemical structures and their interactions improves the understanding of pathology and treatment of microbiome-related diseases such as diabetes, obesity and cancer (Yang, Karr, Watrous, & Dorrestein, 2011). Exposome is another integral component of organism that corresponds to the totality of environmental exposure (non-genetic) over the lifetime of an organism. Exposure to chemical and physical stressors, biologicals, psychological and social stress, as well as cumulative exposures can lead to many non-communicable diseases including metabolic diseases and obesity, cancers, neurodevelopmental and neurodegenerative disorders, allergies and autoimmune diseases, pulmonary and cardiovascular diseases. Exposomics will provide better understanding of the role of environmental risk factors in respiratory disease and other chronic pathologies (Barouki, Audouze, Coumoul, Demenais, & Gauguier, 2018).

2. Rise of the omics

The generation of large datasets in genomics, transcriptomics and proteomics approaches necessitate analysis and integration strategies for interdisciplinary data for better understanding of biological systems (Manzoni et al., 2018). Some of these omics are enumerated in Fig. 1.

2.1 Genomics

Genomics is the study of complete set of genes of an organism, which focuses on structure, function, and evolution of genomes. The new experimental technologies have resulted in a steady stream of larger and more complex genomic datasets flowing into public databases, revolutionizing the study

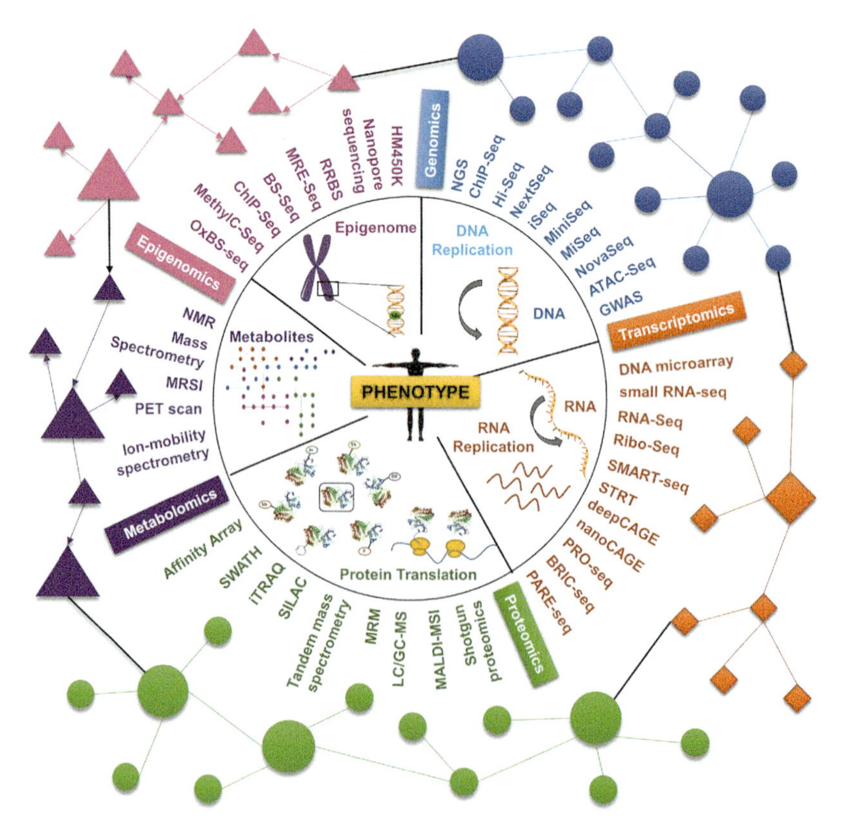

Fig. 1 An overview of omics techniques and approaches for studying biomolecules across different layers of omics and their semantic connections to form the complete connectome which can be studied in health and disease.

of virtually all life processes. Advancement in genetics, comparative genomics, high-throughput biochemistry and bioinformatics are providing biologists with a dramatically improved repertoire of research tools to examine and comprehend the functioning of organisms in healthy and diseased state at molecular level (Collins, Green, Guttmacher, Guyer, & Institute, 2003). Genomics includes identification of structural and functional components, understanding the heritable variation in the human genome, elucidating the mechanism behind evolutionary variation across species, understanding of genes and pathways to develop new therapeutic approaches to disease (Collins et al., 2003). It is important to study identification of variants (copy number variants, single nucleotide variations, structural variations) or mutations (like insertion, deletion, inversions) which exist in the genome of an organism. The techniques available for genomics and capturing genetic variation are Sanger sequencing, DNA-microarrays and next generation sequencing (NGS). Currently, sequencing is mostly performed on NGS platforms that use massively parallel sequencing to sequence millions of DNA fragments from a single sample at a time (Behjati & Tarpey, 2013). There are several NGS platforms that provide low-cost, high-throughput sequencing, which include template/library preparation, sequencing/imaging, and data analysis. It involves various techniques with similar basic methodology but different purposes, such as NextSeq, NovaSeq, iSeq, MiSeq, MiniSeq, Hi-Seq, ChIP-Seq, RNA-Seq, MeDIP-Seq, etc.

Genome-wide association studies (GWAS) involve testing of genetic variants (single nucleotide polymorphisms or SNPs) following a case-control design, in which the occurrence of SNPs is compared between healthy and diseased individuals. In GWAS, thousands of single nucleotide polymorphisms (SNPs) are genotyped for association with a disease or trait (Wang, Cordell, & Van Steen, 2019). Assay for Transposase-Accessible Chromatin using sequencing (ATAC-Seq) is a recent high throughput sequencing technique for studying chromatin accessibility across genome and is mainly used for single-cell genomics. The nucleosome positions, gene expression control machinery, the binding sites of transcription factors and accessibility of chromatin between the biological samples can be studied using this technique (Yan, Powell, Curtis, & Wong, 2020).

2.2 Epigenomics

Epigenomics is the study of heritable changes other than those encoded in the DNA sequence. Epigenetics includes any process that alters gene activity without change in the DNA but leads to heritable modifications.

Epigenetics covers modifications of DNA or chromatin, i.e., DNA methylation, modification of cytosine, and post-translational modification of histones that plays an important role in controlling transcription activity (Callinan & Feinberg, 2006). Most commonly studied epigenetic event is DNA methylation specifically cytosine followed by guanine (CpG) dinucleotides found in CpG islands and found to be methylated (Fazzari & Greally, 2010). DNA methylation controls important cellular processes like embryogenesis and carcinogenesis (Gao et al., 2020). Several techniques are available to study epigenetic changes such as, Methylated DNA immunoprecipitation-sequencing (MeDIP-seq), Whole-genome bisulfite sequencing (WGBS) (including MethylC-seq and BS-seq), Reduced-Representation Bisulfite Sequencing (RRBS), etc., to study patterns of genomic methylation. Chromatin immunoprecipitation sequencing (ChIP-Seq), is used to chart the chromatin modification and transcription factor (TF) binding sites across the genome (Park, 2009).

2.3 Transcriptomics

Transcriptomics involves the study of RNA expression from specific tissues, developmental stage or disease. This offers an insight into cell and tissue specific gene expression, predicting protein isoforms using alternate splicing and assessment of genotype on gene expression with expression quantitative trait loci (eQTL). A comprehensive view of how the transcriptome profile changes affect the biological or the disease condition (Manzoni et al., 2018) that includes study of somatic mutations including insertions and can be obtained using transcriptomics. Various methods include Sanger sequencing of EST (Expressed Sequence Tags) or cDNA library, like Serial and Cap analysis of gene expression (SAGE/CAGE), DNA microarrays, etc. (Dong & Chen, 2013; Tarca, Romero, & Draghici, 2006). It helps in the detection of diagnostic biomarkers, understanding of disease pathways, classification of diseases and monitoring of therapeutic response.

Ribosome footprinting, also known as Ribosome profiling, or Ribo-Seq, or ART-Seq (active mRNA translation sequencing) quantitatively measures ribosome occupancy and translation, by sequencing mRNA fragments protected by ribosome. This helps in identifying the exact position of ribosomes, and also reveals the presence of ribosomes on upstream open reading frames, measuring active translation (Ingolia, 2016). This technique integrates mRNA abundance and translational regulation, and accurately demarcates the translated regions to identify full coding potential of the genome. Ribo-Seq protocol consists of following steps: drug treatment and cell harvesting, nuclease footprinting and RPF isolation, library preparation and sequencing, data analysis and downstream analysis (Calviello & Ohler, 2017).

2.4 Proteomics and interactomics

Proteomics is the qualitative and quantitative study of the proteome (complete set of all expressed proteins in a cell, tissue, or organism in the given context). In addition to profiling all proteins, proteomics also encompasses the study of isoforms, modifications, their interactions and protein complexes (Tyers & Mann, 2003). The analysis of shotgun proteomics data is computationally challenging (Patterson & Aebersold, 2003). The proteomic approaches include gel-based separation: one-dimensional and two-dimensional polyacrylamide gel electrophoresis, and gel-free high throughput technologies, including multidimensional protein identification technology, and separation using Liquid chromatography–mass spectrometry (LC–MS). Shotgun proteomics involves the analysis of samples by extraction, digestion, liquid chromatography, followed by tandem mass spectrometry. Mass spectrometers consist of an ion source, the mass analyzer, and an ion detection system. Analysis of proteins by MS occurs as follows: (a) protein ionization and generation of gas-phase ions, (b) separation of ions according to their mass to charge ratio, and (c) detection of ions. Matrix assisted laser desorption/ionization (MALDI) and electrospray ionization (ESI) are the major ionization methods. There are various mass analyzers in use for proteomics applications, i.e., time-of-flight (TOF), ion trap (and Orbitrap), quadrupole, and Fourier transform ion cyclotron resonance (FTICR). The experimentally derived peptide masses are correlated with the theoretical peptide MS/MS of known proteins in the databases using search tools, such as Mascot (Perkins, Pappin, Creasy, & Cottrell, 1999), Sequest (Eng, McCormack, & Yates, 1994), MassWiz (Yadav, Kumar, & Dash, 2011), MSGF+ (Kim & Pevzner, 2014), X!Tandem (Craig & Beavis, 2004), rescored through tools like Percolator (Kall, Canterbury, Weston, Noble, & MacCoss, 2007; Yadav, Kumar, & Dash, 2012) and passed through FDR analysis and protein inference (Aggarwal & Yadav, 2016; Nesvizhskii & Aebersold, 2005; Yadav et al., 2011).

Interactomics deals with measuring interactions between proteins using affinity purification-mass spectrometry (AP-MS), and how these interactions are topologically arranged into networks. Interactions may occur between different biomolecules proteins, nucleic acids, lipids, and carbohydrates. Interactomes may be described as gene interaction networks, protein-protein interaction (PPI) networks and protein-DNA interactions, etc. Interactomics is applied to compare the networks between biological states, times, or species so as to follow information flow, patterns in network variation or rewiring. It can reveal dysfunctional pathways and speed up biomarker discovery (Keskin, Gursoy, Ma, & Nussinov, 2007). Such studies have been performed employing yeast two-hybrid screens (Y2H screens) and related complementation assays, AP-MS, proximity labeling approaches like BioID and APEX, cross-linking mass spectrometry (XL-MS) and protein co-fractionation coupled to mass spectrometry (CoFrac-MS) (Bludau & Aebersold, 2020).

Algorithms have been developed to predict interaction network of proteins, for the characterization of functions and possible targets for drug discovery. The overall integration of large-scale data emanating from different sources, such as proteomics, metabolomics, and protein interactomics can reveal the foundations of molecular interactions in controlling signal flow. Many computational tools and algorithms have been developed to integrate the analysis results, such as STRING, Cytoscape, Ingenuity Pathway Analyses, and Pathway Studio, etc. (Martins-de-Souza, 2014).

2.5 Proteogenomics and PTMs

Proteogenomics includes the integration of genomics/transcriptomics with proteomics, to search tandem mass spectra against nucleic acid databases to identify, annotate and characterize novel as well as known protein-coding genes. Proteogenomic includes validation of translation, determining the correct reading frame, gene and exon boundaries, alternative splicing, along with discovery of novel genes (Castellana & Bafna, 2010). In this approach, sample specific genomic or transcriptomic data is used to create custom protein sequence databases to help identify novel peptides (not present in reference protein sequence databases) using shotgun proteomics data for gene model refinement (Kumar, Yadav, & Dash, 2017). The database may also be generated using RNA-seq and/or ribosome-profiling data and can also identify protein-coding long noncoding RNA (lncRNA) genes and pseudogenes. False positives are controlled using target-decoy approach (Aggarwal & Yadav, 2016; Yadav, Kadimi, Kumar, & Dash, 2013). The peptides identified using proteogenomics search are mapped to known protein reference databases to establish novelty of the peptides (Nesvizhskii, 2014).

Post-translational modifications (PTMs) are covalent addition of a functional group to amino acid side chains that modulate its activity, subcellular localization, turnover, and interactions. The PTMs initiate signaling from membrane receptor, transduce it to cytoplasmic receptors and convey the signal to nuclear transcription factors that regulate gene expression. Proteins can be modified by phosphorylation, ubiquitination, acetylation, methylation, nitrosylation, etc. These modifications can be transient and reversible, or long-lasting, can act as on/off switch or fine tune responses like a rheostat. The crosstalk between different PTMs and pathways modulates important biological functions (Aggarwal, Banerjee, Talukdar, & Yadav, 2020; Aggarwal, Kandpal, Asthana, & Yadav, 2017; Kandpal, Aggarwal, Jamwal, & Yadav, 2017; Nie, Gong, Liu, & Li, 2017; Wu, Huang, & Yuan, 2019). Reversible phosphorylation controls enzyme activity and signaling pathways. Acetylation regulates DNA recognition, protein stability and its interactions. Ubiquitin targets proteins for degradation, and also DNA repair, signal transduction and autophagy. The dysregulation of PTMs is known to be involved in many diseases (Yadav, 2017). The analysis of specific post translational modifications can be performed on a large-scale by PTM enrichment, followed by mass spectrometry (Pieroni et al., 2020). The data is then analyzed by database search algorithms with variable modification search followed by FDR control.

2.6 Metagenomics and metaproteomics

Metagenomics involves the functional analysis of the community microbial genomes (microbiome) in a particular environmental niche. It includes culture-independent genomic analysis of microbial diversity and their functional roles in the niche such as, soil, water body, gastrointestinal tracts of animals and humans. It provides an unbiased assessment of community structure (species richness and distribution) and its associated functional (metabolic) potential (Hugenholtz & Tyson, 2008). Functional metagenomics screens the metagenomics libraries searching for a particular phenotype such as salt tolerance, antibiotic production or enzyme activity, and then identifies the phylogenetic origin of the cloned DNA. In sequencing based method, clones are screened for the identifying the conserved 16S rRNA genes, followed by sequencing of the complete clone which can identify further interesting genes. Environmental metagenomics from soil, marine water, industrial sludge, etc., and the gut microbiome from insects and humans are the focus of large-scale studies on sequencing metagenomes (Sleator, Shortall, & Hill, 2008). The sequences from the metagenome sample contains different species in a single sample and features like microbial genome size, taxonomy, and functional content, GC content are compared between the samples in comparative studies. These studies offer additional understanding of associations between environment and metagenomes, and improve our understanding of symbiosis, enrichment of gene families and environmental virology (Wooley, Godzik, & Friedberg, 2010).

Metaproteomics also aims to understand and determine the main functional components in a microbial ecosystem using proteomics approaches. It is the large-scale characterization of the full protein complement of the microbiota at a given point in time (Wilmes, Heintz-Buschart, & Bond, 2015). The vital role of metaproteomics studies is to highlight the links between genomic and functional diversity of microbial communities. The outcome and success of metaproteomic approaches are largely dependent upon the metagenomics and metatranscriptomics information available for experimental design and analysis as it can inform judicious database construction and functional analysis and correlation across the omics. The mapping of proteins from the environmental proteomics or metaproteomics samples from native microbial communities can reflect the novel functional pathways and associated genes. Metaproteome analysis involves the selection of microbiota samples, extraction of the community proteome, separation of proteins using 2D-gel electrophoresis or 1D-gel electrophoresis with LC-MS/MS, data acquisition, statistical analysis, analysis of community functional organization and finally linking this proteomic functional diversity to the genetic diversity provided by the microbiome composition. Metaproteomics is applied to the problem of dissecting out the functional indicators of a microbial community, to track new genes in complex metabolic pathways that provide functional insights into microbial ecological parameters like resistance, resilience and functional redundancy (Maron, Ranjard, Mougel, & Lemanceau, 2007).

2.7 Metabolomics

Metabolomics is the large-scale study of all metabolites in a cell and provides a functional context to the proteome. While proteome is the regulatory workhorse of a cell, its impact can be seen in the metabolites and thus metabolomics helps in bridging the genotype-to-phenotype lacuna. It magnifies the changes in proteome and reflects the phenotype at molecular level. Metabolomics includes both targeted and non-targeted analysis of endogenous and exogenous metabolites which are generally small molecules with masses <1500 Da. Metabolomics is useful for assessment of diverse types of stress responses—environmental, mutational, genetic manipulation, comparing growth stages or tissues, and natural product discovery. A global profiling of metabolites uses high resolution instrumentation (typically NMR and MS) along with statistical tools like the principal component analysis (PCA) and partial least squares (PLS). This provides an integrated view of metabolism reflecting the diet and lifestyle influences with respect to diseases. The numerous analytical platforms for large-scale metabolomics analysis are NMR, Fourier transform-infrared spectroscopy (FT-IR) and MS coupled to liquid chromatographic separation techniques that includes NMR, GC-MS, LC-MS, FT-MS and UPLC-MS. NMR spectroscopy is mainly suitable for the bulk analysis of metabolites. The volatile organic compounds and derivatized primary metabolites can be best analyzed using GC-MS. The multitude of diverse semi-polar compounds and secondary metabolites are most amenable for analysis using LC-MS. The metabolic and signaling pathways can be measured in a multiplex and broad manner using the UPLC-MS technique (Zhang, Sun, Wang, Han, & Wang, 2012).

2.8 Fluxomics

Most omics datasets provide information on qualitative pathways activity and lack the data on activity of the proteins which can be modulated or altered by protein post-translational modifications, allosteric regulation, etc. Fluxomics is the study of comprehensive flux in the metabolic network of a cell, which can measure and evaluate the rates of reactions (fluxes) for a network of metabolic reactions in an organism. These metabolic fluxes represent the end outcome of the interaction between gene expression, protein abundance, their enzyme kinetics, regulation and metabolite concentrations (thermodynamic driving forces) which combines to constitute the metabolic phenotype (Winter & Kromer, 2013). Several methods are in use for quantification of metabolic flux, such as flux balance analysis and stoichiometric metabolic flux analysis. The most dependable methods use isotope-labeled precursors of metabolic pathways, primarily ^{13}C-labeled substrates. The ^{13}C-labeled substrates are fed to the cells, tissues or animals which results in ^{13}C containing metabolites, which are then measured. The tracer-based metabolomics is used for determination of concentration and isotope distribution (or labeling pattern) of these metabolites to model the flux distribution in the metabolic network (Cascante & Marin, 2008).

2.9 Phenomics

Phenomics is analysis of phenotypic assessment of morphological, physiological, and biochemical characters of an organism in high throughput, and also includes its association with the genetic, epigenetic, and environmental factors (Grosskinsky, Svensgaard, Christensen, & Roitsch, 2015). Phenotypic variations are a result of dynamic interactions between an organism's genotype and its environment. Phenotypic data can help us recognize which gene variants affect a phenotype, pleiotropic effects and to infer the causes of health, crop yields, disease and evolutionary fitness (Houle, Govindaraju, & Omholt, 2010). The phenomic data spans multiple levels like quantitative genetics, evolutionary biology, epidemiology and physiology. Phenome measurements to untangle causes from correlations, can be gene expression profiling, proteomics mass spectrometry, metabolome-wide association studies, imaging, etc. (Houle et al., 2010).

2.10 Exposomics

The "exposome" concept represents the environmental, i.e., non-genetic, drivers of health and disease. The biological samples with molecular measurements, integrated with the coverage of the internal and external contributors to the exposome, like biological perturbations and external chemicals from air, water, or food or from other natural processes contribute to the exposome. The different approaches for systematic mapping of the exposome are mass spectrometry, sensors, wearables, biostatistics, and bioinformatics. High-resolution mass spectrometry (HRMS) has enhanced our analytical capacity to measure well-known metabolites, pollutants as well as other externally derived small molecules like pharmaceuticals, preservatives, pesticides and other microbial metabolites. Chemicals entities in the body are not stagnant and can react in that environment to form secondary metabolites or altered products which can be predicted by computational tools, and in this direction, the recently proposed environment-wide association studies (EWAS) aim to decipher the environmental causes of disease phenotypes. EWAS as the name suggests, was inspired by the analytical methods developed and applied in GWAS, uses a panel of "exposures," similar to genotype variants, to study the phenotype of interest. A multi-layered network framework of organism and its entirety of exposures needs to be developed for illuminating their roles in health and disease (Vermeulen, Schymanski, Barabasi, & Miller, 2020).

2.11 Single-cell omics

Single-cell omics approaches dig deeper into the developmental and communication networks in cells and tissues delineating the heterogeneity at the single cell level in high-resolution and throughput. Single-cell techniques can generate comprehensive cellular maps from different types of cells in healthy and pathogenic conditions. With the development of advanced computational methods, the multimodal omics data generated using these techniques can be now seamlessly integrated and characterized in a semantic manner. This will allow the classification of cell types and phenotypic insights into interactions and spatial organization of cells (Efremova & Teichmann, 2020). Single-cell RNA sequencing (scRNA-seq) involves high throughput measurement of cellular gene expression levels. The technique allows for a thorough, in-depth portrayal of cellular subtypes and states of a tissue. It includes several high-throughput protocols separating single cells, such as Fluidigm C1 platform, cell-specific barcoded complementary-DNA libraries, and droplet microfluidics and microwells. Several methods have added to the unearthing of new and rare cell types and subtypes and their complex interplay as well as biological mechanisms. The single-nucleotide variations (SNVs) and copy-number variations (CNVs) can be compared in different cell types to showcase the cell-to-cell variability involved in disease processes. Even epigenetic regulation at the single-cell can unveil the status of DNA epigenetic modifications, accessibility and chromosome conformation. A variety of single-cell epigenomic sequencing techniques have been developed, which have been combined with single-cell transcriptomics. The single-cell proteomics is a field in its infancy that uses high-throughput method mass spectrometry (SCoPE-MS) to bring the proteome to the mass spectrometer with minimal loss and the concurrent profiling of peptide along with their quantitation. It can discover >1000 proteins in a single cell. Other approaches for cellular proteins detection can use antibodies conjugated to DNA barcodes. These are concomitantly measured with the transcriptome of a cell in a modified scRNA-seq method. With these sophisticated strategies, scRNA protocols can be multiplexed and applied to provide accurate protein quantification (He, Memczak, Qu, Belmonte, & Liu, 2020).

3. Network biology across regulatory layers and across omics

Systems biology aims to model complex biological systems by employing a holistic view on all cellular processes. The biological layers, such as the genome, transcriptome, proteome and metabolome, maintain homeostasis through their molecular interaction networks. Networks constitute the foundation of biological systems, and systems biology attempts to understand these molecular wirings. Interactome is defined as a network of nodes, which represent individual molecules (gene, protein, DNA, etc.) and connections between these nodes (edges) which reflect their relationship in a graph theoretical format. To establish interactions, experimental assays and computational methods can systematically assemble and predict interactions between molecules, so as to build interactomes that can also be integrated across diverse molecular layers. For example, protein-protein interactions (PPI), protein-protein complexes, PTM networks, disease-gene networks, drug-targeted networks, etc. (Hawe, Theis, & Heinig, 2019).

3.1 PPI networks

Protein–protein interactions (PPIs) orchestrate cellular communication and function through the control of signal transduction pathways and regulatory networks. Biological networks can provide insights into the mechanisms that trigger the onset and progression of diseases. Protein interaction networks are important for decoding the relationships between network structure and function, identifying functional modules, conserved interaction patterns and discovering novel protein functions. Protein interaction studies play a major role in the prediction of genotype-phenotype associations through the use of high-throughput methods like X-ray crystallography, fluorescence and atomic force microscopy, NMR spectroscopy, Yeast two-hybrid (Y2H), gene co-expression methods. The computational interaction prediction involves methods predicting protein domain interactions from existing empirical data on PPIs, and methods relying on theoretical information to predict protein-protein or domain-domain interactions (Safari-Alighiarloo, Taghizadeh, Rezaei-Tavirani, Goliaei, & Peyvandi, 2014).

3.2 Signaling and PTM networks

Protein modifications regulate cell signaling events and rapidly reprogram individual protein functions. PTMs are added and removed in a highly dynamic manner and proteins exist in many different mod-forms. PTMs provide the proteome with an enormous capacity for biological diversity and regulate inter- and intra-cellular communication, cell growth, differentiation, and cell-division. Errors in reading and writing PTMs are causal agents of many human diseases. Cellular signaling responses require rapid modification of specific residues in a protein for signal transduction (Theillet et al., 2012) and regulating biological functions.

3.3 Protein-protein complexes

A protein complex is a group of polypeptide chains linked by noncovalent protein-protein interactions (PPIs). Protein complexes play important roles in biological systems and perform numerous functions, such as DNA transcription, mRNA translation, and signal transduction (Xu et al., 2018). Almost all biological processes involve protein-protein interactions and many of those may require multiple protein-protein interactions to form the quaternary structure of multimeric proteins, to form the protein-protein complexes. The understanding of protein-protein interactions and its specificity at atomic detail requires the knowledge of the three-dimensional (3D) structure of protein complexes and protein-protein interfaces. A single protein can be involved in variety of protein complexes. Same complex can perform different functions depending on several factors such as, stage of cell cycle, the nutritional status of the cell, the cellular compartment, etc. Nuclear magnetic resonance (NMR) spectroscopy, cryo-electron microscopy (CryoEM) and X-ray crystallography are the main experimental techniques to study protein-protein complexes. 3D structures of a large number of protein-protein complexes have facilitated the understanding of the recognition processes. Several web servers and databases are also available for structural analysis of the protein-protein complexes, which includes Protein Data Bank (PDB), protein-protein interaction server, database of interacting proteins (DIP), etc. (Bahadur & Zacharias, 2008).

3.4 Disease-gene networks

Identifying the molecular basis of the disease and their phenotypes is valuable in the prevention, diagnosis and treatment of diseases. Human diseases are a consequence of perturbations in the molecular networks due to genetic mutations, epigenetic changes and pathogens. The properties of disease genes in networks have revealed that genes associated with similar type of diseases, tend to exist in the same neighborhood and form functional topological modules. Genes causing similar phenotypes are also observed to be functionally related and are part of a biological module such as a protein complex or pathway. Such genes also possess considerably higher gene ontology (GO) homogeneity and co-expression tendency. The genes causing the same phenotype tend to form topological clusters and can be used to identify functionally similar genes or uncharacterized disease genes. Disease gene prioritization relies on the proximity of candidate genes to known disease genes within the interactome networks using scoring strategies like guilt-by-association, random walk, random walk with restart algorithm and kernelized score functions. The distance between candidate genes and known disease genes in the PPI network can measure pair-wise protein closeness in a network, useful for prioritizing disease genes (Wang, Gulbahce, & Yu, 2011).

3.5 Drug-target networks

To understand drug targets in the context of cellular and disease networks, a combination of drugs and their targets can be systematically interrogated. It involves the analysis of drug-target networks, evaluation of their network based relationships and quantifying the interrelationships between genes and drug targets (Yildirim, Goh, Cusick, Barabasi, & Vidal, 2007). The development of systems biology and network pharmacology has evolved the drug discovery paradigm. Instead of a linear relationship between drug-target and disease, the current network paradigm is of multiple drugs, targets and diseases integrated as a molecular network. The polypharmacological profile (i.e., on-target and off-target effects) of a drug could lead to both desired therapeutic effects and undesired safety problems. Hence, systematic identification of drug-target interactions (DTIs) is essential in drug discovery, which could help maximize therapeutic effects while minimizing safety problems (Wu, Li, Liu, & Tang, 2018). Identification of drug–target interactions (DTIs) plays an important role in drug discovery and development but experimental determination of DTIs is costly and time-consuming, which makes *in silico* or computational approaches necessary to identify potential DTIs for accelerating drug development and drug-repurposing. Several *in silico* approaches, such as structure-based, ligand-based and machine learning-based methods, have demonstrated their potential in predicting DTIs. Most existing methods for DTI prediction are limited to homogeneous networks or bipartite drug–target networks, and cannot be directly extended to heterogeneous, biological networks. In comparison to homogeneous networks, heterogeneous networks naturally assemble more objects and complementary information from drugs, targets/proteins and their associated diseases (Zeng et al., 2020).

3.6 Symptoms-disease networks

The symptoms are the highest level phenotypes crucial for clinical diagnosis and treatment. The wide range of symptoms is interdependent in the homeostatic process, which when perturbed leads to disease development. Symptoms are directly observable characteristics of a disease and form the primary basis of clinical disease classification. The elucidation of the shared connection between symptoms and genes or the protein–protein interactions in two diseases can help in bringing novel solutions for the diseases. The degree of shared symptom similarity and their shared genes or PPIs can be measured by integrating disease–gene association and protein–protein interaction (PPI) data. The link weights between two diseases was used to construct human symptoms-disease network (HSDN), and reflected by the similarity of their symptoms, which were measured for all pairs of diseases. In HSDN, 7,488,851 links with positive similarity between 4219 diseases were discovered to form a dense network with 94% of the nodes connected to >50% of other nodes. The correlation measured between clinical indicators and disease mechanisms can be useful for functional annotation of genes, and reveals consistencies between diverse disease classes (Zhou, Menche, Barabasi, & Sharma, 2014).

3.7 Network rewiring

Rewiring can be defined as intrinsic restructuring of biological relationships due to conditional transition. With the advancement in large-scale genomic and proteomic technologies that revealed interactions and regulatory relationships between biomolecules, many types of biological networks have been constructed. These include protein interaction, genetic interaction, transcription factor-target regulatory, miRNA-target regulatory, kinase-substrate phosphorylation, and metabolic pathways. Biological networks play a central role in speciation, though the evolutionary speed of biological networks is unknown. In cellular systems, biological networks may rewire at various rates during evolution (Shou et al., 2011). To measure rewiring for individual nodes as well as for the entire network, cosine distance or dissimilarity function are used. This provides an understanding of dynamic behavior and adaptation of networks as well as importance of rewiring in disease. For example, rewiring-based analyses can increase the number of driver mutations found in cancer. Rewiring of protein–protein interactions is more useful to understand dynamic cellular changes and it can detect differential gene essentiality between biological conditions (Hu, Thomas, & Brunak, 2016).

4. Data science in big data era

The speed of data generation in multi-omics sciences now far exceeds the speed of analysis. Due to the barrage of data, the analysis tools and approaches lag behind considerably. Despite these challenges, there have been several studies using the big data in biology in novel innovative ways using state of the art technologies and computational workflows. Large-scale analysis requires some working principles highlighted below.

4.1 FAIR principles

The FAIR data principles specify that data should be findable, accessible, interoperable and reusable as an urgent need to improve the infrastructure supporting data reuse (Wilkinson et al., 2016). There are several resources that already engage in various aspects of FAIR association. Machines can exchange interpretable data and metadata using the RDF, the resource description framework (RDF) globally accepted as key machine readable framework for data and knowledge representation. The lack of consensus on the criteria is a key challenge in achieving the objectives of FAIR principles. To fix this problem, a community-driven approach was used to compile standards, repositories, and policies at a single source. The first step in re-using data is to find the data and metadata both by machines and humans. A global unique and persistent identifier can be assigned and registered in resource where it can be reliably searched. Second step is accessibility which means that the user should be able to understand how to access the data they have found, which can be through access credentials, authentication and whether authorization to access data is provided. The eventual step of reusability entails that the metadata is supplied with good documentation for replication or combination with other resources for reuse. Although the biomedical research community embraces these FAIR guidelines, confusions still exist between being FAIR and how to implement or compare with other such standards. A template for applying FAIR principles was complemented with a FAIR metrics that covers most digital objects. It may not fit all the domain specific requirements, which led to development of a FAIRshake toolkit to assess the FAIRness of a digital resource in a systematic manner (Clarke et al., 2019). Various general purpose data repositories are present to follow good data sharing practices like Dataverse, DANS, FigShare, DataHub, Mendeley Data, and EUDat.

4.2 Reproducible research

The rise of the computational science and bioinformatics has led to exciting and fast evolving developments in several scientific areas with dramatically improved ability to analyze complex high dimensional data. Replication is the ultimate standard by which scientific claims are judged, in which independent investigators address a scientific hypothesis and build up evidence for or against it. Lack of reproducibility can be a burden for an individual researcher as well as others trying to validate the findings (Peng, 2011). A workflow for reproducible research involves data acquisition, data processing and data analysis, all of which require the use of software and automation. It enhances the reproducibility of research methods with documentation, like the R Markdown or the Jupyter notebook. The platform and tools to provision reproducible research are provided by the Open Science Framework (OSF). For reanalysis of data through a unified framework such as GenePattern or Galaxy, etc., the software may provide support for analysis and reproducibility rules. One should be able to recreate critical results as well as be able to use the previous work in a practical and efficient manner to increase productivity (Sandve, Nekrutenko, Taylor, & Hovig, 2013).

4.3 Wearables and wellness data

Wellness wearables are non-invasive devices that help measure a wide range of well-being features like temperature, blood pressure, blood oxygen level, physical activity levels, GPS location, sleep pattern, etc., that aim to enhance the wearer's capacity to perform best physically, mentally, and emotionally. These wearables can track and monitor chronic diseases like diabetes, hypertension, high cholesterol, by enabling the wearer to make healthier lifestyle and diet choices. The measurement of physical activity promotes setting personal goals and targets. A wide range of devices available in different colors and shapes makes it fashionable and promotes daily wellness monitoring. However, there is an uncertainty about data security and accessibility, like selling of data to the health analytical companies or insurance companies which can manipulate the future charges in their favor (Uddin & Syed-Abdul, 2020).

4.4 Data repositories

With the advancement in omics research, large amount of data is being produced. It is essential to store the big multi-omics datasets to make those publically available for future research use. Data repositories store multi-omics experimental (raw) and analyzed data in a logical manner, making it available for reuse. There are number of data repositories available now like: Omics Discovery Index, ProteomeXchange, The Gene Expression Omnibus database (GEO), PRoteomics IDEntifications (PRIDE), MassIVE, BioModels

Database, Kinetic Models of biological systems (KiMoSys), PeptideAtlas, ArrayExpress, Database of Interacting Proteins, IntAct, Genomic Expression Archive (GEA), EVA, MetabolomeExpress, NODE, ExpressionAtlas, ENA, EGA, dbGaP, GenomeRNAi, BioStudies, PAXdb, GPMDB, GNPS, MetabolomicsWorkbench, etc. (Perez-Riverol, Alpi, Wang, Hermjakob, & Vizcaino, 2015; Subramanian et al., 2020).

4.5 Data reanalysis for meta research

In this age of system biology, data from different omics such as genomics, transcriptomics, proteomics, etc., can provide important information for understanding the components responsible for complex diseases. Multiomics high-throughput approaches produce huge volumes of data. Original (raw) and analyzed data are then deposited and stored in public resources such as Gene Expression Omnibus, ArrayExpress, PeptideAtlas, PRIDE, MassIVE, ProteomeXchange, etc. As there is rapid increment in omics data deposition in repositories, new approaches are being developed based on the reanalysis or reuse of stored data. These public datasets can be analyzed using different pipelines to discover, confirm, or highlight new biological findings. By using public data in an appropriate manner, one can address a biological problem without generating new data. In this way, large quantity of samples can be integrated from different public datasets for same context. Public data can also be combined with newly generated data to increase the number of samples. Some proteomics resources such as GPMDB and PeptideAtlas reanalyze data, highlighting control of the number of false positives at both peptide and protein level. Some studies also found new PTMs like O-GlcNAc-6-phosphate, (ADP)-ribosylation, etc., after reanalyzing available phosphoproteomics studies. Another popular data reuse is the building of spectral libraries. Several repositories build their own libraries (e.g., PeptideAtlas, PRIDE) that can be used in spectral searches. Not only raw data but the output files or analyzed files from publically available datasets are useful for reanalysis and to develop new tools (e.g., SpliceVista visualization tool). PRIDE Inspector and PeptideShaker, can be used for visualization, reuse, and reanalysis of the data in repositories (Perez-Riverol et al., 2015).

4.6 Bio containers and workflow engines

Large-scale data generation in omics created a challenge for computational biologists to make data analyses reproducible and scalable, i.e., to rerun, combine and share. The volume of data, complex analysis, multi-stage workflows, and use of multiple tools are some major hurdles. In the past, data reproducibility was hindered by arcane desktop software or high-performance computing systems and cloud computing, where tools were closed-source and vendor OS-hardware dependent with only binary format data. The computational fields have now evolved to embrace open-source distributed frameworks to provide scalability, portability and reproducibility. Although, this has also led to increased technical complexity in installation, such intricacies are resolved by software containers such as Docker and Singularity, which isolate the software and its dependencies. The containers can directly execute the enclosed software without any additional installation processes and allow portable bioinformatics software. BioContainers and BioConda are public community-based repositories of bioinformatics containers. Combining tools to create bioinformatics analysis pipelines is still challenging and several workflow systems/engines set up tool execution steps in sequence or parallel by providing abstraction layers handling the tool connection, execution, error-handling, re-execution, etc. Some examples of workflow engines in bioinformatics are Galaxy, Taverna, Nextflow, Cromwell, Toil, WDL, Rubra, Ruffus and Snakemake (Perez-Riverol & Moreno, 2020).

5. Integrated omics

Integrated omics or multi-omics, requires the multitude of multi-omics data to be integrated in a semantic fashion, dealing with the challenges of data cleaning, normalization, dimensionality reduction, statistical validation, data storage, sharing and archiving. As the time and expense to generate these datasets has decreased, omics data integration has created both thrilling opportunities and plethora of challenges for biologists, computational biologists, biostatistician, and biomathematicians. *Trans*-Omics dynamic networks can identify critical components of biological networks in disease biology (Misra, Langefeld, Olivier, & Cox, 2018). The high-quality multi-omics studies require robust experimental design, appropriate sample storage, careful collection of quantitative multi-omics data and meta-data, reproducible analysis and interpretation of the data, followed by deposition of data in a public repository (Pinu et al., 2019).

5.1 Online public databases

A number of context-specific databases and tools have been developed for the integration of omics data from specific animal models, medical and clinical studies, and selected plant species. Many species-specific and omics-specific databases are now also publicly available that include data on the genome, transcriptome, proteome and metabolome of several model organisms (Pinu et al., 2019). There are many omics databases such as, genome-based databases like 3CDB, 4DGenome, ENCODE, 1000 Genome Project, SGD; transcriptome-based databases like NCBI GEO, TCGA, ICGC; epigenome-based databases like miRBase, lncRNAdb, NGSmethDB, MethylomeDB; proteome-based databases like Proteomics DB, PRIDE, dbPTM, Uniprot; metabolome-based databases like ECMDB, ChEBI, HMDB; multi-omics databases like Plant Metabolic Network database, Reactome, RNAactDrug, DriverDBv3, FlyBase and interactome-focused databases like BioGRID, Reactome, KEGG, STRING, etc.

5.2 Methods for data integration

The analysis of multiomics datasets in a synergistic manner requires semantic data integration approaches (Rohart, Gautier, Singh, & Le Cao, 2017). Integrating multi-omics datasets is challenging due to high dimensionality, limited number of patients, heterogeneity of datasets and modeling of interactions between the different types of omics data (Pierre-Jean, Deleuze, Le Floch, & Mauger, 2020). Integrative analysis can be applied to the discovery of molecular mechanisms, the clustering of samples (e.g., individuals) and the prediction of an outcome, such as survival or efficacy of therapy (Bersanelli et al., 2016). The novel approaches for integrating omics data, are divided into: batch effect removal methods before data integration, unsupervised methods (matrix factorization methods, Bayesian methods, network-based methods, correlation-based), supervised methods (network-based, multi-kernel or multivariate and multi-step based methods).

5.2.1 Batch effect removal methods

The omics data deposited in public repositories is available to other researchers for reuse, but may have batch effects when used to increase the sample numbers in analysis, integrating with more public datasets or with in-house data. However, the disparate experimental and data processing conditions, unique to each dataset or batch, influences the expression values and can hide the biological effect of interest. Therefore, it is important to normalize the merged dataset, to avoid the unwanted source of variation that adversely impacts statistical inference, known as *batch effect*. The batch effect correction algorithms (BECAs) like ComBat, RUV, ARSyN and MultiBaC methods, are used to remove such systematic biases which are first detected by PCA analysis (Ugidos, Tarazona, Prats-Montalban, Ferrer, & Conesa, 2020).

5.2.2 Unsupervised methods

Unsupervised data integration refers to the cluster of methods that draw an inference from input datasets without labeled response variables. Unsupervised methods are used to reduce data dimensions, to highlight underlying factors within the data or to identify clusters on the basis of similarity. The various approaches under unsupervised data integration includes: matrix factorization methods, Bayesian methods, network-based methods and correlation-based (Huang et al., 2017) method.

Matrix factorization (MF) methods are used to discover biological knowledge from multiomics data integration. The tabular matrix format with individual biomolecules in rows (genes, proteins, metabolites, etc.) and their samples in columns is a conventional omics data representation. MF decomposes the data into an amplitude matrix and a pattern matrix (Stein-O'Brien et al., 2018). Several MF approaches are discussed here. Joint Non-negative Matrix Factorization (NMF) decomposes a non-negative matrix into non-negative loadings and non-negative factors:

$$min \|X - WH\|^2, W \geq 0, H \geq 0$$

where X is the matrix of omics data with $M \times N$ dimensions, W is the common factor for $M \times K$ dimension matrix and H is the $K \times N$ dimension coefficient matrix. NMF is a time and memory consuming method that requires appropriate normalization of input datasets. iCluster assumes a regularized joint latent variable similar to W in NMF, but there are no non-negative constraints on it. The loading factor H is the coefficient matrix, with imposed sparsity with penalty functions. The error term is E and the decomposition equation is:

$$X = WH + E$$

The upgraded version is iCluster+, which can incorporate varied data types with diverse modeling assumptions. Another method Joint and Individual Variation Explained (JIVE) factors the input data matrix into two portions- shared factor W^c, and data-specific factor W^s with respective dependent coefficient matrices H^c and H^s (Huang et al., 2017).

Bayesian method is an integrative modeling method for the *unsupervised* analysis of multiomics datasets. Bayesian methods can deal with different types of data with various distributions, even with correlation among datasets (Huang et al., 2017). These methods include approaches, such as

Multiple Dataset Integration (MDI), Patient-specific data fusion (PSDF), Bayesian consensus clustering (BCC) (Subramanian et al., 2020), etc.

Network-based methods can identify network modules and are representative depictions of the disease-associated mechanisms. The nodes represent genes while the edges are the links between two interacting genes. In unsupervised integration, the network-based methods are applied for detecting significant genes, sub-clusters, or co-expression network modules. Many network-based approaches are used for multi-omics data analysis, like PAthway Representation and Analysis by DIrect Reference on Graphical Models (PARADIGM) which is a probability based graphical model framework (Huang et al., 2017). Similarity Network Fusion (SNF) is another network fusion based method that integrates multi-omics data. SNF calculates similarity matrix iteratively per sample to create a network for each data type, by constructing sample by sample similarity matrix (Huang et al., 2017). The Network-based integration of multi-omics data (NetICS) method is a framework for disease gene prioritization that can predict the effect of genetic aberrations, epigenetic changes and miRNAs on downstream genes, and protein expression in the interaction network. It derives a population level ranking of genes for all samples by using a per-sample network-diffusion model on a directed interaction network (Subramanian et al., 2020).

Correlation-based approach for integrating multi-omics dataset includes CNAmet. It is a software for integrative analysis of copy number changes, DNA methylation, and gene expression data. The algorithm calculates weight which connects the expression values to copy number and methylation.

5.2.3 Supervised data integration

Integration of multi-omics data can also be accomplished by supervised methods that utilize known labels from the training omics data. The phenotype labels of samples (disease or normal) are considered and machine learning approaches are trained to evaluate the models. This approach includes network-based, multi-kernel or multivariate and multi-step based methods.

Network-based methods include various approaches for multi-omics datasets integration. A neural network approach with a supervised model like Analysis Tool for Heritable and Environmental Network Associations (ATHENA), can be used for prognosis. ATHENA utilizes grammatical evolution neural networks (GENN) to train the model with selected less noisy features. Several such individual models are combined eventually into a final integrative model, which can be utilized for diagnosis and prognosis. Genes connected in a network tend to have correlated expression, which can be inferred from their molecular interactions through a Cytoscope plug-in jActiveModules. The plug-in calculates highest-scoring sub-network for interesting biological discoveries.

Multiple Kernel Learning or multivariate method uses a decision function to construct a classifier. The decision function depends on diverse input data types (gene expression and copy number variation) by mean of pathway-based kernels. Using Semidefinite Programming/Support Vector Machine (SDP/SVM), it represents each dataset as a kernel function to calculate the similarity between two objects. Different kernels are used for a different transformation of the input data, which provides a precise type of information from each dataset. For example, Fast Fourier Transform (FFT) kernel is specific for recognizing the membrane protein using hydrophobicity. Feature Selection Multiple Kernel Learning (FSMKL) is another supervised learning method that uses the multiple kernel learning (Huang et al., 2017). Multiple kernels are used to measure the similarity between datasets to identify features for disease progression. A linear combination of kernels is used to create a base kernel encoding for each dataset. The coefficients of the kernel define the measure of weight of the several datasets used in the final decision function. When the regulatory features are larger in number than the samples measured, thresholding singular value decomposition (T-SVD) regression method is used to identify the mechanisms of regulation between the two omics datasets, utilizing the sparsity limitation. Sparse multi-block partial least squares (sMBPLS) is a method employing a sparse version of PLS to decompose the multi-omics datasets into small "multi-dimensional regulatory modules" (MDRMs). Partial least squares regression helps in detecting the relationship between input and response variables. Sparse multi-block partial least squares method allows an input with multiple omics datasets (e.g., data from CNV, DNA methylation, and miRNA expression) that regulates the gene expression. This method attempts to optimize the covariance between input and response data. Lasso penalization can also be used to apply the sparsity constraint so as to convert the negligible coefficients to zero (Subramanian et al., 2020).

Multi-Step Analysis method integration of multi-omics data contains two-stage models. One is a mechanistic regression model that segments gene expression data into small segments, each with a principal component. In the second stage, binary result and clinical survival information can be modeled as the response of joint regression, from the previous regression factors. A multi-step analysis method, Multiple Concerted Disruption

(MCD), can perform the integration of CNV, DNA methylation and allelic data for identification of key nodes in a pathway to compare observed against the expected changes. This can reveal functionally important gens with disrupted mechanism and change in expression (Huang et al., 2017).

5.3 Analysis tools and visualization

Integration of multi-omics data provides cohesive information on functional roles of biomolecules from various omics layers. Apart from multi-omics data integration, numerous computational software and visualization tools have been developed to understand the complex network architecture and explore the information in multi-view data and assess their outcome. Numerous databases and computational tools that target the multi-omics data integration challenge, have been developed. The analysis and visualization tools for this purpose are 3Omics, BioCyc/MetaCyc, BiofOmics MADMAX, Gaggle, MassTrix, MetaboAnalyst, MapMan, Omix visualization tool, mixOmics (R package), PaintOmics, CellML, COBRA, Escher, OmicsAnalyzer, IMPaLA, INMEX, IOMA, KaPPA-View, MarVis-Pathway, MetScape 2, ProMeTra, Recon3D, VitisNet, VANTED, PathVisio 3, etc. (Pinu et al., 2019).

5.4 Precision health and longitudinal profiling

This approach involves the use of an individual's personal-data for the treatment. Here, each patient is an individual case, incorporating data from medical history, omics, environment, lifestyle, etc. The two critical aspects of precision health are: data generation and modeling. The information on treated subjects is collected in profiling health care approach to infer treatment response (Daniels & Normand, 2006). A longitudinal profiling study focused on diabetes which followed 109 individuals for 2.8 years (median), measured omics profiles and clinical parameters quarterly as well as monitoring health with wearable devices (Vogt, Green, Ekstrom, & Brodersen, 2019). The Integrated Personal Omics Profiling (iPOP) was one of the first studies that comprehensively integrated multi-omics data for assessment and prediction of health (Li-Pook-Than & Snyder, 2013). A longitudinal study on about 100 people, aimed at setting the groundwork for precision personalized medicine by allowing for unparalleled deep biochemical profiling of healthy people, was designed to understand the differences in biochemical and physiological profiles of healthy and diseased people on individual level. Over several years, the samples and data (diet, stress level, activity level, personal and family medical history) were collected at regular intervals, both in good health and at periods of illness and severe stress for each participant, using whole genome sequencing (Li-Pook-Than & Snyder, 2013).

The few omics-based approaches for analysis of data for precision medicine are data-driven modeling, statistical models, regression analysis, machine learning, clustering, classification, rule-based and logical models, flux control analysis, etc. Some challenges for developing precision medicine are, the influx of new data, integration of heterogeneous data types, data accessibility, delivering multiplex information to patients, integration with clinicians and healthcare systems, complete dependency on heterogeneous, static, and incomplete data. To overcome these challenges, we need better modeling approaches to clear the diagnostic standards of regulatory firms, which are not common in system biology (Duffy, 2016).

6. Ethics of big data sharing and reuse

Ethics is a system of moral principles that incorporates four main bioethics principles: non-maleficence (not communicating non-actionable incidental findings), beneficence (communicating actionable incidental findings), justice (data protection and access management) and autonomy (data sharing preferences) (Mann, Treit, Geyer, Omenn, & Mann, 2021). Big data in health incorporates information from a single person to large cohorts, collecting information about the biological, environmental, lifestyle and clinical aspects concerning human health at one or multiple time points. Ethics framework helps in guiding toward accountable and pertinent usage of the data in government as well as the public sectors. Many hypotheses and interpretations are embedded within the structure of ethical framework, with overarching considerations such as transparency, need to respect persons, to take account of community expectations, and to consider issues of vulnerability that can arise in uses of big data. The need for an ethics framework is due to the sole reliability on data masking techniques, de-identification and the shrinking role of informed consent. There are three kinds of disclosures that lead to the re-identification of masking data, i.e., identity, attribute, and inferential disclosure. Anonymization of data with privacy models does not realize complete reliability as it is developed for static datasets instead of big data. Data linkability, composability and low computation are some features that can enhance the efficacy of privacy models. Reliance on consent is becoming increasingly infeasible in the big data context because data might be linked and used within and across ecosystems that are far removed from the source of information. To resolve this issue, a properly secured framework needs to be implemented (Laurie, 2019). The regulatory bodies managing big data in the health sector are: General Data Protection Regulation (GDPR), the Health Insurance Portability and Accountability Act (HIPAA), CIOMS (Council for International Organizations of Medical Sciences), GA4GH (global alliance for genomics and health), OECD (organization for economic co-operation and development) that work in favor of the protection and privacy of data in health (Kalkman, Mostert, Gerlinger, van Delden, & van Thiel, 2019).

7. Future perspectives, challenges and conclusion

Multiomics research will be more common in the near future, and promising new high-throughput methods such as limited proteolysis for unbiased and proteome-wide profiling of protein conformational changes, proteomics of proteoforms, single-cell omics measurements, or small-molecule interactions will be more routinely added to the repertoire. Such rich multiomics datasets will inspire the development of more complex mechanistic models that can take advantage of the new data. For example, genome-scale metabolic models have already begun to routinely integrate transcriptomic and proteomic data, and will likely have more layers of information added in the near future. This will be increase the demand for data integration approaches, where machine learning holds great promise and is already being used intensively in biomedical applications. The advancement in molecular biological experiments is producing huge amount of data related to genome and RNA sequence, protein and metabolite abundance, protein-protein interaction, etc. It is important to handle these huge data efficiently and scientifically to understand the cell as a system and to develop new applications in biotechnology and biomedical fields. The application of network theory and algorithms can facilitate the analysis and integration of big data (Altaf-Ul-Amin, Afendi, Kiboi, & Kanaya, 2014). One main challenge that persists is the heterogeneity of data formats delivered by the different techniques and reproducibility of analysis. Integration of more than two different omics data formats is still not routine and requires optimized software tools together with well-trained scientists to generate replicable, reproducible and comprehensible analysis workflows (Dihazi et al., 2018). Data governance will be critical for FAIR access to data and ethical frameworks are required to prevent maleficence and increase beneficence. As new methods and machine learning approaches are developed and adopted by biologists, they are likely to become an essential part of multi-omics data analysis. Hybrid techniques that take advantage of the vast computational capabilities while maintaining the underlying mechanistic model are now beginning to emerge and will hopefully gain popularity. It remains to be seen whether a generalized algorithm that can handle all the different types of omics data and network-based information will one day become the default tool for biological data analysis (Heckmann et al., 2018).

Acknowledgments

A.K.Y. is supported by DBT-Big Data Initiative grant (BT/PR16456/BID/7/624/2016). This grant also supports P.T. Translational Research Program (TRP) at THSTI funded by DBT also supports A.K.Y. and S.A.

Conflict of interest

The authors declare no competing interests.

References

Aggarwal, S., Banerjee, S. K., Talukdar, N. C., & Yadav, A. K. (2020). Post-translational modification crosstalk and hotspots in sirtuin interactors implicated in cardiovascular diseases. *Frontiers in Genetics*, *11*, 356. https://doi.org/10.3389/fgene.2020.00356.

Aggarwal, S., Kandpal, M., Asthana, S., & Yadav, A. K. (2017). Perturbed signaling and role of posttranslational modifications in cancer drug resistance. In G. Arora, A. Sajid, & V. C. Kalia (Eds.), *Drug resistance in bacteria, fungi, malaria, and cancer* (pp. 483–510). Cham: Springer International Publishing.

Aggarwal, S., & Yadav, A. K. (2016). False discovery rate estimation in proteomics. *Methods in Molecular Biology*, *1362*, 119–128. https://doi.org/10.1007/978-1-4939-3106-4_7.

Altaf-Ul-Amin, M., Afendi, F. M., Kiboi, S. K., & Kanaya, S. (2014). Systems biology in the context of big data and networks. *BioMed Research International*, *2014*, 428570. https://doi.org/10.1155/2014/428570.

Alyass, A., Turcotte, M., & Meyre, D. (2015). From big data analysis to personalized medicine for all: Challenges and opportunities. *BMC Medical Genomics*, *8*, 33. https://doi.org/10.1186/s12920-015-0108-y.

Bahadur, R. P., & Zacharias, M. (2008). The interface of protein-protein complexes: Analysis of contacts and prediction of interactions. *Cellular and Molecular Life Sciences*, *65*(7–8), 1059–1072. https://doi.org/10.1007/s00018-007-7451-x.

Barouki, R., Audouze, K., Coumoul, X., Demenais, F., & Gauguier, D. (2018). Integration of the human exposome with the human genome to advance medicine. *Biochimie*, *152*, 155–158. https://doi.org/10.1016/j.biochi.2018.06.023.

Behjati, S., & Tarpey, P. S. (2013). What is next generation sequencing? *Archives of Disease in Childhood. Education and Practice Edition*, *98*(6), 236–238. https://doi.org/10.1136/archdischild-2013-304340.

Bersanelli, M., Mosca, E., Remondini, D., Giampieri, E., Sala, C., Castellani, G., et al. (2016). Methods for the integration of multi-omics data: Mathematical aspects. *BMC Bioinformatics*, *17*(Suppl. 2), 15. https://doi.org/10.1186/s12859-015-0857-9.

Bludau, I., & Aebersold, R. (2020). Proteomic and interactomic insights into the molecular basis of cell functional diversity. *Nature Reviews. Molecular Cell Biology*, *21*(6), 327–340. https://doi.org/10.1038/s41580-020-0231-2.

Callinan, P. A., & Feinberg, A. P. (2006). The emerging science of epigenomics. *Human Molecular Genetics*, *15*, R95–101. https://doi.org/10.1093/hmg/ddl095. Spec. No. 1.

Calviello, L., & Ohler, U. (2017). Beyond read-counts: Ribo-seq data analysis to understand the functions of the transcriptome. *Trends in Genetics*, *33*(10), 728–744. https://doi.org/10.1016/j.tig.2017.08.003.

Cascante, M., & Marin, S. (2008). Metabolomics and fluxomics approaches. *Essays in Biochemistry*, *45*, 67–81. https://doi.org/10.1042/BSE0450067.

Castellana, N., & Bafna, V. (2010). Proteogenomics to discover the full coding content of genomes: A computational perspective. *Journal of Proteomics*, *73*(11), 2124–2135. https://doi.org/10.1016/j.jprot.2010.06.007.

Clarke, D. J. B., Wang, L., Jones, A., Wojciechowicz, M. L., Torre, D., Jagodnik, K. M., et al. (2019). FAIRshake: Toolkit to evaluate the FAIRness of research digital resources. *Cell Systems*, *9*(5), 417–421. https://doi.org/10.1016/j.cels.2019.09.011.

Collins, F. S., Green, E. D., Guttmacher, A. E., Guyer, M. S., & US National Human Genome Research Institute. (2003). A vision for the future of genomics research. *Nature*, *422*(6934), 835–847. https://doi.org/10.1038/nature01626.

Craig, R., & Beavis, R. C. (2004). TANDEM: Matching proteins with tandem mass spectra. *Bioinformatics*, *20*(9), 1466–1467. https://doi.org/10.1093/bioinformatics/bth092.

Daniels, M. J., & Normand, S. L. (2006). Longitudinal profiling of health care units based on continuous and discrete patient outcomes. *Biostatistics*, 7(1), 1–15. https://doi.org/10.1093/biostatistics/kxi036.

Dihazi, H., Asif, A. R., Beissbarth, T., Bohrer, R., Feussner, K., Feussner, I., et al. (2018). Integrative omics—From data to biology. *Expert Review of Proteomics*, *15*(6), 463–466. https://doi.org/10.1080/14789450.2018.1476143.

Dong, Z., & Chen, Y. (2013). Transcriptomics: Advances and approaches. *Science China. Life Sciences*, *56*(10), 960–967. https://doi.org/10.1007/s11427-013-4557-2.

Duffy, D. J. (2016). Problems, challenges and promises: Perspectives on precision medicine. *Briefings in Bioinformatics*, *17*(3), 494–504. https://doi.org/10.1093/bib/bbv060.

Efremova, M., & Teichmann, S. A. (2020). Computational methods for single-cell omics across modalities. *Nature Methods*, *17*(1), 14–17. https://doi.org/10.1038/s41592-019-0692-4.

Eng, J. K., McCormack, A. L., & Yates, J. R. (1994). An approach to correlate tandem mass spectral data of peptides with amino acid sequences in a protein database. *Journal of the American Society for Mass Spectrometry*, *5*(11), 976–989. https://doi.org/10.1016/1044-0305(94)80016-2.

Fazzari, M. J., & Greally, J. M. (2010). Introduction to epigenomics and epigenome-wide analysis. *Methods in Molecular Biology*, *620*, 243–265. https://doi.org/10.1007/978-1-60761-580-4_7.

Flores, M., Glusman, G., Brogaard, K., Price, N. D., & Hood, L. (2013). P4 medicine: How systems medicine will transform the healthcare sector and society. *Personalized Medicine*, *10*(6), 565–576. https://doi.org/10.2217/pme.13.57.

Gao, J., Shao, K., Chen, X., Li, Z., Liu, Z., Yu, Z., et al. (2020). The involvement of post-translational modifications in cardiovascular pathologies: Focus on SUMOylation, neddylation, succinylation, and prenylation. *Journal of Molecular and Cellular Cardiology*, *138*, 49–58. https://doi.org/10.1016/j.yjmcc.2019.11.146.

Grosskinsky, D. K., Svensgaard, J., Christensen, S., & Roitsch, T. (2015). Plant phenomics and the need for physiological phenotyping across scales to narrow the genotype-to-phenotype knowledge gap. *Journal of Experimental Botany*, *66*(18), 5429–5440. https://doi.org/10.1093/jxb/erv345.

Hasin, Y., Seldin, M., & Lusis, A. (2017). Multi-omics approaches to disease. *Genome Biology*, *18*(1), 83. https://doi.org/10.1186/s13059-017-1215-1.

Hawe, J. S., Theis, F. J., & Heinig, M. (2019). Inferring interaction networks from multi-omics data. *Frontiers in Genetics, 10*, 535. https://doi.org/10.3389/fgene.2019.00535.

He, X., Memczak, S., Qu, J., Belmonte, J. C. I., & Liu, G. H. (2020). Single-cell omics in ageing: A young and growing field. *Nature Metabolism, 2*(4), 293–302. https://doi.org/10.1038/s42255-020-0196-7.

Heckmann, D., Lloyd, C. J., Mih, N., Ha, Y., Zielinski, D. C., Haiman, Z. B., et al. (2018). Machine learning applied to enzyme turnover numbers reveals protein structural correlates and improves metabolic models. *Nature Communications, 9*(1), 5252. https://doi.org/10.1038/s41467-018-07652-6.

Houle, D., Govindaraju, D. R., & Omholt, S. (2010). Phenomics: The next challenge. *Nature Reviews. Genetics, 11*(12), 855–866. https://doi.org/10.1038/nrg2897.

Hu, J. X., Thomas, C. E., & Brunak, S. (2016). Network biology concepts in complex disease comorbidities. *Nature Reviews. Genetics, 17*(10), 615–629. https://doi.org/10.1038/nrg.2016.87.

Huang, S., Chaudhary, K., & Garmire, L. X. (2017). More is better: Recent progress in multi-omics data integration methods. *Frontiers in Genetics, 8*, 84. https://doi.org/10.3389/fgene.2017.00084.

Hugenholtz, P., & Tyson, G. W. (2008). Microbiology: Metagenomics. *Nature, 455*(7212), 481–483. https://doi.org/10.1038/455481a.

Ingolia, N. T. (2016). Ribosome footprint profiling of translation throughout the genome. *Cell, 165*(1), 22–33. https://doi.org/10.1016/j.cell.2016.02.066.

Kalkman, S., Mostert, M., Gerlinger, C., van Delden, J. J. M., & van Thiel, G. (2019). Responsible data sharing in international health research: A systematic review of principles and norms. *BMC Medical Ethics, 20*(1), 21. https://doi.org/10.1186/s12910-019-0359-9.

Kall, L., Canterbury, J. D., Weston, J., Noble, W. S., & MacCoss, M. J. (2007). Semi-supervised learning for peptide identification from shotgun proteomics datasets. *Nature Methods, 4*(11), 923–925. https://doi.org/10.1038/nmeth1113.

Kandpal, M., Aggarwal, S., Jamwal, S., & Yadav, A. K. (2017). Emergence of drug resistance in mycobacterium and other bacterial pathogens: The posttranslational modification perspective. In G. Arora, A. Sajid, & V. C. Kalia (Eds.), *Drug resistance in bacteria, fungi, malaria, and cancer* (pp. 209–231). Cham: Springer International Publishing.

Kelkar, D. S., Kumar, D., Kumar, P., Balakrishnan, L., Muthusamy, B., Yadav, A. K., et al. (2011). Proteogenomic analysis of *Mycobacterium tuberculosis* by high resolution mass spectrometry. *Molecular & Cellular Proteomics, 10*(12), M111.011627. https://doi.org/10.1074/mcp.M111.011445.

Keskin, O., Gursoy, A., Ma, B., & Nussinov, R. (2007). Towards drugs targeting multiple proteins in a systems biology approach. *Current Topics in Medicinal Chemistry*, 7(10), 943–951. https://doi.org/10.2174/156802607780906690.

Kim, S., & Pevzner, P. A. (2014). MS-GF+ makes progress towards a universal database search tool for proteomics. *Nature Communications, 5*, 5277. https://doi.org/10.1038/ncomms6277.

Kumar, D., Mondal, A. K., Yadav, A. K., & Dash, D. (2014). Discovery of rare protein-coding genes in model methylotroph *Methylobacterium extorquens* AM1. *Proteomics, 14*(23–24), 2790–2794. https://doi.org/10.1002/pmic.201400153.

Kumar, D., Yadav, A. K., & Dash, D. (2017). Choosing an optimal database for protein identification from tandem mass spectrometry data. *Methods in Molecular Biology, 1549*, 17–29. https://doi.org/10.1007/978-1-4939-6740-7_3.

Kumar, D., Yadav, A. K., Jia, X., Mulvenna, J., & Dash, D. (2016). Integrated transcriptomic-proteomic analysis using a proteogenomic workflow refines rat genome annotation. *Molecular & Cellular Proteomics, 15*(1), 329–339. https://doi.org/10.1074/mcp.M114.047126.

Kumar, D., Yadav, A. K., Kadimi, P. K., Nagaraj, S. H., Grimmond, S. M., & Dash, D. (2013). Proteogenomic analysis of *Bradyrhizobium japonicum* USDA110 using GenoSuite, an automated multi-algorithmic pipeline. *Molecular & Cellular Proteomics*, *12*(11), 3388–3397. https://doi.org/10.1074/mcp.M112.027169.

Laurie, G. T. (2019). Cross-sectoral big data: The application of an ethics framework for big data in health and research. *Asian Bioethics Review*, *11*(3), 327–339. https://doi.org/10.1007/s41649-019-00093-3.

Li-Pook-Than, J., & Snyder, M. (2013). iPOP goes the world: Integrated personalized omics profiling and the road toward improved health care. *Chemistry & Biology*, *20*(5), 660–666. https://doi.org/10.1016/j.chembiol.2013.05.001.

Mann, S. P., Treit, P. V., Geyer, P. E., Omenn, G. S., & Mann, M. (2021). Ethical principles, constraints and opportunities in clinical proteomics. *Molecular & Cellular Proteomics*, 100046. https://doi.org/10.1016/j.mcpro.2021.100046.

Manzoni, C., Kia, D. A., Vandrovcova, J., Hardy, J., Wood, N. W., Lewis, P. A., et al. (2018). Genome, transcriptome and proteome: The rise of omics data and their integration in biomedical sciences. *Briefings in Bioinformatics*, *19*(2), 286–302. https://doi.org/10.1093/bib/bbw114.

Maron, P. A., Ranjard, L., Mougel, C., & Lemanceau, P. (2007). Metaproteomics: A new approach for studying functional microbial ecology. *Microbial Ecology*, *53*(3), 486–493. https://doi.org/10.1007/s00248-006-9196-8.

Martins-de-Souza, D. (2014). Proteomics, metabolomics, and protein interactomics in the characterization of the molecular features of major depressive disorder. *Dialogues in Clinical Neuroscience*, *16*(1), 63–73.

Misra, B. B., Langefeld, C. D., Olivier, M., & Cox, L. A. (2018). Integrated omics: tools, advances, and future approaches. *Journal of Molecular Endocrinology*, *62*(1), R21–R45. https://doi.org/10.1530/JME-18-0055.

Nesvizhskii, A. I. (2014). Proteogenomics: Concepts, applications and computational strategies. *Nature Methods*, *11*(11), 1114–1125. https://doi.org/10.1038/nmeth.3144.

Nesvizhskii, A. I., & Aebersold, R. (2005). Interpretation of shotgun proteomic data: The protein inference problem. *Molecular & Cellular Proteomics*, *4*(10), 1419–1440. https://doi.org/10.1074/mcp.R500012-MCP200.

Nie, Q., Gong, X. D., Liu, M., & Li, D. W. (2017). Effects of crosstalks between sumoylation and phosphorylation in normal cellular physiology and human diseases. *Current Molecular Medicine*, *16*(10), 906–913. https://doi.org/10.2174/1566524016666161223105555.

Park, P. J. (2009). ChIP-seq: Advantages and challenges of a maturing technology. *Nature Reviews. Genetics*, *10*(10), 669–680. https://doi.org/10.1038/nrg2641.

Patterson, S. D., & Aebersold, R. H. (2003). Proteomics: The first decade and beyond. *Nature Genetics*, *33*(Suppl), 311–323. https://doi.org/10.1038/ng1106.

Peng, R. D. (2011). Reproducible research in computational science. *Science*, *334*(6060), 1226–1227. https://doi.org/10.1126/science.1213847.

Perez-Riverol, Y., Alpi, E., Wang, R., Hermjakob, H., & Vizcaino, J. A. (2015). Making proteomics data accessible and reusable: Current state of proteomics databases and repositories. *Proteomics*, *15*(5–6), 930–949. https://doi.org/10.1002/pmic.201400302.

Perez-Riverol, Y., & Moreno, P. (2020). Scalable data analysis in proteomics and metabolomics using BioContainers and workflows engines. *Proteomics*, *20*(9), e1900147. https://doi.org/10.1002/pmic.201900147.

Perkins, D. N., Pappin, D. J., Creasy, D. M., & Cottrell, J. S. (1999). Probability-based protein identification by searching sequence databases using mass spectrometry data. *Electrophoresis*, *20*(18), 3551–3567. https://doi.org/10.1002/(SICI)1522-2683(19991201)20:18<3551::AID-ELPS3551>3.0.CO;2-2.

Pieroni, L., Iavarone, F., Olianas, A., Greco, V., Desiderio, C., Martelli, C., et al. (2020). Enrichments of post-translational modifications in proteomic studies. *Journal of Separation Science*, *43*(1), 313–336. https://doi.org/10.1002/jssc.201900804.

Pierre-Jean, M., Deleuze, J. F., Le Floch, E., & Mauger, F. (2020). Clustering and variable selection evaluation of 13 unsupervised methods for multi-omics data integration. *Briefings in Bioinformatics*, *21*(6), 2011–2030. https://doi.org/10.1093/bib/bbz138.

Pinu, F. R., Beale, D. J., Paten, A. M., Kouremenos, K., Swarup, S., Schirra, H. J., et al. (2019). Systems biology and multi-omics integration: Viewpoints from the metabolomics research community. *Metabolites*, *9*(4), 76. https://doi.org/10.3390/metabo9040076.

Rohart, F., Gautier, B., Singh, A., & Le Cao, K. A. (2017). mixOmics: An R package for 'omics feature selection and multiple data integration. *PLoS Computational Biology*, *13*(11), e1005752. https://doi.org/10.1371/journal.pcbi.1005752.

Safari-Alighiarloo, N., Taghizadeh, M., Rezaei-Tavirani, M., Goliaei, B., & Peyvandi, A. A. (2014). Protein-protein interaction networks (PPI) and complex diseases. *Gastroenterology and Hepatology from Bed to Bench*, 7(1), 17–31.

Sandve, G. K., Nekrutenko, A., Taylor, J., & Hovig, E. (2013). Ten simple rules for reproducible computational research. *PLoS Computational Biology*, *9*(10), e1003285. https://doi.org/10.1371/journal.pcbi.1003285.

Shou, C., Bhardwaj, N., Lam, H. Y., Yan, K. K., Kim, P. M., Snyder, M., et al. (2011). Measuring the evolutionary rewiring of biological networks. *PLoS Computational Biology*, 7(1), e1001050. https://doi.org/10.1371/journal.pcbi.1001050.

Sleator, R. D., Shortall, C., & Hill, C. (2008). Metagenomics. *Letters in Applied Microbiology*, *47*(5), 361–366. https://doi.org/10.1111/j.1472-765X.2008.02444.x.

Stein-O'Brien, G. L., Arora, R., Culhane, A. C., Favorov, A. V., Garmire, L. X., Greene, C. S., et al. (2018). Enter the matrix: Factorization uncovers knowledge from omics. *Trends in Genetics*, *34*(10), 790–805. https://doi.org/10.1016/j.tig.2018.07.003.

Subramanian, I., Verma, S., Kumar, S., Jere, A., & Anamika, K. (2020). Multi-omics data integration, interpretation, and its application. *Bioinformatics and Biology Insights*, *14*, 1177932219899051. https://doi.org/10.1177/1177932219899051.

Tarca, A. L., Romero, R., & Draghici, S. (2006). Analysis of microarray experiments of gene expression profiling. *American Journal of Obstetrics and Gynecology*, *195*(2), 373–388. https://doi.org/10.1016/j.ajog.2006.07.001.

Theillet, F. X., Smet-Nocca, C., Liokatis, S., Thongwichian, R., Kosten, J., Yoon, M. K., et al. (2012). Cell signaling, post-translational protein modifications and NMR spectroscopy. *Journal of Biomolecular NMR*, *54*(3), 217–236. https://doi.org/10.1007/s10858-012-9674-x.

Tyers, M., & Mann, M. (2003). From genomics to proteomics. *Nature*, *422*(6928), 193–197. https://doi.org/10.1038/nature01510.

Uddin, M., & Syed-Abdul, S. (2020). Data analytics and applications of the wearable sensors in healthcare: An overview. *Sensors (Basel)*, *20*(5). https://doi.org/10.3390/s20051379.

Ugidos, M., Tarazona, S., Prats-Montalban, J. M., Ferrer, A., & Conesa, A. (2020). MultiBaC: A strategy to remove batch effects between different omic data types. *Statistical Methods in Medical Research*, *29*(10), 2851–2864. https://doi.org/10.1177/0962280220907365.

Vermeulen, R., Schymanski, E. L., Barabasi, A. L., & Miller, G. W. (2020). The exposome and health: Where chemistry meets biology. *Science*, *367*(6476), 392–396. https://doi.org/10.1126/science.aay3164.

Vogt, H., Green, S., Ekstrom, C. T., & Brodersen, J. (2019). How precision medicine and screening with big data could increase overdiagnosis. *BMJ*, *366*, l5270. https://doi.org/10.1136/bmj.l5270.

Wang, M. H., Cordell, H. J., & Van Steen, K. (2019). Statistical methods for genome-wide association studies. *Seminars in Cancer Biology*, *55*, 53–60. https://doi.org/10.1016/j.semcancer.2018.04.008.

Wang, X., Gulbahce, N., & Yu, H. (2011). Network-based methods for human disease gene prediction. *Briefings in Functional Genomics*, *10*(5), 280–293. https://doi.org/10.1093/bfgp/elr024.

Wilkinson, M. D., Dumontier, M., Aalbersberg, I. J., Appleton, G., Axton, M., Baak, A., et al. (2016). The FAIR guiding principles for scientific data management and stewardship. *Scientific Data*, *3*, 160018. https://doi.org/10.1038/sdata.2016.18.

Wilmes, P., Heintz-Buschart, A., & Bond, P. L. (2015). A decade of metaproteomics: Where we stand and what the future holds. *Proteomics*, *15*(20), 3409–3417. https://doi.org/10.1002/pmic.201500183.

Winter, G., & Kromer, J. O. (2013). Fluxomics—connecting 'omics analysis and phenotypes. *Environmental Microbiology*, *15*(7), 1901–1916. https://doi.org/10.1111/1462-2920.12064.

Wooley, J. C., Godzik, A., & Friedberg, I. (2010). A primer on metagenomics. *PLoS Computational Biology*, *6*(2), e1000667. https://doi.org/10.1371/journal.pcbi.1000667.

Wu, Z., Huang, R., & Yuan, L. (2019). Crosstalk of intracellular post-translational modifications in cancer. *Archives of Biochemistry and Biophysics*, *676*, 108138. https://doi.org/10.1016/j.abb.2019.108138.

Wu, Z., Li, W., Liu, G., & Tang, Y. (2018). Network-based methods for prediction of drug-target interactions. *Frontiers in Pharmacology*, *9*, 1134. https://doi.org/10.3389/fphar.2018.01134.

Xu, B., Liu, Y., Lin, C., Dong, J., Liu, X., & He, Z. (2018). Reconstruction of the protein-protein interaction network for protein complexes identification by walking on the protein pair fingerprints similarity network. *Frontiers in Genetics*, *9*, 272. https://doi.org/10.3389/fgene.2018.00272.

Yadav, A. K. (2017). Commentary: Deep phosphoproteomic measurements pinpointing drug induced protective mechanisms in neuronal cells. *Frontiers in Physiology*, *8*, 174. https://doi.org/10.3389/fphys.2017.00174.

Yadav, A. K., Bhardwaj, G., Basak, T., Kumar, D., Ahmad, S., Priyadarshini, R., et al. (2011). A systematic analysis of eluted fraction of plasma post immunoaffinity depletion: Implications in biomarker discovery. *PLoS One*, *6*(9), e24442. https://doi.org/10.1371/journal.pone.0024442.

Yadav, A. K., Kadimi, P. K., Kumar, D., & Dash, D. (2013). ProteoStats—A library for estimating false discovery rates in proteomics pipelines. *Bioinformatics*, *29*(21), 2799–2800. https://doi.org/10.1093/bioinformatics/btt490.

Yadav, A. K., Kumar, D., & Dash, D. (2011). MassWiz: A novel scoring algorithm with target-decoy based analysis pipeline for tandem mass spectrometry. *Journal of Proteome Research*, *10*(5), 2154–2160. https://doi.org/10.1021/pr200031z.

Yadav, A. K., Kumar, D., & Dash, D. (2012). Learning from decoys to improve the sensitivity and specificity of proteomics database search results. *PLoS One*, 7(11), e50651. https://doi.org/10.1371/journal.pone.0050651.

Yan, F., Powell, D. R., Curtis, D. J., & Wong, N. C. (2020). From reads to insight: A hitchhiker's guide to ATAC-seq data analysis. *Genome Biology*, *21*(1), 22. https://doi.org/10.1186/s13059-020-1929-3.

Yang, J. Y., Karr, J. R., Watrous, J. D., & Dorrestein, P. C. (2011). Integrating '-omics' and natural product discovery platforms to investigate metabolic exchange in microbiomes. *Current Opinion in Chemical Biology*, *15*(1), 79–87. https://doi.org/10.1016/j.cbpa.2010.10.025.

Yildirim, M. A., Goh, K. I., Cusick, M. E., Barabasi, A. L., & Vidal, M. (2007). Drug-target network. *Nature Biotechnology*, *25*(10), 1119–1126. https://doi.org/10.1038/nbt1338.

Zeng, X., Zhu, S., Hou, Y., Zhang, P., Li, L., Li, J., et al. (2020). Network-based prediction of drug-target interactions using an arbitrary-order proximity embedded deep forest. *Bioinformatics*, *36*(9), 2805–2812. https://doi.org/10.1093/bioinformatics/btaa010.

Zhang, A., Sun, H., Wang, P., Han, Y., & Wang, X. (2012). Modern analytical techniques in metabolomics analysis. *Analyst*, *137*(2), 293–300. https://doi.org/10.1039/c1an15605e.

Zhou, X., Menche, J., Barabasi, A. L., & Sharma, A. (2014). Human symptoms-disease network. *Nature Communications*, *5*, 4212. https://doi.org/10.1038/ncomms5212.

CHAPTER FIVE

Proteome analysis using machine learning approaches and its applications to diseases

Abhishek Sengupta, G. Naresh, Astha Mishra, Diksha Parashar, and Priyanka Narad*
Amity Institute of Biotechnology, Amity University Uttar Pradesh, Noida, India
*Corresponding author: e-mail address: pnarad@amity.edu

Contents

Abstract

With the tremendous developments in the fields of biological and medical technologies, huge amounts of data are generated in the form of genomic data, images in

Advances in Protein Chemistry and Structural Biology, Volume 127
ISSN 1876-1623
https://doi.org/10.1016/bs.apcsb.2021.02.003

medical databases or as data on protein sequences, and so on. Analyzing this data through different tools sheds light on the particulars of the disease and our body's reactions to it, thus, aiding our understanding of the human health. Most useful of these tools is artificial intelligence and deep learning (DL). The artificially created neural networks in DL algorithms help extract viable data from the datasets, and further, to recognize patters in these complex datasets. Therefore, as a part of machine learning, DL helps us face all the various challenges that come forth during protein prediction, protein identification and their quantification. Proteomics is the study of such proteins, their structures, features, properties and so on. As a form of data science, Proteomics has helped us progress excellently in the field of genomics technologies. One of the major techniques used in proteomics studies is mass spectrometry (MS). However, MS is efficient with analysis of large datasets only with the added help of informatics approaches for data analysis and interpretation; these mainly include machine learning and deep learning algorithms. In this chapter, we will discuss in detail the applications of deep learning and various algorithms of machine learning in proteomics.

1. Introduction to proteomics

The term proteome denotes the complete set of proteins in a cell. It is known as the large-scale characterization of entire protein of whole cells, tissues and organisms. It deals with the study of protein function, localization and protein-protein interaction (Aslam, Basit, Nisar, Khurshid, & Rasool, 2017). As nucleotides deal with the genotype of the organisms, proteins deal with phenotypes. It is not possible to fully understand the mechanisms underlying diseases with solely studying the genome (Feist & Hummon, 2015). Only with the study of protein functions and their modifications, targets for drug molecules can be identified. The main aim of proteomics is to understand the localization of proteins by creating a three-dimensional map of a cell. The dynamic nature of proteomics helps to determine the external and internal cues of the biochemical machinery associated with the cells (Aslam et al., 2017). It also helps us to study post-translational modifications, synthesis, and degradation. It is also known as taking the snapshot of protein environment at a given time (Dhingraa, Gupta, Zhen, & Fu, 2005). It determines the 3D structure all proteins in a genome. It helps to determine the function of proteins. It also helps to characterize post-translational modifications on proteins (Dhingraa et al., 2005). Now, proteins are one of the most essential and informational molecules that part take in biological processes (Ong & Mann, 2007). They are involved both structurally and functionally in the workings of a cell, like help form cell scaffold, gene regulation, aid in bio-signaling, transportation of solute-solvent across the cell membranes, protein synthesis, look out for any functional abnormalities in cells, regulation of proteins (which is very informative in study of disease pathology), and many other metabolic processes of the cells (Padula et al., 2017). In biomedical research, to identify the disease, understanding and studying the behavior of our proteome is crucial. In transcriptomics data, mRNA molecules are to be found in abundance. The traditions techniques prove insufficient and lack the capacity to analyze more than a few proteins at a time from the huge set of complex biological data samples (Padula et al., 2017). Thus, Mass spectrometry is of the modern techniques used to perform the task mentioned above in an integrated fashion (Lin, Shaler, & Becker, 2006). To establish a relationship between a disease and the phenotypic behavior the type of progression of disease, later development of the disease and its regulatory mechanism in body must be studied (Lin et al., 2006). Through Mass Spectrometry, we are able to formulate an inventory of proteins that are easily identifiable through available information (Lin et al., 2006). These proteins extracted from the different types of tissues are used to generate the differentially expressed proteins' lists from the cellular samples. When we perform proteome analysis using Mass Spectrometry, the dataset is differentiated into groups of two based upon the size of the proteins (Patterson, 1995). The MS approaches include—the top down approach and the bottom up approach (Patterson, 1995). Further, different Bioinformatics tools are used for data abstraction of the large datasets thus created; using the deep neural networks formed using DL algorithms. Simply put, this helps us make sense of the data in sound and image recognition and also, text analysis (Tran et al., 2019). The DL neural networks usually have two properties—firstly, they are made up of layers on layers of multi-layered structures of nonlinear processing units, and secondly, at each of these layers different types of supervised and unsupervised algorithms are used to learn features of the particular dataset (Tran et al., 2019). The applications of deep learning are vast and versatile from techniques like NLP (natural language processing), drug discovery, ASR (automatic speech recognition), field of bioinformatics and biomedical data analysis including—genome sequencing, protein structure predictions and analysis, medical imaging data analysis, etc. (Tran et al., 2019). Proteomics is basically the study of such proteins, their structures, features, properties and so on; which help in quantification, identification, their analysis, proteome classification and proteins expressions differential comparison obtained from biological samples under conditions characteristic to them (Smith, Kelleher, & Consortium for Top Down Proteomics, 2013). As a form of data science, Proteomics has helped us progress excellently in the field of genomics technologies. Based on qualitative and quantitative aspects of proteomics data, network analysis helps in formulation of new hypothesis for diagnosis (Smith et al., 2013). In this chapter, we will cover the methods and tools used for dispensing raw data, spectral data analysis produced through mass spectrometry technique which further helps in identification of peptides, quantification of proteins or peptides, and further downstream data analysis, network inference or transformation of data through deep learning and machine learning algorithms (Swan, Mobasheri, Allaway, Liddell, & Bacardit, 2013).

1.1 Overview of proteins

Proteins are important biomolecules which are comprised of long chains of amino acids bonded together by peptide bonds. An amino acid is comprised of a carboxylic group (COOH), a functional group (R) and an amino group (NH2) (Aslam et al., 2017). There are 20 various forms of R groups that make different amino acids. All these amino acids give rise to the structure and function of a protein (Aslam et al., 2017). The R groups are stabilized by various interactions such as dipole-dipole, covalent and non-covalent. Proteins are synthesized by the two-step process of transcription of DNA and translation of mRNA. After translation, the folding of polypeptide chains occurs (Padula et al., 2017). Proteins occur in various forms in our body, depending upon the diversity of their structures (Padula et al., 2017). All the biological reactions in living organisms are catalyzed by special class of proteins called enzymes. They bind firmly with their complementary substrates and increase the speed of the reaction (Dhingraa et al., 2005). Membrane channels and pumps are other classes of proteins which are involved in the regulation of small molecules and fluxes of ions to and from the membranes of cells. Proteins are also present in the immune system as antibodies that help in recognition and destruction of foreign particles and antigens (Dhingraa et al., 2005).

1.2 Protein structures

1.2.1 Primary structure

Proteins occur in L-α form of amino acids in humans. The term peptide is used for sequences fewer than 50 amino acids while polypeptide is used for longer sequences. The primary structure of proteins is comprised of linear chain of joining the end of C-terminus with the N-terminus (Smith et al., 2013).

1.2.2 Secondary structure

Depending upon the hydrogen bonding, the strands of proteins have various forms of structural confirmations. The secondary structure of proteins is formed by two types of forms—the α-helix and the β-sheet. The α-helix occurs in the form of right handed coiled strand (Smith et al., 2013). The side chain amino acids extend to outside in α-helix. Hydrogen bonds are formed between C and O C=O and N—H bond between N and H. In case of β-sheet, the hydrogen bonds are formed between rather within strands. It occurs in the form of a sheet formed by side by side (Smith et al., 2013). The O- from one strand is bonded with H- of adjacent strand. The strands are formed anti-parallel and parallel to each other. It is more stable bond due to well aligned hydrogen bonds (Smith et al., 2013).

1.2.3 Tertiary structure

The 3D structure of a protein is formed by the tertiary structure. A stabilized structure is formed by bending and twisting by achieving lowest energy state (Padula et al., 2017). It occurs as random and irregular folding stabilized by interactions between side chains of amino acids. The disulfide bridges in cysteine forms strong and stabilized tertiary structure due to oxidation of sulfhydryl groups (Padula et al., 2017). This holds various parts of protein together. Besides these, ionic interactions between negatively and positively charged ions form stabilized tertiary structure of protein (Padula et al., 2017).

1.2.4 Quaternary structure

Most often the proteins are made up of combination of multiple polypeptide chains. These are known as protein subunits (Feist & Hummon, 2015). When the subunits are formed by same units, they are called homodimer and that of different subunits are called heterodimer. The complex protein structure is formed by quaternary structure through interactions between the subunits (Feist & Hummon, 2015). Various interactions such as ionic interactions, disulfide bridges and hydrogen bonding form the stabilized and final shape of the protein complex (Feist & Hummon, 2015).

1.3 Different subdivisions of proteomics

1.3.1 Structural proteomics

It refers to map out the entire nature and structure of protein complexes in a particular cell rather than comparing the normal cell or tissue and diseased state. It aims to characterize all protein-protein interactions and identify protein complexes (Anderson & Anderson, 1998). The architecture of the protein assembly can be obtained by isolation of specific cellular organelle by purification (Ong & Mann, 2007). The estimation of cellular states and their static differences is done through the microarray experimentation and SILAC technology is used to measure the kinetics of biological processes for various points in time in proteomics, post which isotopic labeling of the three states cellular is performed (Blackstock & Weir, 1999). The proteasome inhibition on the smaller molecules initiates upon the appearance of changes in nucleolar proteome. The kinetics between these groups of proteins is observed to be quite similar (Colinge & Keiryn, 2007). Similar profiles that are categorized together into one group are shared among temporal profiles; these groups share functional relationships and thus make data biologically relevant to study (Colinge & Keiryn, 2007).

1.3.2 Expression proteomics

Expression proteomics refer to the quantitative analysis of protein expression within the samples. The expression of entire proteome or sub-proteome can be compared using this approach. This helps in the identification of disease specific proteins (Blackstock & Weir, 1999). For instance, tumor samples and similar samples from normal individual can be compared together (Blackstock & Weir, 1999). Techniques such as 2D gel electrophoresis, mass spectrometry, etc., can be used. The data obtained could be compared with the microarray data. Correlation of the expression among proteins and mRNA is a part of proteome data quantitative analysis (Anderson & Anderson, 1998). Effect of translational regulation is observed in protein expression. Cellular and physiological states differences are studied in phenotypic proteomics (Anderson & Anderson, 1998). Spatial distribution, which can also be termed temporal kinetics performed over mediating pathways and cell organelles or other processes of cell can be partnered among proteins for their interaction (Colinge & Keiryn, 2007). For course, this diverse and large amount of proteomics data is subjected to machine learning tools and algorithms for its proper analysis (Colinge & Keiryn, 2007). mRNA aided gene expression quantitatively in proteins is in the form of transcriptional and translational gene regulations. This analysis is done at transcriptomics level (Colinge & Keiryn, 2007). The KEGG pathway database is used to study reaction pathway of expression map and functional phenotype in biological processes GO are identifies through this analysis. Cell line shift templates are used to study cellular phenotypes (Chen, Hou, Tanner, & Cheng, 2020).

1.3.3 Interaction proteomics

Peptide identification and proteins interaction help realize our cellular processes and their uses. Analysis of molecular networks, identification of the participating molecules is possible through interaction proteomics (Pandey & Mann, 2000). Proteins interactions data inundation through ever advancing technologies has led to cumulation of interaction databases and interaction networks (Pandey & Mann, 2000). Mathematical analysis of networks is used in cell and molecular biology. The behavior of interactomes networks is studied with the help of network biology (Pandey & Mann, 2000). Mass spectrometry, based on the quantitative side of proteomics, provides reliable information on protein interaction. Proteins sub cellular localization has become a prerequisite for study of cell function and thus is mapped using this approach (Ong & Mann, 2007). The inventory proteins are used to help create organelle in organelles proteomics. Detection of contaminations in the sample which preparing pure organelles can also be identified through MS because of their high sensitivity (Ong & Mann, 2007). The proteins are then allocated to different endomembrane system for the organelles prediction, these include plasma membrane, Golgi apparatus, ER, etc. (Ong & Mann, 2007).

Fig. 1 shows the use of all three different types of proteomics.

1.4 Major proteomics methods

There are various techniques used in proteomics. Some of them are discussed below. Table 1 shows the major techniques used in proteomics.

1.4.1 Gel electrophoresis

2D gel electrophoresis is one of the most widely used techniques for the study of proteomics. It has high resolution power and with the use of IPG strips, the stability of the pH increases (Bjellqvist et al., 1982). It uses ampholytes, surfactants and chaotropes for the complete alkylation and reduction of cysteine bonds, thereby removing the proteins from the sample (Bjellqvist et al., 1982). The procedure includes use of 8 M chaotrope urea and 4% zwitterionic surfactant CHAPS. In case of bacterial and plant samples, sample disruption is required using disruption methods (Marc, Christian, Ron, et al., 1996). Due to increased surface area, it allows higher extraction of proteins. A combination of increasing solubilization power is used for protein extraction such as PBS or Tris-HCL (Marc et al., 1996). This is then subjected to the action of surfactant and mixture of chaotrope

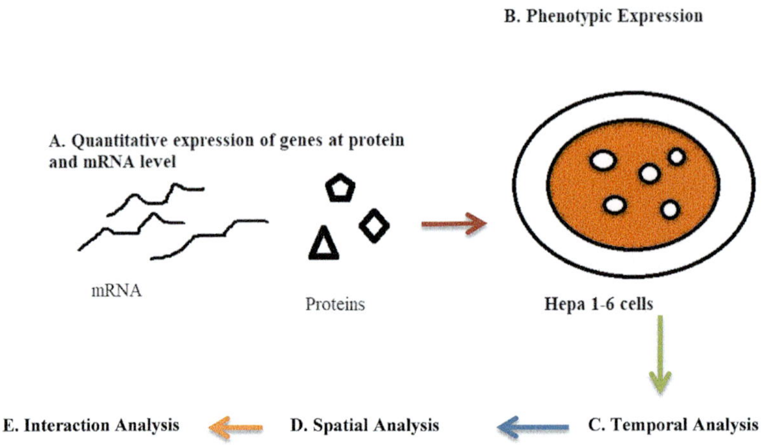

B. Phenotypic Expression

A. Quantitative expression of genes at protein and mRNA level

mRNA

Proteins

Hepa 1-6 cells

E. Interaction Analysis

D. Spatial Analysis

C. Temporal Analysis

Fig. 1 Different types of proteomics involved in analysis of expression of genes in epigenetic level.

such as thiourea, urea and zwitterionic surfactant CHAPS. The remaining insoluble material is boiled in SDS to reduce complexity of the proteome (Marc et al., 1996). It is necessary to do thorough alkylation and reduction of cysteine residues to produce single proteoforms. Phosphine based reducing agents are widely used since they have affinity only toward disulfide bonds. They don't react with alkylating reagents such as thiol based reducing agents (Marc et al., 1996). After separating proteoforms according to their molecular size, isoelectric focusing is used to resolve all proteoforms according to their isoelectric point (Issaq & Veenstra, 2008). It is the point at which the net charge of a protein is zero. It is carried out in IPG strips. The IPG strips are subjected to high field strengths or voltages upto 10,000 V and at extremely low currents such as 1 μA. Hence, 7–9 M of urea with 2 M of thiourea is used to maintain protein stability and disrupt association of lipids during IEF (Rabilloud & Lelong, 2011). Urea disrupts the protein structure by disturbing the water structure, making hydrophobic interactions less compact by weakening them, disrupting ionic interactions and Vander Waal forces (Rabilloud & Lelong, 2011). Surfactants such as CHAPS, amidosulfobetaine, etc., with greater solubilizing power are comprised of two different regions—an ionisable group interacting with solvent molecules such as water and hydrophobic region interacting with the protein (Issaq & Veenstra, 2008). They are used in the form of zwitterion in IEF, i.e., a positively charged group amine and a negatively charged group sulfoxide (Issaq & Veenstra, 2008). They interact typically with the solvents

Table 1 Some of the important methods used in proteomics.

S. No.	Techniques	Advantages	Disadvantages
1.	Two Dimensional Electrophoresis (2DE)	Gives information about post-translational modifications (PTMs). Used for protein separated and quantitative expression analysis	Separates basic, acidic and hydrophobic proteins poorly
2.	Isotope-coded affinity tag (ICAT)	Used in quantitative proteomics. Highly reproducible and sensitive. Can detect peptides with low detection levels	Acidic proteins and non-cysteine residues cannot be detected
3.	Isobaric tags for relative and absolute quantitation (iTRAQ)	Can be multiplexed. Provides high throughput results	Complexity of sample increases. Fractionation of peptides is required
4.	Differential Gel Electrophoresis (DIGE)	Highly reproducible and sensitive. Gives information about post-translational modifications (PTMs). Used for protein separated and quantitative expression analysis	Quite expensive due to use of fluorophores. Unable to label proteins without lysine
5.	Stable Isotope Labeling with Amino acids in Cell culture (SILAC)	High degree of labeling using direct isotope	Tissue samples cannot be labeled
6.	Protein Array	Highly sensitive, reproducible, multiplexed, high throughput, sample consumption is low. Used in quantification of proteins and biomarker detection	Amount of protein produced is limited. Expression level is low. Requires large amount of antibodies
7.	Multidimensional Protein Identification Technology (MUDPIT)	Can identify large proteins and protein-protein interactions. High degree of separation	Cannot be used for quantitative analysis. Difficulty increases with isoforms and huge data set
8.	Shotgun	Covers large number of proteomes	Difficulty in obtaining proteomic information
9.	Bait-prey affinity isolation	Robust, can detect protein-protein interaction	Large precipitation of samples. Can sometimes give false positive results
10.	Ligand Blotting	Used for protein-protein interaction studies	Requires high optimization

and cancel out the charge on them thereby reducing net charge on the molecule to zero (Westermeier, 2014). They enhance the solubility of the protein molecule with solvents bringing to their isoelectric point so that the protein cannot function. This is caused due to neutral charge on amino acids thereby altering their solvent interactions (Westermeier, 2014). Dithiothreitol (DTT) is considered as the most effective reagent for reduction of disulfides. It is used at a high concentration of ~20 mM. It is thiol reagent used to alkylate protein thiols (Westermeier, 2014). Alkylation and reduction of proteins is done in a two-step process with DTT. First, proteins are treated with DTT for 30 min and after that they are treated with iodoacetamide at double the concentration of DTT. This step is mandatory to ensure that the thiols are alkylated (Westermeier, 2014). Phosphine based reducing agents are also used to react with disulfide bonds. They perform both reduction and alkylation in a single step (O'Farrell, 1975). Fractionation of the protein samples is done to load samples in IPGs using 7 M urea, surfactant such as 1% C7BzO and 2 M thiourea. The IPG strips are then mounted inside a series of chambers filled with chaotropic solution. After diffusion, the strips are subjected to SDS PAGE (O'Farrell, 1975).

1.4.2 SDS PAGE

It is a widely used method to identify, separate and purify polypeptides (Ramos, Garcia, Perez-Riverol, Leyva, et al., 2011). Polyacrylamide gels are formed by extreme crosslinking of the acrylamide using a cross linking agent N,N′ methylenebisacrylamide. The polymerization step is carried out with the help of an initiator ammonium persulfate and a catalyst N,N,N′, N′-tetramethylethylenediamine (TEMED) (Ramos et al., 2011). It is widely applicable in forensic science, biotechnology, molecular biology, biochemistry and genetics to segregate proteins according to their electrophoretic mobility (Ramos et al., 2011). Sodium dodecyl sulfate (SDS) is an anionic detergent used to denature secondary and tertiary structure of the proteins (Shevchenko, Wilm, Vorm, & Mann, 1996). It also imparts negative charge to the protein molecules with respect to their mass. The sample is then heated at 60 °C for further denaturation (Shevchenko et al., 1996). Tracking dye is used to track the progress of protein through the gel. For high molecular weight molecules, lower percentage of gel is used and vice versa for low molecular weight molecules (Wittig & Schagger, 2009). The gels are then stained with Coomassive Brilliant Blue stain for detection under autoradiography. The size of a protein can be estimated by comparing known molecular weight ladder and the distance migrated by proteins (Wittig & Schagger, 2009).

1.4.3 HPLC

High performance liquid chromatography (HPLC) is a commonly used technique in proteomics and other fields to segregate biological compounds (Xie, Liu, Qian, Petyuk, & Smith, 2011). HPLC provides effective analysis of complex trace analytes and generates highly sensitive results. It is an improved version of column liquid chromatography (Xie et al., 2011). Solvents are introduced into the column under higher pressure upto 400 atm (Xie et al., 2011). The segregation takes place in a separate column under the reaction between a mobile phase and a stationary phase (Xie et al., 2011). The stationary phase is comprised of a diatomaceous earth, silica, polymer gels, etc., with tiny pores in a separate column (Wright et al., 2014). The column is generally made up of stainless steel or glass. The mobile phase consists of a solvent forced at high pressure into the column (Wright et al., 2014). The sample is injected into the mobile phase via syringe through a loop. The sample molecules travel through the column at different rates due to varying affinities with the stationary phase. A detector is used to detect the individual substances leaving the column (Wright et al., 2014). The results are displayed in the form of a chromatogram using suitable HPLC software on the computer. It helps to identify and quantify the substances (Wright et al., 2014).

1.4.4 Mass spectrometry

Mass Spectrometry is a tool efficient and powerful enough for this purpose as it is useful in identification of proteins along with the chemical alternations of proteins post the post-translational modifications (Yates, 2011). It is a highly reliable, robust label-free analysis technique enables detection of different proteins from a complex mixture in an acceptable timeframe (Yates, 2011). It can be used for the detection of a range of molecular substances along with proteins like glycans, peptides, metabolites, lipids, therapeutics, etc. Primarily, mass spectrometry is not quantitative, however, today it is used to produce large amounts of data because of the application of stable isotopes onto the molecules being analyzed (Yates, 2011). These stable isotopes are used for labeling the molecules (Yates, 2011). Currently many different MS techniques are available, but, the two most efficient techniques are MALDI and the other, Desorption electrospray ionization (Yocum & Chinnaiyan, 2009). There are three parts of a mass spectrometer: it has an ion source, mass to charge ratio (M/Z) of ionized analytes measured using mass analyzer and also the detector that can register the number of ions at each of the mass by charge values (Yocum & Chinnaiyan, 2009). Both figuratively and literally mass analyzer is central to this apparatus. During sample preparation (proteomics data) in a typical experiment, on the section surface, the analysts on each x-y coordinate are ionized by the mass spectrometer (Yocum & Chinnaiyan, 2009). This results in an ordered array of mass spectra. Ion density maps for various M/Z values are reconstructed by statistical analysis of data on various software available for the task (Yocum & Chinnaiyan, 2009). Next step is the identification of the selected M/Z values, determined through mass comparison with the relevant databases for known ones, further, fragmentation using Tandem Mass spectrometry technique (MS/MS) of particular ions that will reveal particulars about the chemical structure of precursor ion previously hidden (Patterson, 1995). Currently, four basic mass analyzers are majorly used for research purposes, time of flight (TOF), ion trap, Fourier transform ion cyclotron analyzers, quadrupole, all build differently with unlike designs and have varying performance relatively (Patterson, 1995).

MS is used to accurately determine the molecular weight of biopolymers, chemical compounds and all the other molecules whether small or large (Patterson, 1995). The different sections of a mass spectrometer are illustrated in Fig. 2:-

- First the biological samples are stored in the spectrometer as a system
- Ionization chamber
- The mass analyzer
- Ions detector
- And software components for storage and analysis of the spectra thus produced

MS procedure is widely accepted as compared to NMR because of its ability to detect metabolites at very low concentrations even at femto to atto molar range (Gygi & Aebersold, 2000). It is used to identify and quantify

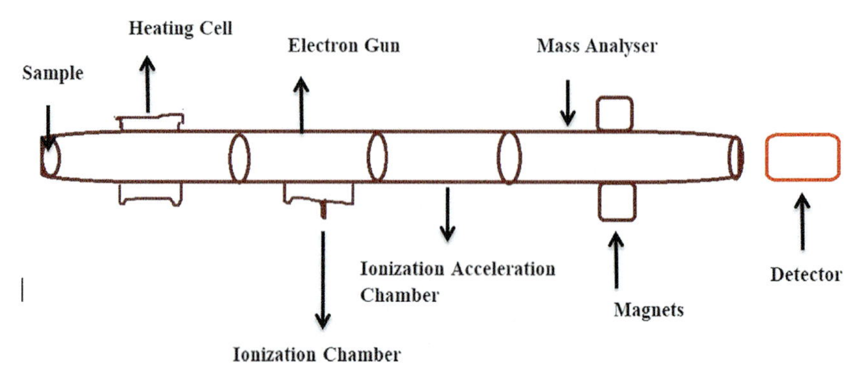

Fig. 2 A diagram showing structure of a typical mass spectrometer.

metabolites with high sensitivity, resolution and dynamicity (Gygi & Aebersold, 2000). The following are the steps involved in MS based analysis of metabolites—preparation of sample, separation of the sample by chromatographic, capillary electrophoresis and extraction, ionization of sample, conversion and detection of metabolites with respect to their mass to charge ratio (m/z) (Gygi & Aebersold, 2000).

- *Sample Preparation*: Biological samples such as cell culture, tissue, and biofluids such as saliva, blood, bile, urine, seminal fluid, etc., can be used for metabolome analysis (Aebersold & Mann, 2003). It is crucial to do sample preparation carefully along with its management in the form of bio-banking and biorepositories with proper labeling. It is required to generate high throughput, reproducible, optimum output with enriched metabolite coverage and recovery (Aebersold & Mann, 2003).
- *Extraction*: The prepared samples are extracted using nonpolar and polar compounds based on targeted and untargeted approaches (Diamandis, 2004a, 2004b). Methanol-water-chloroform is an approach widely used to extract hydrophobic and hydrophilic compounds (Diamandis, 2004a, 2004b). The mixture is then subjected to centrifugation which yields a biphasic mixture of lower (organic) and upper (aqueous) layers. Other polar organic solvents such as mixture of methanol or acetonitrile with water are also used. Dichloromethane is used to perform organic extraction (Diamandis, 2004a, 2004b).
- *Separation*: Separation of the two phases is done using variety of techniques such as gas chromatography, liquid chromatography coupled with MS system. Sometimes the mixture is directed injection into real time MS (Yates, 1998). For polar metabolites, capillary electrophoresis coupled with MS is used for separation and profiling in biological samples (Yates, 1998). For nonpolar metabolites, reversed phase LC with C18 columns is used (Yates, 1998). GC/MS or LC/MS is preferably used for targeted and untargeted metabolomics which produces high chromatographic resolution (Yates, 1998). GC/MS is used to analyze volatile organic compounds (VOCs) in biological samples such as organic acids and fatty acids (Yates, 1998). VOCs can be separated by solid phase microextraction technique (Wilkins et al., 1999). It separates organic compounds from solid, aqueous and gaseous materials (Wilkins et al., 1999). While GC-MS is used to analyze less polar biomolecules such as perfumes, lipids, waxes, essential oils, etc., LC-MS is used to analyze more polar biomolecules such as nucleotides, organic acids,

polyamines, etc. (Wilkins et al., 1999). Both GC-MS and LC-MS are used to analyze amino acids, steroids, fatty acids, alcohols, etc. (Wilkins et al., 1999).

- *Ionization*: The samples are ionized using positively or negatively charged ions in gas phase. It is done by injecting dry nitrogen and heat followed by evaporation (Yates, Ruse, & Nakorchevsky, 2009). The samples are then converted into positively and negatively charged ions. Depending upon the polarity of the molecules, various types of ionization sources are used such as electrospray ionization sources (ESI), matrix assisted laser desorption ionization (MALDI), chemical ionization (CI), etc. (Yates et al., 2009).
- *Detection*: The ions are detected by high resolution mass analyzers depending on the mass to charge ratios of the fragmented ions (Yates et al., 2009). Ion trap, time of flight (TOF), quadrupole time of flight (QTOF), etc., are some the mass analyzers widely used. It produces mass spectrum with high resolution, sensitivity and accuracy (Yates et al., 2009).
- *Analysis and Identification*: The enormous amount of raw data extracted from MS contains specific metabolic signals which can be analyzed using specialized software for data interpretation and identification of metabolite of interest (Lin et al., 2006). Free available software and tools are used for processing of data, assessment and their quantification (Lin et al., 2006). Pre-processing of raw data such as retention time correction, peak detection, spectral filtering, peak alignment, noise elimination and peak normalization are required (Lin et al., 2006). The data is then prepared for integrity checking, identifying compound name and normalization using clustering, statistical, and multivariate and univariate analysis (Lin et al., 2006). Following all these steps, processes such as enrichment analysis, pathway mapping, pathway analysis and functional interpretation are performed (Shiio & Aebersold, 2006). Tools such as MetaCore, 3Omics, Progenesis and Metaboanalyst can be used for processing raw mass spectrum data and connecting with the databases such as Chemical Entities of Biological Interest (ChEBI), In Vivo/In Silico Metabolites Database (IIMDB), Madison Metabolomics Consortium Database (MMCD), Human Metabolome Database (HMDB), Kyoto Encyclopedia of Genes and Genomes (KEGG), Metabolite and Tandem MS Database (METLIN), Reactome, BioCyc, ChemSpider, PubChem, MetaCyc, MetaboLights, etc. (Shiio & Aebersold, 2006).

1.5 Applications of proteomics approaches to diseases

1.5.1 Biomarker identification

Protein profiling is invaluable to biomarker discovery. The intrinsic genetic coding of the body and its effects on the environment around are reflected by the proteins in our systems (Diamandis, 2004a, 2004b). Proteins are specialized in regulating the biochemical processes of the body, such as metabolism, thus, to be able to study these processes protein identification, characterization and quantification is extremely important (Diamandis, 2004a, 2004b). Expression of proteins and their functions are highly subjective due to the modifications brought about by translational events, post-translational events, through transcription. The subtle changes in the intricate protein molecular organization can be biomarkers in it (Diamandis, 2004a, 2004b). The techniques being developed and employed in biomarker discovery have the following purposes in aim: first being to improve upon the currently available biomarkers, secondly, to develop different kinds of approaches that enable us to discovered unconventional biomarkers and bestow greater specificity and sensitivity in the validation of novel biomarkers (Wagner, Verma, & Srivastava, 2004). Thirdly, to employ combinations of biomarkers by putting together a panel of biomarkers currently available or using other techniques in identifying protein peak patterns that could be used in disease detection, etc. (Wagner et al., 2004).

In this report, SELDI TOF generated raw spectral data will be used in the processing. SELDI TOF stands for Surface-enhanced laser desorption/ ionization (time-of-flight) (Tyagi et al., 2010). This technique produces proteomics data that is highly relevant for biomarker discovery as its technology combines mass spectrometry and retentate chromatography wonderfully (Tyagi et al., 2010). A high throughput technique, an extension to the original MALDI TOF mass spectrometry method, SELDI TOF is a refinement to the MALDI TOF technique, its suitable for proteins with low molecular weight (Tyagi et al., 2010). However, the Protein Chip, the resulting matrix focuses on purification along with ionization/desorption step. Proteins chips are a kind of Protein array that are ligand binding, solid phase assay system formed by immobilized proteins on a surface like cellulose membrane or glass or micro/nano-particles or microbeads (Tyagi et al., 2010). In this technique, chromatographic separation allows the proteins to be differentiated on the basis of their physical and chemical characteristics like acidic affinity, basic affinity, hydrophobic or hydrophilic, metal affinity (Tyagi et al., 2010).

1.5.1.1 Biomarkers

A Biological Marker or Biomarker is simply the measure of a bodily characteristic or detected molecule that indicates a biological response associated with pharmacological reaction to a medical treatment, a pathogenic process, etc., often, measured objectively, one or a combination of these biomarkers are used to evaluate the health of an individual, during clinical assessment like cholesterol level, blood sugar, heart rate and so on (Veenstra et al., 2005). Today, a highly diverse range of biomarkers are used that are specific and accurate to our different organ systems such as circulatory, nervous, immune, metabolic system (Veenstra et al., 2005). The Biomarkers discovered are only ideal if they could be easily detected in body wastes or easily extracted like urine and blood, cost efficient to be tested multiple times, variable with treatment, harmless, relatively regular across different populations (Veenstra et al., 2005). Expression of biomarkers during disease is shown in Fig. 3.

Biomarker is a powerful and dynamic tool for clinical investigation with its applications ranging from diagnosis to prognosis, form screening to randomized trials, from observational to analytical epidemiology, to name a few (Veenstra et al., 2005). These qualities have made biomarker research an avenue for tremendous work and opportunity. Biomarkers can be categorized in two, relevant to the intended purpose of the biomarker: firstly, predictive, that can be used in predictive deduction of the response of an illness to a particular treatment, and secondly, diagnostic, which incorporates the

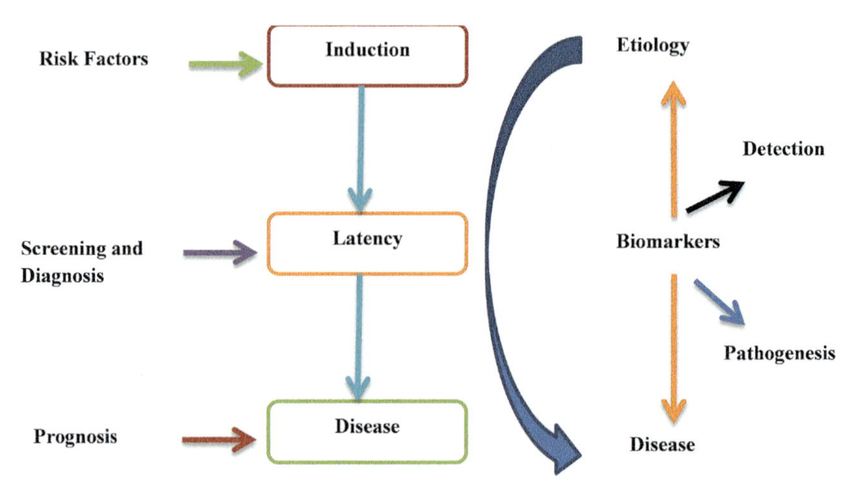

Fig. 3 Flowchart showing the expression of biomarkers during disease.

disease diagnosis, further course of the conditions like progression, recurrence, fatality, cause, regression, etc. (Veenstra et al., 2005). Application of new high throughput techniques has led to the discovery of different powerful biomarkers, such as molecular biomarkers that could even predict the susceptibility of an individual to a disease. However, their application in clinical assessment is scare due the cost and risk factors (Tyagi et al., 2010).

1.5.1.2 Data classification and biomarker identification

Biomarker discoveries have become a sequential affair, meaning it follows a highly sequential approach that lead to findings of novel biomarkers. First step, namely the identification of potential biomarkers from a list of biological constituents that might be the relevant ones (Wagner et al., 2004). Their relevance is tested based upon their impact or correspondence to the disease being studies. The better the connection between the two, the higher the potential for those biological constituents to be next biomarkers for the disease (Wagner et al., 2004). Next step in this sequential approach is validation of the previously shortlisted biological constituents (Diamandis, 2004a, 2004b). They are cross referenced, their correspondence to the disease tested against more specific means, re-estimation even for their indication to the outcome of the disease or its status. Classification is assigning each of the spectrums to a group having similar spectra, so the similar samples are then grouped together and other dissimilar ones separated into another group (Diamandis, 2004a, 2004b). A large set of plausible biomarkers are used to build the classifier in classification algorithms. They include KNN (k-nearest neighbor), RF (random forest), SVM (support vector machines), etc., as mentioned above in the report (Swan et al., 2013).

Picking out best candidate biomarkers from the large number of biological elements is the main focus of identification studies. In terms of proteomics, selecting the plausible biomarker equates selecting the candidate proteins that differential one sets of samples from another group (Veenstra et al., 2005). There are various ways to go about doing this, by facing the problem of dimensionality that causes the number of peak intensities (the variables) to exceed the total number of input samples itself. Therefore, reducing the dissension of regression space is necessary (Tyagi et al., 2010). The first approach of reducing dimension is redefining regression space to only the selectively considering important variables in the matrix that correspond to our features of interest (Tyagi et al., 2010). This will reduce the dimension effectively. Another approach involves the subspace redefined by the components. The first approach is focused toward identification of those features that are responsible for the differences in intensities between these groups (Tyagi et al., 2010). The second approach, however, evaluates how important are these features embedded in an algorithm for classification (Tyagi et al., 2010).

1.5.1.3 Biomarkers validation

In the process of internal validation, the sample data is split into two sets, learning and a test set. Then the training of the classifier is done on this learning set, the classifier is then validated using the test set (Wagner et al., 2004). To avoid any overfitting, this type of internal validation is a crucial and necessary evaluation procedure so as to avoid using learning set characteristics which might not be of interest, corresponding to disease status but rather particular to the current set (Wagner et al., 2004). The test set is to be exempt from the entire primary processing including pre-processing procedures and peak detection and identification steps, when the identification study's statistical analysis includes selecting differential biological constituents followed by an algorithm meant for classification (Veenstra et al., 2005). This is a case of misuse and misemployment of the techniques of cross validation as it leads to an over estimation of biomarker performance artificially. Therefore, the best suggestion is to avoid any pre-processing of entire dataset before splitting it into training set and test set appropriately (Diamandis, 2004a, 2004b). Along with internal validation of the biomarkers, external validation is still crucial for reliable information. One aspect of biomarkers application that has been over-shadowed and mistaken for one another for a long time is the difference between their proficiency in classifying different subjects and the estimation by odds-ratio of strength of association between disease status/outcome and biomarkers levels (Wagner et al., 2004). This must be included in the validation procedure of biomarkers, and in the estimation of their proficiency. ROC curves (receiver operating characteristic curves) and an extension to these curves, PROC curves (predictive receiver operating characteristic curves) are used to estimate the biomarkers' classification performance and second one predicting positive and negative values along with the usual estimation (Veenstra et al., 2005).

1.5.2 Identification of peptides and proteins

The instrument of mass spectrometer produces its results in form of fragmentations (spectra), which are then input into other analysis tools for the identification of proteins present in the sample (Baldwin, 2004). Through this the proteins sequences are also obtained. But our main focus lies in extracting the list of significant peaks, as each of the peaks thus acquired correspond to the different peptides and proteins present in the given biological sample. Databases can also be utilized for this purpose of recording the protein sequences (Baldwin, 2004). While the spectrum is measured with calculation of the match score between the theoretical spectrum and the experimental spectrum with the help of swissProt like query databases (Baldwin, 2004). For protein identification, the peptide match that scores the highest is given the most due consideration. The approaches used for protein identification include PSM (however, its score is often considered false), target decoy or FDR (Gevaert et al., 2003). For posterior error in probability other tools like maxquant can also be utilized. In order to yield a greater number of proteins from the sample, data analysis id performed multiple times on the sample (Gevaert et al., 2003). Form the fragmented spectrum, identification of peptides is performed through the (de novo) sequencing of peptide technique that customizes the MS mass by ratio vs intensity results to perform protein identification. For better protein identification, along with de novo tools other algorithms are also applied (Gevaert et al., 2003).

Through the spectrum input, method of fragment characterization and the analysis of mass differences between the peaks, the identification of peptides is carried out (Baldwin, 2004). Mass spectrometry also helps in studying proteins' post translation modifications through the method of peaks PTM among other things (Baldwin, 2004).

1.5.2.1 Protein inference

Sequences of peptides are reconstructed into their original proteins; this is termed as protein inference. Following this, a list of proteins is obtained from these peptides (Mann & Pandey, 2001). Earlier, the proteins from this list are digested and segmented into their constituent peptides and intensify these peptides (Mann & Pandey, 2001). Their features are unique to the proteins they constitute, therefore, the longer the peptides the more information we have about original proteins. Various tools and algorithms are used to assemble these peptides into their parent proteins (Mann, Hendrickson, & Pandey, 2001). The probabilistic and statistical are the two models of proteins inference. Justifiably smaller sets of proteins for the detected peptides are present in the probabilistic model (Mann et al., 2001). However, the statistical model is further sub-divided into the two sub-models as the Bayesian inference model and the hierarchical spectral model. The PSM score also affects the protein inference model performance (Mann et al., 2001). The reduced number of repeated proteins is the FDR's, and proteomics datasets that are very large are segmented into FDR's during identification of proteins. Null distribution is generated in the FDR's estimation approach; this is done by combining the model of logistic regression and the permutation method (Mann & Pandey, 2001). A list of proteins is generated as the result of protein inference. Also, if any further proteins are present in the list of grouped proteins that are individually distinguishable are grounded on the peptides observed (Mann et al., 2001).

1.5.2.2 Quantification of the abundance of protein

Through quantitative analysis of the proteome, protein levels are athwart in the profusion of the proteome sample (Riffle & Eng, 2009). As the amount of data multiplied with high performance quantitative proteome analysis, the demand for better analysis method in bioinformatics surged (Rabilloud & Lescuyer, 2014). Two methods have been categorized under the quantitative proteomics approaches, these are as follows:

(1). Relative Quantification method—the ratios are considered in this relative quantification method. Any changes in the relative concentration of proteins in the sample are recorded (Rabilloud & Lescuyer, 2014).

(2). Absolute Quantification method—the number copies of proteins concentration across the sample are defined in the absolute quantification method (Rabilloud & Lescuyer, 2014).

On basis of means of their quantification, the above methods are divided into further categories: labeled method and labeled-free method (Riffle & Eng, 2009).

1.5.2.3 Labeled method

Two quantification approaches in the labeled method include:

(1) *MS1 based labeling*

This is often termed as first stage MS. The spectra of MS1 produces various isotopic patterns for all the different sample on the basis of the labeling (Riffle & Eng, 2009). These sample can be:

- *In vitro samples*—in vitro experiments involve strategies like dimethyl based labeling, isotope tags, isotope protein labeling.
- *In vivo samples*—the culture of tissue in in vivo experiments involve of isotope labeling with nitrogen (isotope 15) metabolic labeling, amino acids (Riffle & Eng, 2009).

In bioinformatics, MaxQuant method is used in MS1 labeling. SILAC labeling is used to design this method in order to obtain high resolution data.

The raw results produced through MaxQuant method with framework help is utilized in the downstream statistical analysis (Rabilloud & Lescuyer, 2014).

(2) *MS2 based labeling*

This also known as isobaric labeling. MS2 based labeling detects the quantification signals from mass range of extremely low magnitude. In MS2, isobaric labeling is often used strategy (Rabilloud & Lescuyer, 2014). So even the molecules with identical mass where differentiated with the help of heavy isotopes. Protein pilot software is used to process the raw data produced from MS2 based labeling with instrumentational aid (Rabilloud & Lescuyer, 2014).

1.5.2.4 Labeled-free method

The diverse spectra of samples processed in labeled-free method is obtained with the help of technique of LC-MS (liquid chromatography-mass spectrometry). Variations in the dataset are addressed to intensity normalization for correction (Riffle & Eng, 2009). MaxLFQ algorithm of the MaxQuant method aids labeled-free method sample quantification. In the field of clinical studies, imaging mass spectrometry (IMS) technique has emerged as the quantification tool (Riffle & Eng, 2009).

1.6 Major proteomics resources

Some of the important proteomic tools and resources are discussed below and shown in Table 2:

1.6.1 PX consortium

PX consortium is comprised of MS based proteomics repositories. It collects information from the primary resources such as PRIDE and PASSEL and secondary resources such as PeptideAtlas (Chen, Zhao, Ma, & Zhu, 2015). It provides consistent, user friendly and harmonized data. The data is stored in format of PX XML and can be accessed using a universal and unique PXD identifier (Chen et al., 2015). It retrieves MS data from PRIDE and SRM data from PASSEL. It is advantageous that the submitted data remains private and facilities such as manuscript submission are provided to allow reviewers and editors to access the data. In order to access all PX datasets, user can utilize ProteomeCentral repository (Chen et al., 2015). It provides the searching facility of metadata associated with PASSEL and PRIDE (Vizcaino, Cote, Csordas, Dianes, et al., 2013).

Table 2 Some of the important resources for analysis of proteomics data.

S. No.	Resources	Description
1.	iProX	Provides facility for raw data storage and submission. Helps in data analysis
2.	Yale Protein Expression Database (YPED)	Bioinformatics based repository for proteomic research
3.	Tranche	Provides facility for raw data storage and submission. Helps in data analysis
4.	PHOSIDA (phosphorylation site database)	Provides facility for raw data storage and submission. Helps in phosphoproteomic MS data storage
5.	Phospho.ELM	Helps in phosphoproteomic MS data storage
6.	PhosphoSitePlus	Provides facility for raw data storage and submission. Provides information of PTMs from MS data
7.	Reactome	Database for pathways, reactions and biological processes
8.	PANTHER Pathway	Database for data analysis and tools for gene ontology
9.	Search Tool for the Retrieval of Interacting Genes/Proteins (STRING)	Database for protein-protein interaction analysis
10.	Kyoto Encyclopedia of Genes and Genomes (KEGG)	Database for diseases and pathways
11.	Pathway Commons	Database for biological pathways
12.	Signaling Network Open Resource (SIGNOR)	Database for biological entities relationships
13.	Covariance inverse (COVAIN)	Tool for correlation network analysis, statistics and time series
14.	IKAP	Tool for analysis of kinase activities from phosphoproteomics data
15.	Neglog	Software for study of protein-protein interaction networks in humans
16.	ProLoc-GO	Tool for the study of Gene Ontology and protein subcellular localization

Table 2 Some of the important resources for analysis of proteomics data.—cont'd

S. No.	Resources	Description
17.	Human Integrated Protein-Protein Interaction Reference (HIPPIE)	Tool for protein-protein interaction networks
18.	Interaction Network GO Annotator (INGA)	Tool for study of protein interaction networks, sequence similarity and domain assignments
19.	Pathview	Contains R package for pathway visualization and data integration
20.	PP2A	Tools consisting of integrated flowchart for human interaction proteome

1.6.2 MassIVE

The MassIVE is an important data repository developed by Center for Computational Mass Spectrometry, University of California, San Diego (Chen et al., 2020). It provides global exchange of MS data. It helps researchers to access public files and raw datasets. It provides a suitable social platform of networking for researchers (Chen et al., 2020). It provides users to access, browse, download and comment on datasets. The data can be submitted to ProteoSAFe (Vizcaino, Deutsch, Wang, Csordas, et al., 2014). The dataset files of MassIVE are categorized as follows:

- License files—specify conditions for downloading datasets.
- Spectrum files—consist of mass spectrum files.
- Result files—contain output of search engine.

It provides fields for data submission such as PTMs, species, contact information and instruments (Vizcaino et al., 2014). It provides a list of all public datasets in tabular form and allows users to access, sort, filter and share data. It provides password protected access to the users to access the datasets (Vizcaino et al., 2014). The results can be reanalyzed through various tools such as discovery of unexpected modifications using MODa, top-down protein identification using MS-Align+, database search using MixDB and proteogenomics searches using ENOSI against transcriptomics and genomics sequences (Vizcaino et al., 2014).

1.6.3 PASSEL

PASSEL provides submission of SRM datasets such as experimental results and raw data. mQuest is used to submit raw data (Farrah, Deutsch, Kreisberg, Sun, et al., 2012). It provides access to original files and results.

It utilizes a web interface for data submission process (Farrah et al., 2012). It contains information of submitter, instrument, sample source, information about targeted and peptide ions (Farrah et al., 2012). All these information are available in TraML format in a tab separated file. Besides these, MS files are avaible in mzML, .wiff, .d and .raw format (Farrah et al., 2012).

1.6.4 Chorus

Chorus is cloud based application software used to store, analyze and share MS data in spite of original raw file format. It is freely accessible and openly accessed (Martens, 2011). It provides cloud based computing environment globally through which data can be easily analyzed and customized in Map Reduce format (Martens, 2011). Algorithms such as distributed and parallel can be used for large datasets. It utilizes MS and chromatographic data for protein sequence identification (Martens, 2011).

1.6.5 PRIDE

The PRIDE database is used to store experimental data. It stores information about biological metadata, mass spectra, PTMs and protein expression values (Cote, Griss, & Dianes, 2012). The data can be submitted in two ways—partial and complete. For complete submission, the data has to be in PRIDE XML or mzIdentML format (Cote et al., 2012). In case complete submission, tools such as PRIDE inspector can be used (Cote et al., 2012). It provides a tool called PRIDE Converter 2 tool suite which is a platform independent and open source for converting several search engine files to PRIDE XML (Wang, Fabregat, Rios, Ovelleiro, et al., 2012; Wang, Weiss, Simonovic, Haertinger, et al., 2012). The tool also supports four various applications—PRIDE XML filter, PRIDE mzTab, PRIDE Converter 2 and PRIDE XML Merger. All these files support graphical user interface (Wang, Fabregat, et al., 2012; Wang, Weiss, et al., 2012). PRIDE Converter 2 supports data from MASCOT (.dat), OMSSA (.csv) and X!Tandem (.xml). Partial submission enables data from proteomics, MS, etc. (Wang, Fabregat, et al., 2012; Wang, Weiss, et al., 2012). PRIDE PX submission tool is used to submit data which is an open source platform with graphical user interface facility (Vizcaino et al., 2013). PRIDE Inspector is also used to browse and visualize MS proteomics data. It is an open source standalone tool used by researchers to edit and review manuscript (Vizcaino et al., 2013). It supports all kinds of formats such as mzData, mzML, mgf, ms2, etc. It is used to visualize experimental data, metadata and spectrum information (Griss, Foster, Hermjakob, & Vizcaino, 2013). It retrieves information from UniProt, NCBI and Ensembl. PRIDE Cluster is used to obtain PRIDE data (Griss et al., 2013). It retrieves information of peptide identification and spectrum data (Griss et al., 2013).

1.6.6 PeptideAtlas

PeptideAtlas is a data reprocessing resource and contains data of SRM related tools. It is one of the well curated and biggest protein expression data resources. It contains information of 27% of human genes from Ensembl (Deutsch, Lam, & Aebersold, 2008). The tool has been developed into two more versions—PeptideProphet and ProteinProphet. They provide accurate procedure to curate data of protein/peptide identification. PeptideAtlas contains 14,000 different entries from Swiss-Port and UniProt database (Deutsch et al., 2008). Data in PeptideAtlas are organized into various build for reprocessing that includes data from a single sub-proteome or proteome. The builds are generated from data deposited on other public repositories such as PRIDE containing MS/MS spectra (Farrah, Deutsch, Hoopmann, Hallows, et al., 2013). These data are mapped with a reference protein database such as Ensembl, UniProt, etc. The data are then annotated with sequence alignments and genome mapping and finally linked to databases such as Human Protein Atlas (Farrah et al., 2013). This contains information about protein-peptide mappings, protein abundances and observances, cross reference to databases such as UniGene and RefSeq (Farrah et al., 2013). All these results are finally processed and loaded into systems biology experiment analysis management system (SBEAMS). Moreover, PeptideAtlas can be used a web search interface for protein search using peptide sequence in FASTA format, gene name, protein accession number and keyword (Farrah et al., 2013). Protein view page is used to give all the information of a specific protein (Deutsch et al., 2008). Basic information about the protein is provided in the top section. Sequence motifs would provide the information of peptide coverage of protein (Deutsch et al., 2008). It also contains information of MS data of all peptides. PeptideAtlas also provides users with Cytoscape plug in. It also works with Chromosome Explorer and HUPO chromosome based Human Proteome Project and provides information about protein identifications per chromosome (Deutsch et al., 2008). PeptideAtlas supports various resources such as SRMAtlas, ATAQS and TIQAM. TIQAM stands for targeted identification for quantitative analysis by MRM (Farrah et al., 2013). It is a desktop application used to select peptide sequences and transitions. It is further categorized into TIQAM-viewer, digestor and PeptideATlas

(Farrah et al., 2013). SRMAtlas is a resource to quantify and detect proteins using SRM based proteomics workflow (Farrah et al., 2013). ATAQS stands for automated and targeted analysis with quantitative SRM tool. It is software which contains the modules and tools to manage, validate, design and analyze MRM assays (Deutsch et al., 2008). It connects with web services such as TIQAM for generating in silico peptides of a protein, PABST (peptide atlas based SRM transition for generating optimal transitions) and PIPE2 for generating lists of proteins and designing MRM assay, etc., using FireGoose (Deutsch et al., 2008).

1.6.7 MaxQB

It is a database developed to display and store collections of large amount of data of proteomics projects (Schaab, Geiger, Stoehr, Cox, & Mann, 2012). It widely employed as an analysis platform and a generic repository for high resolution bottom up experiments. It contains information of fragment spectra, label-free derived intensities, protein and peptide identifications (Schaab et al., 2012). It facilitates data analysis and reduces data heterogeneity. MaxQuant is used to integrate new datasets into MaxQB. It also links data with other proteomic databases such as Ensembl, UniProt, etc. (Schaab et al., 2012). It allows users to retrieve information from the database using different fields such as organism and gene name. It also gives information about the expression of proteins (Schaab et al., 2012).

1.6.8 GPMDB

It is the most well renowned protein expression database. It provides information about MS/MS data from other repositories such as X!Tandem and PX (Keshava Prasad, Goel, Kandasamy, Keerthikumar, et al., 2009). The information is stored in XML files and allows protein and peptide identifications. It supports other formats such as .pkl, mzData, .dta and mzXML, etc. (Keshava Prasad et al., 2009). Besides these, users can also use X!Hunter for analyzing data. All the results are compared with Ensembl genome database (Keshava Prasad et al., 2009). GPMDB is a web-based interface and provides users with quick accession by accession number, sequence search using keywords, gene names, protein identifiers from other databases such as KEGG, GO, BRENDA, etc. (Keshava Prasad et al., 2009). The protein table lists all matched proteins when users input through protein accession number.

Protein view displays all peptide identifications of a specific protein. GPMDB is also linked to Ensembl and HUGO genome nomenclature committee (Keshava Prasad et al., 2009).

1.6.9 MOPED

MOPED is a repository for protein expression information from several model organisms and humans (Kolker, Higdon, Haynes, Welch, et al., 2012). It contains information about Meta data, quantitative data and protein level expression data. Users can access the data through keywords, source organism, and accession number (Kolker et al., 2012). It provides facility of reanalyzing MS data from databases such as systematic protein investigative research environment (SPIRE) (Montague, Stanberry, Higdon, Janko, et al., 2014). SPIRE is connected with other tool such as OMSSA and X!Tandem. MOPED is used to identify protein absolute expression and estimate concentration values of spectrum. The information of the identified protein is linked with various pathway and protein databases such as KEGG, Gene Cards, Reactome and UniProt (Montague et al., 2014). It contains information of organisms like *C. elegans*, *S. cerevisiae*, mouse and human (Montague et al., 2014). The web interface of MOPED provides three important panels—protein relative expression, protein absolute expression and gene relative expression (Higdon, Stewart, Stanberry, Haynes, et al., 2014). All of these panels contain search box for queries using keywords. The expression data is displayed in protein ID and expression summary tab (Higdon et al., 2014). The expression summary table is used to display expression information in a tabular form containing information such as sequence coverage, protein accession, spectral count, description, gene name and concentration (Higdon et al., 2014).

1.6.10 ProteomicsDB

It is a human protein expression database developed to store information about peptide identifications and protein quantification values (Yasset, Emanuele, Rui, Henning, & Juan, 2015). It was developed joinly by SAP and Technical University of Munich. It is comprised of more than 300 experiments, 18,000 human genes, and 62 projects representing 90% of human proteome and 70 million spectra from human tissues, body fluids and cancer cell lines (Yasset et al., 2015). It provides facilities of real time analysis, rapid data mining and visualization. It is used for data submission and reviewing data (Yasset et al., 2015). The data is submitted at three levels: first, experimental data containing description, name and scope; second, projects containing experiment summary, title and publication; and third, experimental files (Yasset et al., 2015). It links to other databases such as MaxUant, PRIDE, PeptideAtlas and MASCOT. It is a widely used user friendly web interface that contains information of human proteome, protein expression and function (Yasset et al., 2015). Protein expression tab gives the expression of a given protein and human body map in more than 30 tissues, body fluids and organs. Chromosome view gives the information of chromosome region, peptide length, description, PSMs and sequence coverage (Yasset et al., 2015).

1.6.11 PaxDb

PaxDb is a database which contains information about absolute protein abundance, deep coverage of proteome and data post-processing (Wang, Fabregat, et al., 2012; Wang, Weiss, et al., 2012). It contains data from PeptideAtlas and PRIDE. It matches protein sequence with STRING database. The results are displayed as molar concentrations and protein abundance (Wang, Fabregat, et al., 2012; Wang, Weiss, et al., 2012). It allows multiple protein requests simultaneously using an ad hoc query approach, as well as comparing and browsing complete datasets (Wang, Fabregat, et al., 2012; Wang, Weiss, et al., 2012). It lists information from UniProt in protein view. Protein sequences of a specific organism can be accessed through the navigation panel. All PaxDb data are freely available (Wang, Fabregat, et al., 2012; Wang, Weiss, et al., 2012).

1.6.12 Human proteinpedia

It is a public protein repository developed for human proteome information, protein expression in different human tissues and cell lines, PPI (i.e., protein-protein interactions), enzyme substrate relationships, PTMs and also subcellular localization (Goel, Harsha, Pandey, & Prasad, 2012). As a database that is annotation oriented HPRD (Human Protein Reference database) contains Western blotting data; MS based experiments, immunohistochemistry, fluorescence, microarrays and co-immunoprecipitation (Goel et al., 2012). Users can annotate data by using web forms to upload experimental evidences, uploading data in a batch mode, sending data through mail to the support team and by setting up laboratories for data upload (Goel et al., 2012). It contains more than 249 laboratories containing information of more than 2 million peptides, 2906 subcellular localizations, 15,000 proteins, 5 million spectra and 2710 distinct experiments (Kandasamy, Keerthikumar, Goel, Mathivanan, et al., 2009). It not only contains MS derived data but also proteomics data from HPRD (Kandasamy et al., 2009). Users can access the query though gene symbol, accession number, protein name and type of protein feature (Kandasamy et al., 2009).

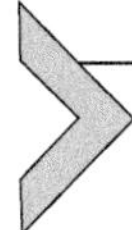

2. Overview of machine learning algorithms

2.1 Definition

Machine learning is a method which allows systems to make decisions automatically without the need of any external support (Breiman, 2001). It allows systems to learn and understand from the data given. It returns the outcome through pattern matching and analysis (Breiman, 2001). It involves generating algorithms and programs that improve the performance of the task. With the ever evolving technology, mass spectrometry techniques and instrumentation have also become highly applicable and popular in the fields of clinical practice and pharmacology (Breiman, 2001). But the data generated through MS has become more intricate and surplus (in terms of complexity of data, its amount, and its dimensionality) due to hyped accuracy, sensitivity, mass resolution, spatial resolution, acquisition speed (Breiman, 2001). Now, the problem arises of how to efficiently and comprehensively analyze enormous amounts of data and extract meaningful information (chemical, spatial or otherwise) (Dayhoff & DeLeo, 2001). Application of Informatics approach fueled with specialized ML algorithms (machine learning) as helped tackle this problem as it processes and provides relevant insight about the dataset (Dayhoff & DeLeo, 2001). Machine learning algorithms add dimension to the proteomics data making it translatable and versatile for real life application by detecting patterns, structures within the data. Mostly, Unsupervised and supervised machine learning algorithms are use in data clustering and data classification, respectively (Dayhoff & DeLeo, 2001).

2.2 Types of algorithms

2.2.1 Supervised ML algorithms

Pattern prediction and predicting the features of a dataset which has labels is mostly done using the supervised learning algorithms (Edwards, Wu, & Tseng, 2009). These algorithms mostly work in three steps: first being selection of the labeled training data, model is then processed to validation or optimization and third, this model is used to predict a new set of unlabeled data (Edwards et al., 2009). To be able to discriminate among sets of samples, supervised algorithms address the problems of classification. The two most widely used algorithms in clinical purposes are random forest (i.e., RF) and support vector machines (i.e., SVM) (Edwards et al., 2009). Through the introduction of kernel function, support vector machines can go past computational limitations in linear classification. Moreover, random forests are good at processing larger datasets and efficiently eliminate overfitting problems (Hoopmann & Moritz, 2013).

The two of these algorithms have been delivering promising results in processing a wide range of biological samples, for example, in processing tissue samples of cancer patients (Hoopmann & Moritz, 2013). The two approaches have been utilized in differentiating different kinds of cancer, that includes breast cancer, thyroid cancer, liver cancer and colon cancer, done so by analyzing the MS dataset and ascertaining tumorigenesis origin regardless of their metastatic sites (Swan et al., 2013). In sarcomas and metastatic melanoma, identification of recurrence and survival associated proteins and protein signatures have been identified (Swan et al., 2013).

2.2.2 Unsupervised ML algorithms

In order to categorize the dataset, unsupervised algorithms have no need of samples being previously labeled (Swan et al., 2013). Therefore, in processing unfamiliar data, unsupervised algorithms have proved to be invaluable in providing tentative results and figured unknown patterns in the datasets previously hidden (Hoopmann & Moritz, 2013). They group together parts of data that share similar features and characteristics, this produces clustered or segmented data, and further samples can be similarly fashioned into the clusters based on their likeliness that is spectral similarity (Edwards et al., 2009). Most widely used unsupervised algorithms are hierarchical clustering, k-means; density based spatial clustering of noise applications and portioning around medoids (i.e., PAM). Though sensitive to anomalous data outliers and data points, k-means is a popular algorithm (Hoopmann & Moritz, 2013). PAM, on the other hand, has proved robust while processing datasets with anomalous points and outliers. Clustering dendrogram interactive analysis gives additional advantage to hierarchical clustering over k-means but require much more space (Hoopmann & Moritz, 2013). Outliers and noisy data are best processed effectively by density based spatial clustering algorithm. Thus, heterogeneity in morphologically similar nearby cells is best recognized by unsupervised clustering algorithms (Hoopmann & Moritz, 2013).

2.2.3 Semi-supervised ML algorithms

Semi-supervised ML falls in between supervised ML and unsupervised ML. As discussed in the previous two types, there is a possibility of two cases—first with no labels for all observations and second with labels for all observations (Swan et al., 2013). It is comparatively expensive due to cost to label and requirement of skilled human experts. Semi-supervised algorithms are considered to be the best candidates for building models. They carry essential information about group parameters (Swan et al., 2013).

2.2.4 Reinforcement ML algorithms

Reinforcement learning deals with artificial intelligence that allows machines to work independently without any manual interpretation to reach their goals and to interact with dynamic environment (Edwards et al., 2009). Hence, software and machines can attain ideal behavior. By getting instructions from the feedback, it learns how to behave thus improves it in longer run. This feedback is called reinforcement signal (Edwards et al., 2009). The actions are taken based on the current which is quite different from supervised learning. In case of supervised learning, the instruction is given to learn the correct answer. Whereas, in the case of reinforcement learning, there is no such kind of answer key (Swan et al., 2013). The machine learns from its own experience. It works by self-observation from the environment that would minimize the risk and maximize the reward (Edwards et al., 2009). It works in an iterative manner and learns continuously. Reinforcement algorithms follow Markov Decision Process to produce intelligent programs and works as follows:

- Agent provides input state.
- Action is performed by decision making function provided by the agent.
- The environment provides reward to the agent.

Algorithms based on Reinforcement ML are deep adversarial networks, Q-learning and Temporal Difference (TD) (Swan et al., 2013).

Some of the important machine learning algorithms are discussed in Table 3.

2.3 Applications of ML to proteomics data

Proteome investigation including the protein modification and protein expression analysis in large datasets is performed with the help of mass spectrometry which is one of the most capable tools available to us for quantitative proteomics data analysis to help understand cellular mechanisms,

Table 3 Important machine learning algorithms.

S. No.	Algorithms	Advantages	Disadvantages
1.	Artificial Neural Networks (ANNs)	Helps to solve complex functions using multilayer perceptron	Results cannot be read and time consuming
2.	Random Forest	Can handle large number of datasets and attributes	Less sensitive
3.	Support vector machines (SVMs)	Helps to solve complex functions using kernels	Time consuming and requires multiple parameters
4.	Naïve Bayes	Easiest and fastest method. Useful in case of datasets with missing values	Wrong assumption of independent nature of attributes
5.	Rule based classifiers	Easily readable and helpful in identifying putative biomarkers	Chances of overfitting
6.	Decision Trees	Easy interpretation of outputs. Useful in case of datasets with missing values	Complexity increases with large datasets and algorithms used

disease progression or genotypic-phenotypic relations (Allmer, 2011). Mass Spectrometry also aids feature detection, protein sequence identification. In MS, we measure the mass by charge ratio of the protein molecules. Identification of proteins cataloging models of diseases or in the treatment or those diagnostically applicable is the main aim of such proteome investigations (Allmer, 2011). Biomarker is a powerful and dynamic tool for clinical investigation with its applications ranging from diagnosis to prognosis, form screening to randomized trials, from observational to analytical epidemiology, to name a few (Wagner et al., 2004). Biomarker is simply the measure of a bodily characteristic or detected molecule that indicates a biological response associated with pharmacological reaction to a medical treatment, a pathogenic process, etc., often, measured objectively, one or a combination of these biomarkers are used to evaluate the health of an individual, during clinical assessment like cholesterol level, blood sugar, heart rate and so on (Veenstra et al., 2005). Application of new high throughput techniques has led to the discovery of different powerful biomarkers, such as molecular biomarkers that could even predict the susceptibility of an individual to a disease (Allmer, 2011). Protein profiling is invaluable to for biomarker discovery. The intrinsic genetic coding of the body and its effects on the environment around are reflected by the proteins in our systems (Geer et al., 2004). Mass Spectrometry is a tool efficient and powerful enough for this purpose as it is useful in identification of proteins along with the chemical alternations of proteins post the post-translational modifications (Kinsinger, Apffel, Baker, Bian, et al., 2012). It is a highly reliable, robust label-free analysis technique enables detection of different proteins from a complex mixture in an acceptable timeframe (Kinsinger et al., 2012). It can be used for the detection of a range of molecular substances along with proteins like glycan, peptides, metabolites, lipids, therapeutics, etc. (Kinsinger et al., 2012). Currently, four basic mass analyzers are majorly used for research purposes, time of flight (TOF), ion trap, Fourier transform ion cyclotron analyzers, quadrupole, all build differently with unlike designs and have varying (Kinsinger et al., 2012). In this report, SELDI TOF generated raw spectral data will be used in the processing. SELDI TOF stands for Surface-enhanced laser desorption/ionization (time-of-flight). This technique produces proteomics data that is highly relevant for biomarker discovery as its technology combines mass spectrometry and retentate chromatography wonderfully (Loo et al., 1992). In the early times, the proteomics actually is a sort of a qualitative discipline. In the experiment of proteomics, the identification of proteins is accomplished in the protein complexes, tissues and cell organelles, etc. the digestion of enzymes which is enzymatic digestion this is the stage which occurs during the procurement of the mixtures of the proteins and proteins identification, there are large and heavy collection of peptides which we obtained as a result and these peptides are proteolytic in nature and after sometime they get analyzed via proteomics that's why stated as per shotgun (Loo et al., 1992). The PSTM which is known as post-translational modifications which is mentioned in phosphorylation which is a peptide inventory. In the proteomics which is qualitative, two important things which have more focus are correctness and depth analysis. In the output which we get by performing the experiment we obtained a list which has proteins (Loo et al., 1992). For the prediction and classification, it needs the patterns of the residues which are required for the aminoalkanoic acid and are further coupled with the approaches of synthetic intelligence like machine learning (Guyon & Elisseeff, 2003). There are the sequences of the peptides and there mapping is done or finished into the positions of genomes which are also meant for their coding (Tran & Doucette, 2008). From the peptide sequences we get to find that the genes which are expressed by them possess the similar property as the pseudogenes (Tran & Doucette, 2008). As the world of bioinformatics is rising, there is a new field which comes under it and it is emerging at high rate, basically it works with the MS (mass spectrometry) based proteomics sets of data throughout there is a conjugation which is present between the annotation of genome or proteome and DNA sequences datasets (Tran & Doucette, 2008). As peptides undergo post-translational modifications so their analysis takes place with the help of mass spectrometry as a result the peptides which are unmodified got improved or advanced in contrast with the peptides which are modified (Tran & Doucette, 2008). There are peptides which are PTM they are known as PTM because they are unmodified peptides. They have functional sites which are exposed to the proteins that are why they are of boundless biological interests (Tran & Doucette, 2008).

Application of Informatics approach fueled with specialized ML algorithms (machine learning) processes and provides relevant insight about the dataset (Geer et al., 2004). Machine learning algorithms add dimension to the proteomics data making it translatable and versatile for real life application by detecting patterns, structures within the data (Geer et al., 2004). Mostly, Unsupervised and supervised machine learning algorithms are use in data clustering and data classification, respectively. There are two ways to practically use deep learning to proteomics data which is MS based are: first, mass spectral peaks directly and second, to the identified proteins which got identified with the help of the sequence searching from the database (Geer et al., 2004).

Machine learning plays a key role in analyzing proteomics data. It makes extraction of essential information from huge amount of proteomics data easily (Guyon & Elisseeff, 2003). Supervised learning is used to quantify proteomics data and build new models. Labels used in supervised learning include regression for continuous numeric values and classification for discrete categories (Guyon & Elisseeff, 2003). It is used to analyze quantitative data from MS and annotate the data experimentally or manually. Unsupervised learning is used for downstream analysis in proteomics (Swan et al., 2013). Common models used in supervised learning include Random Forest, Bayesian Classifiers, Support Vector Machines (SVM), Decision trees and Artificial Neural Networks (Swan et al., 2013). These models are useful in subtype specific therapeutics development and studies related to protein subcellular localization (Guyon & Elisseeff, 2003). Dimensionality reduction techniques such as Linear Discriminant Analysis (LDA), principal component analysis (PCA) and t-Distributed Stochastic

Neighbor Embedding (t-SNE) are used in building models due to high dimensionality for proteomics data (Dayhoff & DeLeo, 2001). Regularization is used to reduce requirement of large number of features and model complexity. It adds penalties to models having more parameters (Dayhoff & DeLeo, 2001). Deep learning is another approach used in proteomics to extract useful information from large number of features. It is also used in cellular signaling systems and improves model accuracy and overcomes overfitting problem (Tran et al., 2019). Hierarchical Clustering is technique of unsupervised learning used to assess MS data. Other approaches of unsupervised learning include—Peptide identification Arbiter by machine learning to detect peptide associated with MS spectrum and ProtVec to represent biological sequences. Unsupervised learning is used less frequently in proteomics as compared to supervised learning (Tran et al., 2019).

2.3.1 Example of machine learning in proteomics[a]

We have processed the dataset and trained a model on Jupyter Notebook and used four machine learning algorithms in the process—logistic regression, decision tree, knn, and random forest algorithms (Supplementary Material 1 in the online version at https://doi.org/10.1016/bs.apcsb.2021.02.003). Our first step was to load all the libraries required from *numpy*, *pyplot* to *seaborn* and learning_curve. Next, we have written the codes for the performance functions and plotting functions. The performance functions that will evaluate the model include accuracy, specificity, sensitivity, precision, recall and F1 score. And the plotting functions will help us plot three graphs for each algorithm—plot learning curve, Plot n_samples vs fit_times for scalability of the model and Plot fit_time vs score for performance of the model. We used the codes to load the samples while simultaneous concatenating frames, adding labels. We have then divided the dataset into dependent and independent variables, split the dataset into training set and test set, and then feature scaling of dataset. We then performed logistic regression, decision tree, k-nearest neighbor and random forest algorithms.

3. MS in proteomics

The technique Mass spectrometry (MS) plays a significant part in the examining of the biological samples and biology related data (Bantscheff, Lemeer, Savitski, & Kuster, 2012). There result contains the estimations

[a] Stepwise screenshots of the code are attached as Supplementary Material 1.

of mass to charge proportion and there relative abundance which is called as relative intensity. When we do the analysis the mass spectrometry-based analysis is utilized, the MS based analysis is done in three stages which are: initially is the stage of pre-processing of the data is take place, the stage second is the stage in which the querying of the database is done, the last stage of analysis where the data is analyzed (Bantscheff et al., 2012). The instruments of mass spectrometry generate the output which has sequences they are managed in the form of flat flies in which the result of these instruments are stored (DeSouza & Siu, 2013). Mass spectrometer is the device which is used for the production of data in the experiment but that data got affected by various errors which are instruments saturation, distortion in the instruments and in graphs the peaks becomes broaden, etc., for reducing the errors and improvising we use the algorithms which are preprocessed (DeSouza & Siu, 2013). Databases are used with new technologies in the enhancement of the data of raw spectra comprises of few parameter with the biological result and are known as Meta information and in the analysis of huge volume of data (Bantscheff et al., 2012). The tandem MS is comparable to the mass spectrometry. The process of extracting in which the information is extracted from the spectral graph or from spectral data becomes easy or simplified with the usage of mass spectrometry (Bantscheff et al., 2012). The SwissProt database, various databases of proteins, queried with the software of protein which is Mascot is used in the proteins and peptides identification. The database querying is done which is used in the proteins and peptides identification with the help of algorithm proICAT (Tran & Doucette, 2008). There are various others softwares to which they are associated that are used in the analysis of the mass spectrometry data like SEQUEST is one of the software but in contrast the obtained output is analogous to the protein quantification output (Tran & Doucette, 2008). The development of algorithms take place which are required for the spectral data manipulation it is manipulated more effectively and efficiently with the use of algorithms the algorithms are pep3D, msInspect (Smith et al., 2013). These are used in contrasting and comparison of the result of experiments and in finding the peaks location in mass spectrometry also they are in association with peptides and proteins (Smith et al., 2013).

3.1 Workflow of the MS proteomics

In several experiments they used so many types of variants and there is requirement of number of steps but for the proteomics experimental data analysis eight steps as shown in Fig. 4 are required which are as follow (Domon & Aebersold, 2006):-

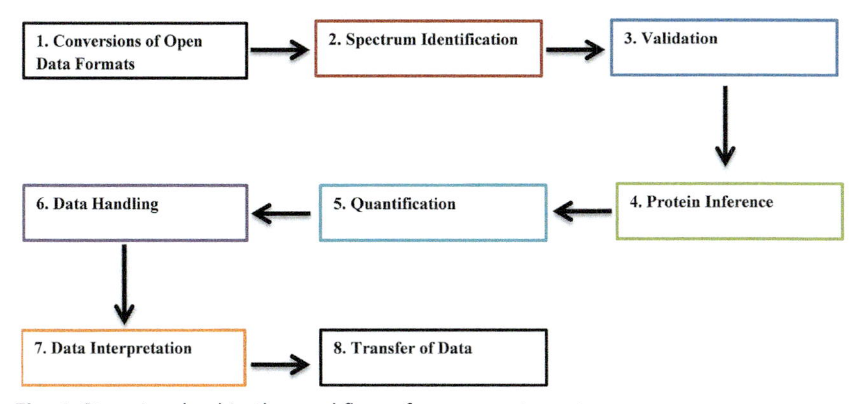

Fig. 4 Steps involved in the workflow of mass spectrometry.

(1). *Processing and conversion by open data formats*—In this there are the data which is present in the XML formats. There is the software which is developed for the formats like XML formats so we can be able to use these formants and can-do work on them (Domon & Aebersold, 2006). There is another format which is mgf format; in this the spectral peak fragments can be extracted. These are process in which XML files works in the association of search engines (Domon & Aebersold, 2006). They also provide us with the raw data collection which is obtained from the mass spectrometry software (Domon & Aebersold, 2006).

(2). *Spectrum identification by search engine*—The identification of peptides is done by one of the methods of bioinformatics in which the FASTA format sequences are extracted (Shiio & Aebersold, 2006). When we get the list of sequences of proteins and the fragmented ionized or the theoretical spectra are created with the help of protein sequences (Shiio & Aebersold, 2006). The theoretical spectra matching are done with observed ion spectra with the help of search engine which is of sequencing. MASCOT and SEQUEST are the name of search engines which are used in this method (Shiio & Aebersold, 2006).

(3). *Validating putative identification*—The spectrum determination is done with the help of search engines and using method like top scoring that are check for the accuracy and the identification done right for the results validation and the way by which searched done take place when FDR is assigned to the threshold (which is already given) (Shiio & Aebersold, 2006). There is a technique in which the sequences of protein which are present in the databases of protein only got inverted and this technique is referred to as decoy searching. There comparison is done with respect to their frequencies and decoy proteins (reversed) exposed to affix the proteins which are targeted (Shiio & Aebersold, 2006).

(4). *Data management system organizations*—The data storage is necessary for its annotation it is stored in the system of management and in the management system of local lab info and from these we can be able to access them easily and are secured and protected in the databases (Shiio & Aebersold, 2006).

(5). *Interpretation of the protein list*—The conclusion about their responses which are biological and their nature is carried out by the Interpretation of protein lists. Data interpretation is done by so many ways and methods but we applied the approach in which the subsets and protein list are taken and the analysis is done which is based on the proteins functional classification (Shiio & Aebersold, 2006). For the visualization of the proteins in an organized manner like according to the groups which are specific on the basis of their protein family, molecular information and about the various biological processes with the help of the web interface they get uploaded to the Panther and David (Domon & Aebersold, 2006).

(6). *Data is transferred to the data repositories (public)*—When the data analysis is done it is then depositing to the repositories. In proteomics the data is accessible to all because it is deposited in the databases which are public (Domon & Aebersold, 2006). These databases contain two types of data the data which is raw and the data which is processed which is obtained by the experiments (Domon & Aebersold, 2006).

3.1.1 Mass spectrometry techniques

The samples are inserted directly into the source of ionization for this process the separation technique is used. There are various types of separation techniques which are LC, chromatography with gas and electrophoresis via capillary (Sleno, 2012). The technique of ionization is used for the mass spectrometry-based analysis of the biomolecules which large and polar. MALDI and ESI are the ionization which is involved in this technique and ESI comprise of LC which has performance of high level (Sleno, 2012). From the technique of ionization which is MALDI we obtained different kinds and types of spectra. MALDI has the mass analyzer known as (TOF) time of flight. SELDI which is a surface enhanced laser desorption ionization is another ionization technique (Wilhelm et al., 2014). The mass spectrometry produces the samples which are collected and undergo scanning at different span of time (Wilhelm et al., 2014). At the time of scanning they select precursor ions one or more than two then fragmentation of them takes place and spectrum is achieved as a result. When ions are ionized they undergo fragmentation, the ions got fragmented into two or more fragments these fragmented parts are known as fragment ion (Wilhelm et al., 2014). After fragmentation, there is a deflection region in which these ions deflected on basis of their charges and size the ion with higher charge will be deflected more as comparison to the ion with smaller charge and in this injecting of samples takes place (Wilhelm et al., 2014). After deflection there is detection region in which the MS data graph is obtained from detection of ions. In mass spectrometry we can obtain the data of different type which have meanings but different from other spectrum with the help of a single spectrum (Wilhelm et al., 2014). They have meaning which is different because there process of analysis which is performed is also different (Wilhelm et al., 2014).

3.1.2 Spectra data handling

The spectral data and its Meta information (Sleno, 2012). This data is stored in the two separate file and file system managed these data and information (Sleno, 2012).

The steps which are required for the handling of the spectra data are:-

- Data is stored and modeled.
- Methods and data integration is done.
- Data pre-processing.
- Querying of Meta data is done with the use of experiments.

For mass spectrometry analysis there are many software and tools which are available like LIMS which is the tool used in the data analysis and in querying of the data with the sequences identification of proteins or peptides known as laboratory information management systems (Sleno, 2012). Data integration of patients to the digital records obtained from bio-clinical instruments. The XML based format is introduced by the community of proteomics and it is introduced for standardizing the spectra representation (Sleno, 2012). The spectra also contain the information of metadata. We can access the spectra and it is sharable but mass to charge ratio and the relative intensity are not directly accessible (Sleno, 2012). Experiment data, contents of the spectra and its data manipulation information, etc., for the description of spectrum the XML schema is used (Sleno, 2012). In the database which has spectra data like the mass to charge ratio and relative intensity we can model them. Databases should be efficient in spectra data storing and it should be manageable in order of data retrieval, the files import and export should be easy and it should be updated in a timely manner (Sleno, 2012).

3.1.3 Example of data processing of samples from MS data[b]

3.1.3.1 Pre-processing raw mass spectrometry data

3.1.3.1.1 Loading of samples In order to load the sample files onto MATLAB, we added the path of the files on its directory using the function *addpath* function of the program. Then we have used the code visible to load the dataset samples, two control csv files and two cancer csv files. We have assigned variables to the files as s1, s2, s3 and s4 and referenced the two columns of each file as MZ1, MZ2, MZ3, MZ4 and Y1, Y2, Y3, Y4 variables for mass by charge ratio and intensity values, respectively. Using the plot function, we can inspect and make sure that the spectrograms have been loaded properly without errors in the process (Supplementary Material 2 in the online version at https://doi.org/10.1016/bs.apcsb.2021.02.003). There are various features available to inspect the spectrograms more closely by zooming in, changing the colors of the plots (we have used red, blue, neon green, pink to differentiate among the samples clearly), the lines could be changed to various others structures (we have used dashes in the pink color sample), placing the cursor the plot points can be viewed for any particular region of the graph, etc.

3.1.3.1.2 Baseline correction We have run the code for baseline correction for each of the samples.

3.1.3.1.3 Spectral alignment of profiles To standardize the mass by charge values, we have used the *msalign* function. Sometimes, there are mass spectrometers that can have miscalibration that results in variations b/w the observed mass by charge ratio vector and the ions' true time of flight values. So, then the appearance of systematic shifts occurs in repetitive experiments. We have performed the same alignment procedure for other two samples as well.

3.1.3.1.4 Normalization of spectra Here, we have used the *msnorm* function to normalize the samples. Between ionized and desorbed proteins, a systematic difference can occur due to repeated experiments. This function helps normalize and standardize these differences.

[b] Stepwise screenshots of the code are attached as Supplementary Material 2.

3.1.3.1.5 Noise reduction We have successfully extracted the main signal from the noise by using the function *mssgolay* which also helps retain the sharpness of main peaks signals with high frequency. Now, the samples have been pre-processing thoroughly following all the necessary steps. The spectrograms have been smooth with a second order polynomial filter.

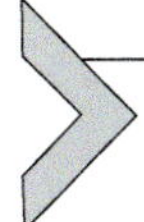

4. Case study 1: Classification and biomarker prediction using proteomics data

As you see in Fig. 5 about mass spectrometry flowchart, proteins or peptides sequences identification done by the experiments in mass spectrometry with the aid of various software and tools and all these are helpful in extracting their mass-charge ratio and there relative intensity information. When protein identification is completed we get the spectra as output. After identification, next step is quantification in which the theoretical and the experimental data are compared. PSTM modification takes place.

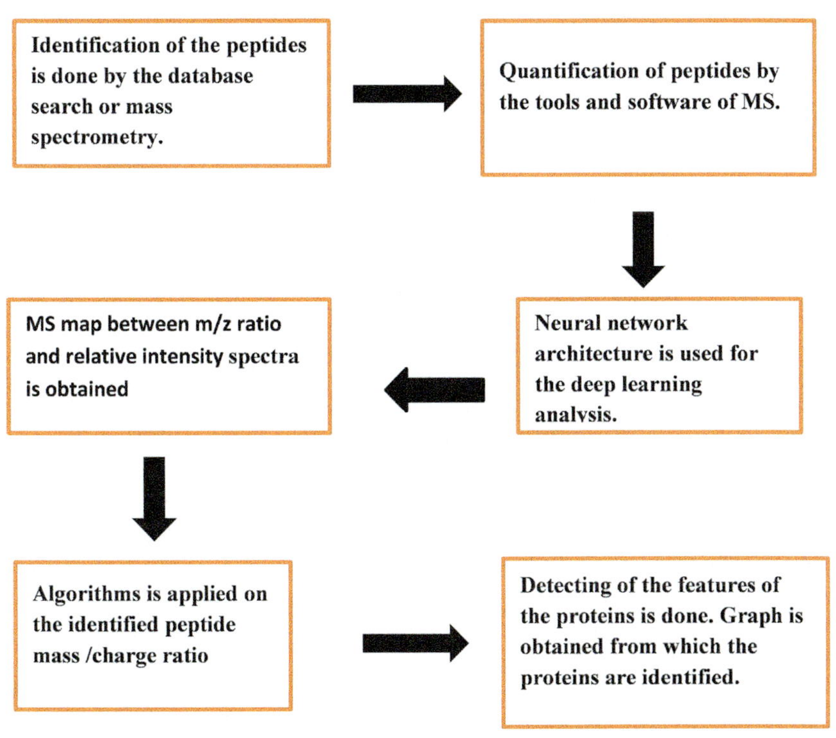

Fig. 5 Flowchart showing the steps of methodology involved in MS based proteomics data.

There are datasets on which the deep learning algorithm with their architecture is applied. These datasets we get from the detection or from the analysis of peptides features and intensity.

4.1 Study requirements

4.1.1 About the dataset

The dataset we used in the case study was from Centre for Cancer Research website under *Clinical Proteomics Program*. It is available to public for downloading and using for personal projects. The dataset consists of 162 ovarian cancer samples and 92 controls. This data was produced using the WCX2 protein-chip where an upgraded PBSII SELDI TOF mass spectrometer was used in the production of the spectra. However, in the case study we have processed data from two control files and two cancer sample data files, as my focus is to evaluate the processing on these file which can finally lead to novel biomarker discovery rather than identifying a biomarker by processing the entire file in the dataset. These will be our test files in a study where we will follow all the steps that typically are undertaken in biomarker discovery.

4.1.2 About software applications

For the pre-processing of our dataset files, we will be using an interactive programming platform, MATLAB, as it is highly specialized by scientific programming, computing. It is widely used by academics for analytical programming like data analysis, pattern recognition, visualization, developing algorithms, experimentation, and so on. According to the requirements of particular fields, MATLAB provides toolboxes that contain libraries of functions specific to the field. For example, we will be using the Bioinformatics Toolbox for this study, specifically the libraries for Mass spectrometry and Bioanalytics. These toolboxes are also properly documented as to guide the user through the process. While bioanalytical techniques like mass spectrometry are essential in the identification of biomolecules as well as their quantification. In this toolbox, there are functions that will help in improving the quality of the datasets, like normalizing it, baseline correction of the peak values, etc. It is a very successful, hassle-free working environment because of its concise, intuitive syntax with very few exceptions, is another one of the reasons for choosing to process the datasets on it.

For the next step, where we will be applying the machine learning algorithms on the pro-processed data, we will do so with the help of Jupyter Notebook application. This is a versatile, powerful tool for coding for data analysis; also it is an open-source application. We can run the code and visualize its results on the application itself. Data analysis programs like statistical modeling, data cleaning, training, building machine learning models, data visualization all can be performed efficiently on this platform. The document processed can also be converted into a number of different formats output. Sharing of the codes and other documents is also fairly easier and support well by this application, however, this feature would not be of much importance in the current project. Since it is based on Python, its syntax is neat and user friendly, making it best opting for coding machine learning algorithms for this project.

4.1.3 Workflow

It is a strenuous and long process that involves use of high throughput technology to generate the dataset first of all as shown in Fig. 6. As iterated above will all the reasoning as to importance of proteomics data produced through mass spectrometry being highly relevant to biomarker research (Veenstra et al., 2005).

4.1.4 Mass spectrometry on samples

The intensity is relative to the mass by charge ratio and can be represented in form of peaks or as a spectrum with equal if not greater accuracy (Walther & Mann, 2010). Mass spectrometry being a diagnostic tool that gauges the ion's mass-to-charge ratio the outcomes are regularly introduced as a mass spectrum; it plots the intensity as a function to the mass to charge ration. Mass spectrometry is utilized in a wide range of fields and is applied to unprocessed samples just as mixtures that are complex (Walther & Mann, 2010). The spectrum is plotted in was that ion signal or intensity is represented against mass to charge ratio. These spectra are utilized to decide the sample's isotopic or elemental signature, the masses of molecules, and to clarify the chemical character or structure of particles and other chemical compounds as well as their chemical characteristics (Walther & Mann, 2010).

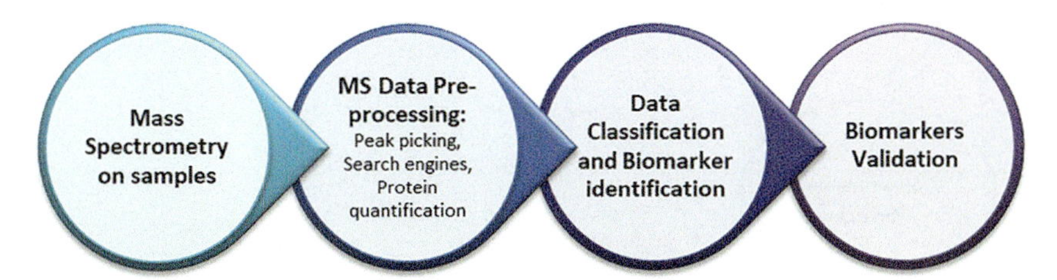

Fig. 6 The flowchart shows simplified phases of biomarker discovery.

4.1.5 MS data pre-processing

The data thus obtained is called raw mass spectrometry data as because technical variations results in the contamination of data, it contains lots of irregularities, outlier values (DeSouza & Siu, 2013). Fig. 7 shows important steps involved in pre-processing of MS data. Directly applying ML algorithms to this unprocessed raw data will not yield true, reliable results, therefore, the datasets are pre-processed to remove these pollutants and irregularities; also referred to as experimental noise. First off, noise filtration is performed on the raw data through the use of the method of wavelet transform (DeSouza & Siu, 2013). The smaller detail coefficients are shrieked to zero, relinquishing any noise from our dataset. The dataset now called as Denoised spectrum as it is marked in Fig. 7 (DeSouza & Siu, 2013). Baseline in the spectrum is basically the signal that is left over after the features of interest to the user have been completely extracted or it can also be the underlying smooth curve in the spectrum (DeSouza & Siu, 2013).

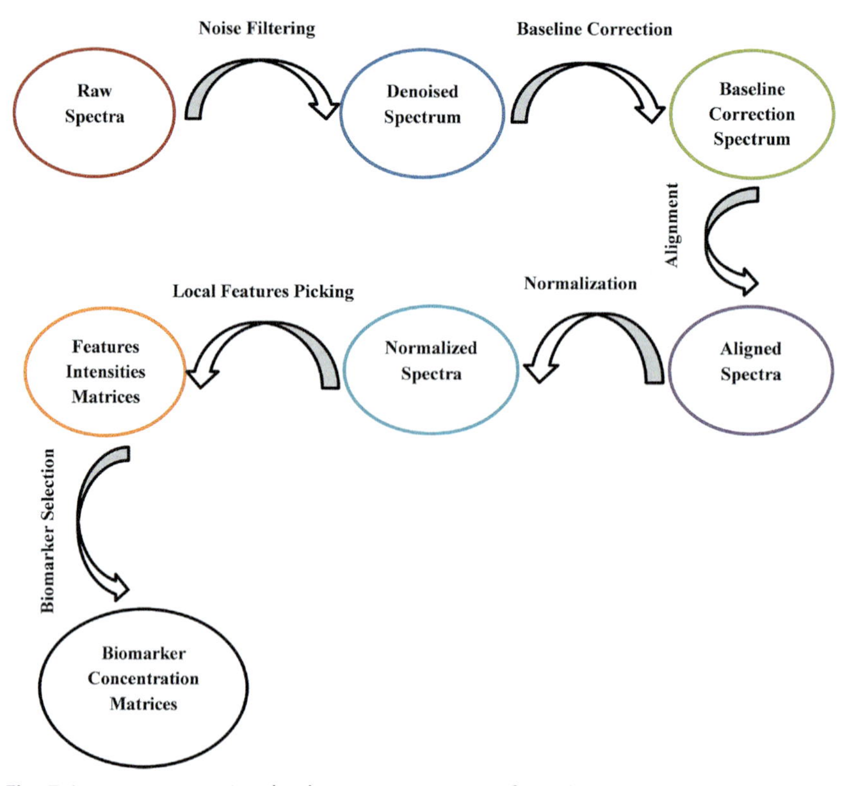

Fig. 7 Important steps involved in pre-processing of MS data.

So, baseline correction is simply the removal of this undesirable waste signal that is masking and hiding the features of interest in the spectrum from us. At this point, the dataset is referred to as the baseline corrected spectrum to differentiate it to the level of processing it belongs to (DeSouza & Siu, 2013). The disturbance in the alignment of the spectrum arises simply due the principles of mass spectrometry (Bantscheff et al., 2012). This causes the mass by charge axis to shift. When the axes of time of flight are adjusted properly, this disturbance in alignment is ridden (Bantscheff et al., 2012). This called alignment of the spectra and the dataset will now be called aligned spectra for ease. To be able to compare the data from different sample normalization of the dataset is essential (Bantscheff et al., 2012). It makes samples comparable by regarding the total amount of protein to be approximately same among all the samples. The ion current total is equivalent to the sum total of protein (Walther & Mann, 2010). The output by the mass spectrometer is dependent upon the number of ions that collide with the detector (Walther & Mann, 2010). Now, the intensities of each of these are divided by the total ion current (TIC). This is a highly efficient and robust method of normalization; however, newer methods are being developed as the approach is a little intuitive (Walther & Mann, 2010). The dataset, now pronounced normalized spectra, has been properly polished through the rigorous pre-processing and can be used for analysis (Walther & Mann, 2010).

Peak Detection: After the pre-processing, when a clear signal has been obtained, next aim ahead is to locate and isolate regions of mass by charge spectra that correspond to a peptide and then quantifying the corresponding intensity values too (Wilhelm et al., 2014). There are several approaches to peak detection on normalized spectrum. First strategy dictates that the maxima in the plot to be isolate from the other neighboring noise around them, extracting just the peaks that respond to the peptides (Wilhelm et al., 2014). Other method is considering this the problem of wavelet coefficient space, and the region with highest clustering of these wavelet coefficients at similar area with different altitudes/scales, then that is considered corresponding to peaks (Wilhelm et al., 2014). The third method uses criterion model based, so the models are used to fit required peaks. These peaks, thus requited from the normalized dataset, are aligned with peaks of other samples to find the ones that are similar or coincide meaning that point to similar biological state input (Wilhelm et al., 2014).

Finding of features of interest: here the features of interest being the peak detected in the previous step, the next step is naturally to estimate the intensities of these peaks (Walther & Mann, 2010). Now the simplest route to doing this seems to measure the heights of the peaks and equate it to their intensity as these peaks are just 2D shapes having more prominent height than width in most cases (Walther & Mann, 2010). However, it is known that the height of the peaks corresponds to the lower mass by charge but as these mass by charge ratio values increase the peak becomes wider. This settles that the area under the peak as to be calculated in order find the intensity (Tran & Doucette, 2008). The problem we face now is that most of these peaks overlap and are too close to be measure distinctly and accurately, so distribution mixtures based new methods have been developed to help deconvolve the features, increase the resolution of each and every peak allowing for distinct reading to be done (Tran & Doucette, 2008). Every sample now has its own characteristic set of common number of peaks. On analyzing the association between the peaks and factors of interest, relevant peaks are isolated as the next step toward biomarker discovery (Tran & Doucette, 2008).

4.1.6 Identifying features of interest and classifying the protein profiles

The feature of interest marked by circles as significant features. This process involved several steps as described in the methodology above. The first step was to rank all the key features that might me candidates to the main features of interest. Using the function *rankfeatures*, that assumes that each mass by charge ration value is independent and thus computes a two way *t*-test, this is also a filtering approach, so does not employ any learning algorithm. At this point, when all the possible features of interest have been identified, we used the function of *cvpartition* to use this available data and classify the samples into normal and cancer ones. Since, we used only four sample files in total we have run the cross validation with upto 20% holdout, this gives a better estimation of the performance of the classifier. Using the principal component analysis, PCA, we have processed the datasets to reduce their dimensionality. The final step was to identify the features of interest from all the available candidates for which we have used the *randfeatures* function that uses the k-nearest neighbor algorithm classifier upon the features subsets that were randomized previously. The appearance of features in output, is clumped together at several spaces. Further classification can be performed reveal the most significant features among these.

Through the above analysis of cancer and control samples, we have been able to identify multiple viable features of interest highlighted by circles, that is, proteins that are candidate for being novel biomarkers that could predict the status of disease, etc. So, by processing the entire dataset following this protocol will help us yield meaningful results. The accuracy of the sample data and its relevance can be tested through model training and comparing using machine learning algorithms (Tran & Doucette, 2008). The test set and training set have considerable diversion, thus, data can be used grouped properly by the classifier. The accuracy, sensitivity, specificity, all lie between the range of 75–81 units for decision tree, k-nearest neighbor and random forest. The models' performance has been satisfactory for application in successful biomarker identification and validation (Tran & Doucette, 2008).

Proteomics by means of the computational method has able to mellowed significantly and also coping up with the tremendous amount of knowledge which is obtained by the current technology MS instrumentations (Smith et al., 2013). The identification of sequence of peptides and the process of quantification of the proteins get analyzed in a very steadfast manner. In the down streaming where analysis of the results which are obtained by the quantification of the protein are tested and from them knowledge about the biomedical is perceived (Smith et al., 2013). In the analysis of the proteomics data various tools are used because of them the analysis becomes accurate less chance of failing of the experiment. As we know, the MS technique is used in detection of biomarkers, discovery of the drugs for the biomedical department (Smith et al., 2013).

In the MS experimentation, the protein culture which is extracted from the tissue or cells and in obtaining of the sequence of peptides helps to identify the masses and relative intensity which results into the formation of the peaks (Tran & Doucette, 2008). The databases which are used for the searching process are not very much efficient, therefore the software of MS is able to find the accurate results (Martens, 2011). Quadra pole technique of the MS is used in the experiment for the obtaining the spectrum of MS and the comparisons are made for the accuracy in the data analysis. From the report the analysis of the data is done with the computational approaches in the field of bioinformatics (Martens, 2011).

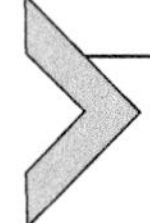

5. Case study 2: Deep learning to large-scale data analysis in mass spectrometry-based proteomics

5.1 Deep learning

As a part of machine learning, artificial intelligence is one of the most tremendously escalating fields out there (Tran et al., 2019). One of the most useful technologies in AI is deep learning. Applications like image recognition, sound recognition, text recognition, are most efficiently performed through deep learning neural networks that enables competent abstraction of data at very large scales (Tran et al., 2019). Following are two important properties of deep learning algorithms:

(1) Multi-layered structure of processing units that are nonlinear.
(2) At each layer feature presentation takes place with the help of unsupervised and supervised learning algorithms (Tran et al., 2019).

Neural networks constitute the framework of deep learning algorithms; these help in drug discovery, NLP (natural language processing), speech recognition and much more as bioinformatics applications (Tran et al., 2019). The versatile algorithms can process data like protein structures and medical imaging data to genomic sequences (Tran et al., 2019).

Hidden layers at each level help in feature extraction in neural networks. Inconsistency in reconstructed data in minimalized through deep learning (Tran et al., 2019). Deep brief networks help perform speech recognition tasks and even NLP. Further, image recognition uses convolutional (Tran et al., 2019).

In deep neural networks, input layer activation takes place first which is followed by the spread of activation across the neural network to the last and final layer of the network for extraction (Tran et al., 2019). These connections are weighted to predict and generate results as they help minimalize difference in real data. For nonlinear layers, activation function, loss function, etc., are part of the deep learning framework. Neural networks and various models form the final architecture (Tran et al., 2019).

5.2 Downstream analysis[c]

5.2.1 Ms datasets

- Mass spectrometry produces the datasets through analysis. The MS dataset used here is sourced ProteomeXchange, from Pride. Databased are identified from them (Supplementary Material 3 in the online version at https://doi.org/10.1016/bs.apcsb.2021.02.003).
- Mascot and SeQuest software used.
- Peptide sequences in FASTA format obtained from NCBI.
- The spectra based on amino acid sequences.

[c] Stepwise screenshots of the code are attached as Supplementary Material 3.

- Spectral data obtained from database, Spectradb.
- The M/Z (mass/charge) ratios plotted against peptide intensities in the graphs.

5.3 Dl algorithm and script

Feature detection performed through recurrent NN (neural network) and convolutional NN.

- R programming language for statistical data analysis.
- Isodetection and isogrouping of proteins.
- *Convolutional neural network*: the features are extracted from the input images. Pooling and convolutional layers make up convolutional NN (Tran et al., 2019). Convolution is performed on the input, followed by designing of feature maps, nonlinear transformation and finally result is obtained through max pooling (Tran et al., 2019).
- *Recurrent neural network*: sequential data is dealt through recurrent NN. The representation of recurrence is hidden (Tran et al., 2019). Activation function can be represented through the following equation:

$$\begin{aligned} h_{(t)} &= g\left(b + Uh_{(t-1)} + Wx_{(t)}\right) \\ o_{(t)} &= c + Vh_{(t)} \end{aligned} \tag{1}$$

$$f_{(t)} = H(W_{io} \cdot X_{it} + W_{hh} \cdot f_{t-1} + b_o) \tag{2}$$

where, H—activation function, W_{io}—weight matrix of the (neurons) connecting layers, W_{hh}—RNN state weight matrix, b_o—at layer o (bias), f_{t-1}—state (previous).

$$f_t = (1 - a_t) \cdot f_{t-1} + a_t \cdot f_t' \tag{3}$$

f_{t-1}—state hidden (previous), f_t'—state (current), a_t—frame current to decision (final).

$$\begin{aligned} f_t' &= H(W_{hh} \cdot f_{t-1} + W_{oh} \cdot X_{ot} + b_h) \\ a_t &= \sigma\left(W_a \cdot f_t' + b_a\right) \end{aligned} \tag{4}$$

X_{ot}—layer o (output), W_{oh}—weight matrix (Xot to RNN) layers, b_h—RNN layer (bias), σ—Sigmoid activation function, W_a—weight matrix, b_a—corresponding (bias).

- The deepISO is a deep learning model in which the two neural networks CNN and RNN are used for the feature extraction. The tensor flow library is used for its implementation (Tran et al., 2019).

5.4 Deep learning output

Isogrouping and isodetecting are the two steps are:-

- Isogrouping
- Isodetecting
  1. From mass spectrometry we obtained the graph which undergo scanning and its relative intensity are detected and for the formation of features the isotopes are obtained is known as Isodetecting (Tran et al., 2019).
  2. When the graph which is obtained undergoes the same process but in this it's mass to charge ratio is detected and feature were detected by the isotopes it is known as Isogrouping (Tran et al., 2019).

On the proteomexchange data, feature detection algorithms are used and there is mass spectrometry map on which running is done with the aid of maxquant (Tran et al., 2019). The features which are so common their detection and extraction are done with the help of neural networks of deep learning (Tran et al., 2019).

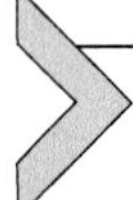

6. Application of deep learning in proteomics

In the field of biomedicines Deep learning applications are: -

i. *In medical images analysis*

- Deep learning has much architecture which is used in the medical image analysis one which is used is the convolutional network which is used for the images of MRI and for the CT scan (Tran et al., 2019).
- For the detection of tumor in the brain or for the detection of mitosis deep neural networks are used (Tran et al., 2019).
- For the ECG based prediction the recurrent neural network is used (Tran et al., 2019).

ii. *In genome sequencing and in the analysis of gene expression*

- The prediction of splicing patterns the architecture of deep neural networks is used and also for the gene expression inference it is used (Tran et al., 2019).
- Protein binding sites binds to the RNA. There prediction is done by the architecture of deep belief networks (Tran et al., 2019).

- The architecture which is used for prediction of non-coding and also denovo functions from the sequences by recurrent neural networks (Tran et al., 2019).

iii. *In structure prediction of proteins*

- The protein structure properties like accessibility of their solvents are predicted by the convolutional neural networks (Tran et al., 2019).
- The architecture of recurrent neural networks is used in the secondary structure prediction (Tran et al., 2019).
- Prediction of proteins which are disorder and protein binding sites are predicted using the deep belief network (Tran et al., 2019).

The field of the MS and proteomics, DL are becoming a powerful tools or techniques in the world of the bioinformaticians. From the small-scale analysis to the large-scale analysis these techniques play an important role (Tran et al., 2019). The data of proteome combining with the technologies is emerged as a new tool in the bioinformatics. Mechanism of post-translational or regulation is becoming significant with the trends of the genes or proteins. The data of these tools of bioinformatics and the databases which are associated need to be updated timely (Tran et al., 2019).

The understanding of the complex biology systems is enhancing with the increasing interest in the field of the computational tools or biology (Tran et al., 2019). The analysis is done only on the proteomics data but we can able to do it in the field of genomics and transcriptomic data. New tools and techniques are emerging for the analysis of the large-scale data for further improvisation in the world of bioinformatics (Tran et al., 2019).

7. Conclusion

Proteomics approach has far reaching capabilities, due to the relentless development of high throughput technologies, study of proteomics and its contribution to biomarkers discovery has spread several folds. However, looking into the process of data analysis for conclusive biomarker discovery, the proper interpretation of data is many times not achieved due to the poor design of study, variations in the samples, their heterogenicity or even lack of enough biological data samples. The field being relatively new faces the preliminary technical and biological challenges that have yet to be solved. These are the challenges that will have to be met with further work in the field like designing studies with greater consideration and accuracy, constructing strict protocols that are to be followed and revised with time to keep them up to date with the present techniques and technologies, validation essays must be made more robust by exploring newer innovative techniques, tools and techniques are to be improvised and innovated to formulate combinations that are much more powerful than the primary, original ones. Interdisciplinary studies must be conducted that will help cope with the ever rising complexity of the proteomics data, simply speaking traditional methodology has to be revolutionized to give better hybrid solutions, those could help unravel the mysteries of highly cryptic diseases like cancers and Alzheimer's disease. The felicitate biomarker discoveries, it is an urgent need to optimize and standardize study design, pre-analytical and analytical assays components, and for stringent validation strategies. Importance of quality indexing for pre-analytical steps is crucial irrespective of the complexity of techniques involved in later procedure. There is a need to eliminate sources of potential bias or other factors confounding in nature. The biomarker studies are also not suitable and applicable to studies at largescale due to reason like several techniques lacking reproducibility. At last, this could be summarized in three points—retrieval of knowledge from proteomics data has the prerequisite of effective pre-processing of raw data; relevant biomarker identification, then validation in only possible through largescale well designed studies; improvisation in methodology has to be done through interdisciplinary approach by collaborations among biologists, biostatisticians, clinicians and computer scientists.

References

Aebersold, R., & Mann, M. (2003). Mass spectrometry-based proteomics. *Nature*, *422*(6928), 198–207.

Allmer, J. (2011). Algorithms for the de novo sequencing of pep-tides from tandem mass spectra. *Expert Review of Proteomics*, *8*, 645–657.

Anderson, N. L., & Anderson, N. G. (1998). Proteome and proteomics: New technologies, new concepts, and new words. *Electrophoresis*, *19*(11), 1853–1861.

Aslam, B., Basit, M., Nisar, M. F., Khurshid, M., & Rasool, M. H. (2017). Proteomics: Technologies and their applications. *Journal of Chromatographic Science*, *55*(2), 182–196.

Baldwin, M. A. (2004). Protein identification by mass spectrometry: Issues to be considered. *Molecular & Cellular Proteomics*, *3*(1), 1–9.

Bantscheff, M., Lemeer, S., Savitski, M. M., & Kuster, B. (2012). Quantitative mass spectrometry in proteomics: Critical review update from 2007 to the present. *Analytical and Bioanalytical Chemistry*, *404*(4), 939–965.

Bjellqvist, B., Ek, K., Righetti, P. G., Gianazza, E., Gorg, A., Westermeier, R., et al. (1982). Isoelectric focusing in immobilized pH gradients: Principle, methodology and some applications. *Journal of Biochemical and Biophysical Methods*, *6*, 317–339.

Blackstock, W. P., & Weir, M. P. (1999). Proteomics: Quantitative and physical mapping of cellular proteins. *Trends in Biotechnology*, *17*(3), 121–127.

Breiman, L. (2001). Random forests. *Machine Learning*, *45*, 5–32.

Chen, C., Hou, J., Tanner, J. J., & Cheng, J. (2020). Bioinformatics methods for mass spectrometry-based proteomics data analysis. *International Journal of Molecular Sciences*, *21*, 2873.

Chen, T., Zhao, J., Ma, J., & Zhu, Y. (2015). Web resources for mass spectrometry-based proteomics. *Genomics, Proteomics & Bioinformatics*, *13*(1), 36–39.

Colinge, J., & Keiryn, L. B. (2007). Introduction to computational proteomics. *PLoS Computational Biology*, *3*(7), 114.

Cote, R. G., Griss, J., & Dianes, J. A. (2012). The PRoteomics IDEntification (PRIDE) converter 2 framework: An improved suite of tools to facilitate data submission to the PRIDE database and the ProteomeXchange consortium. *Molecular & Cellular Proteomics*, *11*, 1682–1689.

Dayhoff, J. E., & DeLeo, J. M. (2001). Artificial neural networks: Opening the black box. *Cancer*, *91*(8 Suppl), 1615–1635.

DeSouza, L. V., & Siu, K. W. (2013). Mass spectrometry-based quantification. *Clinical Biochemistry*, *46*, 421–431.

Deutsch, E. W., Lam, H., & Aebersold, R. (2008). PeptideAtlas: A resource for target selection for emerging targeted proteomics workflows. *EMBO Reports*, *9*, 429–434.

Dhingraa, V., Gupta, M., Zhen, T. A., & Fu, F. (2005). New frontiers in proteomics research: A perspective. *International Journal of Pharmaceutics*, *299*(1–2), 1–18.

Diamandis, E. P. (2004a). How are we going to discover new cancer biomarkers? A proteomic approach for bladder cancer. *Clinical Chemistry*, *50*(5), 793–795.

Diamandis, E. P. (2004b). Mass spectrometry as a diagnostic and a cancer biomarker discovery tool: Opportunities and potential limitations. *Molecular & Cellular Proteomics*, *3*(4), 367–378.

Domon, B., & Aebersold, R. (2006). Mass spectrometry and protein analysis. *Science (New York, N.Y.)*, *312*(5771), 212–217.

Edwards, N., Wu, X., & Tseng, C. W. (2009). An unsupervised, model-free, machine-learning combiner for peptide identifications from tandem mass spectra. *Clinical Proteomics*, *5*, 23–36.

Farrah, T., Deutsch, E. W., Hoopmann, M. R., Hallows, J. L., et al. (2013). The state of the human proteome in 2012 as viewed through PeptideAtlas. *Journal of Proteome Research*, *12*, 162–171.

Farrah, T., Deutsch, E. W., Kreisberg, R., Sun, Z., et al. (2012). PASSEL: The PeptideAtlas SRM experiment library. *Proteomics*, *12*, 1170–1175.

Feist, P., & Hummon, A. B. (2015). Proteomic challenges: Sample preparation techniques for microgram-quantity protein analysis from biological samples. *International Journal of Molecular Sciences*, *16*(2), 3537–3563.

Geer, L. Y., Markey, S. P., Kowalak, J. A., Wagner, L., Xu, M., Maynard, D. M., et al. (2004). Open mass spectrometry search algorithm. *Journal of Proteome Research*, *3*, 958–964.

Gevaert, K., Goethals, M., Martens, L., Van Damme, J., Staes, A., Thomas, G. R., et al. (2003). Exploring proteomes and analyzing protein processing by mass spectrometric identification of sorted n-terminal peptides. *Nature Biotechnology*, *21*, 566–569.

Goel, R., Harsha, H. C., Pandey, A., & Prasad, T. S. (2012). Human protein reference database and human proteinpedia as re-sources for phosphoproteome analysis. *Molecular BioSystems*, *8*, 453–463.

Griss, J., Foster, J. M., Hermjakob, H., & Vizcaino, J. A. (2013). PRIDE cluster: Building a consensus of proteomics data. *Nature Methods*, *10*, 95–96.

Guyon, I., & Elisseeff, A. (2003). An introduction to variable and feature selection. *Journal of Machine Learning Research*, *3*, 1157–1182.

Gygi, S. P., & Aebersold, R. (2000). Mass spectrometry and proteomics. *Current Opinion in Chemical Biology*, *4*(5), 489–494.

Higdon, R., Stewart, E., Stanberry, L., Haynes, W., et al. (2014). MOPED enables discoveries through consistently processed proteomics data. *Journal of Proteome Research, 13*, 107–113.

Hoopmann, M. R., & Moritz, R. L. (2013). Current algorithmic solutions for peptide-based proteomics data generation and identification. *Current Opinion in Biotechnology, 24*, 31–38.

Issaq, H., & Veenstra, T. (2008). Two-dimensional polyacrylamide gel electropho-resis (2D-PAGE): Advances and perspectives. *BioTechniques, 44*(4), 697–700.

Kandasamy, K., Keerthikumar, S., Goel, R., Mathivanan, S., et al. (2009). Human proteinpedia: A unified discovery resource for proteomics research. *Nucleic Acids Research, 37*, 773–781.

Keshava Prasad, T. S., Goel, R., Kandasamy, K., Keerthikumar, S., et al. (2009). Human protein reference database—2009 update. *Nucleic Acids Research, 37*, 767–772.

Kinsinger, C. R., Apffel, J., Baker, M., Bian, X., et al. (2012). Recommendations for mass spectrometry data quality metrics for open access data (corollary to the Amsterdam principles). *Proteomics, 12*, 11–20.

Kolker, E., Higdon, R., Haynes, W., Welch, D., et al. (2012). MOPED: Model organism protein expression database. *Nucleic Acids Research, 40*, 1093–1099.

Lin, S., Shaler, T. A., & Becker, C. H. (2006). Quantification of intermediate-abundance proteins in serum by multiple reaction monitoring mass spectrometry in a single-quadrupole ion trap. *Analytical Chemistry, 78*, 5762–5767.

Loo, J. A., Quinn, J. P., Ryu, S. I., Henry, K. D., Senko, M. W., & McLafferty, F. W. (1992). High-resolution tandem mass spectrometry of large biomolecules. *Proceedings of the National Academy of Sciences of the United States of America, 89*, 286–289.

Mann, M., Hendrickson, R. C., & Pandey, A. (2001). Analysis of proteins and proteomes by mass spectrometry. *Annual Review of Biochemistry, 70*, 437–473.

Mann, M., & Pandey, A. (2001). Use of mass spectrometry-derived data to annotate nucleotide and protein sequence databases. *Trends in Biochemical Sciences, 26*(1), 54–61.

Marc, R. W., Christian, P., Ron, D. A., et al. (1996). From proteins to proteomes: Large scale protein identification by two-dimensional electrophoresis and amino acid analysis. *Nature Biotechnology, 14*(1), 61–65.

Martens, L. (2011). Proteomics databases and repositories. *Methods in Molecular Biology, 694*, 213–227.

Montague, E., Stanberry, L., Higdon, R., Janko, I., et al. (2014). MOPED 2.5—An integrated multi-omics resource: Multi-omics profiling expression database now includes transcriptomics data. *Omics: A Journal of Integrative Biology, 18*, 335–343.

O'Farrell, P. H. (1975). High resolution two-dimensional electrophoresis of proteins. *The Journal of Biological Chemistry, 250*, 4007–4021.

Ong, S. E., & Mann, M. (2007). Stable isotope labeling by amino acids in cell culture for quantitative proteomics. *Methods in Molecular Biology (Clifton, N.J.), 359*, 37–52.

Padula, M. P., Berry, I. J., O'Rourke, M. B., Raymond, B. B., Santos, J., & Djordjevic, S. P. (2017). A comprehensive guide for performing sample preparation and top-down protein analysis. *Proteomes, 5*, 11.

Pandey, A., & Mann, M. (2000). Proteomics to study genes and genomes. *Nature, 405*, 837–846.

Patterson, S. D. (1995). Matrix-assisted laser-desorption/ionization mass spectrometric approaches for the identification of gel-separated proteins in the 5–50 pmol range. *Electrophoresis, 16*, 1104–1114.

Rabilloud, T., & Lelong, C. (2011). Two-dimensional gel electrophoresis in proteomics: A tutorial. *Journal of Proteomics, 74*, 1829–1841.

Rabilloud, T., & Lescuyer, P. (2014). The proteomic to biology inference, a frequently overlooked concern in the interpretation of proteomic data: A plea for functional validation. *Proteomics, 14*, 157–161.

Ramos, Y., Garcia, Y., Perez-Riverol, Y., Leyva, A., et al. (2011). Peptide fractionation by acid pH SDS-free electrophoresis. *Electrophoresis*, *32*, 1323–1326.

Riffle, M., & Eng, J. K. (2009). Proteomics data repositories. *Proteomics*, *9*, 4653–4663.

Schaab, C., Geiger, T., Stoehr, G., Cox, J., & Mann, M. (2012). Analysis of high accuracy, quantitative proteomics data in the MaxQB database. *Molecular & Cellular Proteomics*, *11*(3), 1–10.

Shevchenko, A., Wilm, M., Vorm, O., & Mann, M. (1996). Mass spectrometric sequencing of proteins from silver-stained polyacrylamide gels. *Analytical Chemistry*, *68*, 850–858.

Shiio, Y., & Aebersold, R. (2006). Quantitative proteome analysis using isotope-coded affinity tags and mass spectrometry. *Nature Protocols*, *1*(1), 139–145.

Sleno, L. (2012). The use of mass defect in modern mass spectrometry. *Journal of Mass Spectrometry*, *47*, 226–236.

Smith, L. M., Kelleher, N. L., & Consortium for Top Down Proteomics. (2013). Proteoform: A single term describing protein complexity. *Nature Methods*, *10*, 186–187.

Swan, A. L., Mobasheri, A., Allaway, D., Liddell, S., & Bacardit, J. (2013). Application of machine learning to proteomics data: Classification and biomarker identification in postgenomics biology. *Omics: A Journal of Integrative Biology*, *17*(12), 595–610.

Tran, J. C., & Doucette, A. A. (2008). Multiplexed size separation of intact proteins in solution phase for mass spectrometry. *Analytical Chemistry*, *81*, 6201–6209.

Tran, N. H., Qiao, R., Xin, L., Chen, X., Liu, C., Zhang, X., et al. (2019). Deep learning enables de novo peptide sequencing from data-independent-acquisition mass spectrometry. *Nature Methods*, *16*, 63–66.

Tyagi, S., Raghvendra, R., Singh, U., Kalra, T., Munjal, K., & Vikas. (2010). Practical applications of proteomics-a technique for large-scale study of proteins: An overview. *International Journal of Pharmaceutical Sciences Review and Research*, *3*(1), 87–90.

Veenstra, T. D., Conrads, T. P., Hood, B. L., Avellino, A. M., Ellenbogen, R. G., & Morrison, R. S. (2005). Biomarkers: Mining the biofluid proteome. *Molecular & Cellular Proteomics*, *4*(4), 409–418.

Vizcaino, J. A., Cote, R. G., Csordas, A., Dianes, J. A., et al. (2013). The PRoteomics IDEntifications (PRIDE) database and as-sociated tools: Status in 2013. *Nucleic Acids Research*, *41*, 1063–1069.

Vizcaino, J. A., Deutsch, E. W., Wang, R., Csordas, A., et al. (2014). ProteomeXchange provides globally coordinated proteomics data submission and dissemination. *Nature Biotechnology*, *32*, 223–226.

Wagner, P. D., Verma, M., & Srivastava, S. (2004). Challenges for biomarkers in cancer detection. *Annals of the New York Academy of Sciences*, *1022*, 9–16.

Walther, T. C., & Mann, M. (2010). Mass spectrometry-based proteomics in cell biology. *The Journal of Cell Biology*, *190*(4), 491–500.

Wang, R., Fabregat, A., Rios, D., Ovelleiro, D., et al. (2012). PRIDE inspector: A tool to visualize and validate MS proteomics data. *Nature Biotechnology*, *30*, 135–137.

Wang, M., Weiss, M., Simonovic, M., Haertinger, G., et al. (2012). PaxDb, a database of protein abundance averages across all three domains of life. *Molecular & Cellular Proteomics*, *11*, 492–500.

Westermeier, R. (2014). Looking at proteins from two dimensions: A review on five decades of 2d electrophoresis. *Archives of Physiology and Biochemistry*, *120*, 168–172.

Wilhelm, M., Schlegl, J., Hahne, H., Gholami, A. M., Lieberenz, M., Savitski, M. M., et al. (2014). Mass-spectrometry-based draft of the human proteome. *Nature*, *509*, 582–587.

Wilkins, M. R., Gasteiger, E., Gooley, A. A., Herbert, B. R., Molloy, M. P., Binz, P. A., et al. (1999). High-throughput mass spectrometric discovery of protein post-translational modifications. *Journal of Molecular Biology*, *289*(3), 645–657.

Wittig, I., & Schagger, H. (2009). Native electrophoretic techniques to identify protein-protein interactions. *Proteomics*, *9*, 5214–5223.

Wright, E. P., Partridge, M. A., Padula, M. P., Gauci, V. J., Malladi, C. S., & Coorssen, J. R. (2014). Top-down proteomics: Enhancing 2d gel electrophoresis from tissue processing to high-sensitivity protein detection. *Proteomics*, *14*, 872–889.

Xie, F., Liu, T., Qian, W. J., Petyuk, V. A., & Smith, R. D. (2011). Liquid chromatography–mass spectrometry-based quantitative proteomics. *The Journal of Biological Chemistry*, *286*, 25443–25449.

Yasset, P. R., Emanuele, A., Rui, W., Henning, H., & Juan, A. V. (2015). Making proteomics data accessible and reusable: Current state of proteomics databases and repositories. *Proteomics*, *15*, 930–949.

Yates, J. R. (1998). Mass spectrometry and the age of the proteome. *Journal of Mass Spectrometry*, *33*, 1–19.

Yates, J. R., III. (2011). A century of mass spectrometry: From atoms to proteomes. *Nature Methods*, *8*(8), 633–637.

Yates, J. R., Ruse, C. I., & Nakorchevsky, A. (2009). Proteomics by mass spectrometry: Approaches, advances, and applications. *Annual Review of Biomedical Engineering*, *11*, 49–79.

Yocum, A. K., & Chinnaiyan, A. M. (2009). Current affairs in quantitative targeted proteomics: Multiple reaction monitoring-mass spectrometry. *Briefings in Functional Genomics & Proteomics*, *8*(2), 145–157.

CHAPTER SIX

Network-based strategies for protein characterization

Alessandra Merlotti[a], Giulia Menichetti[b,c], Piero Fariselli[d], Emidio Capriotti[e], and Daniel Remondini[a,*]

[a]Department of Physics and Astronomy, University of Bologna, Bologna, Italy
[b]Center for Complex Network Research, Department of Physics, Northeastern University, Boston, MA, United States
[c]Department of Medicine, Brigham and Women's Hospital, Harvard Medical School, Boston, MA, United States
[d]Department of Medical Sciences, University of Torino, Turin, Italy
[e]Department of Pharmacy and Biotechnology, University of Bologna, Bologna, Italy
*Corresponding author: e-mail address: daniel.remondini@unibo.it

Contents

Abstract

Protein structure characterization is fundamental to understand protein properties, such as folding process and protein resistance to thermal stress, up to unveiling organism pathologies (e.g., prion disease). In this chapter, we provide an overview on how the spectral properties of the networks reconstructed from the Protein Contact Map (PCM) can be used to generate informative observables. As a specific case study, we apply two different network approaches to an example protein dataset, for the aim of discriminating protein folding state, and for the reconstruction of protein 3D structure.

1. Introduction

In the last decade several models describing a protein as a network of interacting residues were used for characterizing the relationship between structure and function (Greene, 2012; Grewal & Roy, 2015; Yan et al., 2014).

Advances in Protein Chemistry and Structural Biology, Volume 127
ISSN 1876-1623
https://doi.org/10.1016/bs.apcsb.2021.05.001

The studies of the protein contact network were used for detecting important residues for protein stability and dynamics (Böde et al., 2007; Brinda & Vishveshwara, 2005; Taylor, 2013), active sites (Amitai et al., 2004) and protein folding kinetics (Bagler & Sinha, 2007). In general, the works in the field focus on the properties arising for the local and global topology of the network (Bollobas, 1998). For the topological analysis, the distribution of the degree of nodes and the shortest paths among nodes are considered the main observables for the description of the protein structure. In this work we analyzed the classical network analysis techniques for the study of protein folding. In particular, we focus on the application of the protein contact network analysis to the reconstruction of the protein three-dimensional structure and to the characterization of the protein folding mechanism.

2. The protein folding problem

Protein folding is the process by which the polypeptide chain reaches its native three-dimensional (3D) structure conformation. The Anfinsens' experiments carried out in the 1970s lead to the conclusion that under favorable conditions, protein will fold consistently into its native structure which is encoded in its amino acid sequence (Anfinsen, 1973). Although this view of the folding mechanism has been challenged by new experimental evidence (Dishman & Volkman, 2018), the large amount of crystallographic data collected in the Protein Data Bank (wwPDB consortium, 2019) reinforce the idea of the uniqueness of the folded conformation. The existence of a stable and kinetically accessible native conformation of the proteins determined by the amino acid sequence enhanced the development of several theoretical models and computational methods for studying the protein folding mechanism (Compiani & Capriotti, 2013; Dill, 1990). The majority of the available models and methods focus on three aspects of the same problem related to the prediction of the native structure, the thermodynamics and the kinetics of the folding process (Dill et al., 2007; Dill & MacCallum, 2012). The prediction of the protein structure from the amino acid sequence is a challenging problem that drew the attention of the scientific community at the end of the 1980s when few hundreds protein structures were made available on the Protein Data Bank (Fariselli et al., 2007).

The seminal work from Chothia and Lesk studying the relationship between protein sequence and structure found that homologous proteins retain the same general fold (Chothia & Lesk, 1986). This observation laid the foundation for the development of computational structure prediction methods which rely on detectable similarity with known protein structures or ab initio methods (Baker & Sali, 2001). During the last two decades the Critical Assessment of Structure Protein (CASP) evaluated the quality of the prediction algorithms tracking the progress in the field (Kryshtafovych et al., 2019). Recently, a dramatic improvement of the performance in the prediction of the protein 3D structure was driven by the successful application of deep learning techniques (Senior et al., 2020). Although the prediction of the native conformation of a protein from its sequence achieved an unprecedented level of performance, the folding mechanism description at thermodynamic and kinetic levels is still incomplete. In the last few years statistical and machine learning algorithms have been developed for predicting the stability of a protein structure and the folding rate. Nevertheless, at the current stage reliable and general models for describing the free energy landscape of the folding process are unavailable. In this context, several methods have been developed for predicting protein stability and folding rate (Chang et al., 2015; Magliery, 2015; Sanavia et al., 2020). These approaches rely on protein 3D structure which is used to identify the interacting residues along the amino acid sequence. Such information is essential for estimating the stability of the native conformation and determining the mechanism of the protein folding.

3. Modeling folding kinetics

The study of the folding kinetics is important for calculating the time by which a protein reaches its native conformation and for identifying the formation of metastable conformations during the folding process. Thus, for the characterization of the folding kinetics were developed several theoretical models based on a simplified representation of the protein structure (Compiani & Capriotti, 2013). Depending on the level of cooperativity in the formation of the native conformation, the models of protein folding were classified in three groups: hydrophobic collapse, nucleation–condensation and framework models. The main differences among these models depend on the role played by the secondary structure in the formation of the native conformation. Among them, the Diffusion-Collision model is one of the first quantitative models for predicting the folding time (Karplus & Weaver, 1976). This hierarchical model represents the protein as a set of partially formed secondary structure elements that reach the native state through stochastic collision events. More recent models based on non-local interactions between residues consider the static 3D of the protein as a proxy for the prediction of the protein folding rate (Gromiha & Selvaraj, 2001; Ivankov & Finkelstein, 2004; Plaxco et al., 1998; Zhou & Zhou, 2002). These methods rely on the observation that the average sequence separation between contacting residues in the native conformation correlates with the folding rate and transition state of single-domain proteins. Thus, the definition of interacting residues assumed an important role in the determination of the folding mechanism, and for the development of more sophisticated methods based on predicted contact maps and machine learning approaches for predicting the folding rate and mechanism (Capriotti & Casadio, 2007; Huang & Gromiha, 2010; Punta & Rost, 2005). In general, all the methods represent the protein as a graph where the nodes are the residues connected by an edge when the distance between two nodes is below a given threshold. Such representation, which is equivalent to a contact map, is used to compute the distribution of non-local interactions among residues.

4. Protein structure representation

In the last two decades the Protein Structure Initiative strongly contributed to the identification of new protein structures (Grabowski et al., 2016). Such information is important for studying the function of a protein that is related to geometrical features defining the secondary structure of the protein and determining its fold (Hrmova & Fincher, 2009). Currently the PDB collects ~177 K structures which are classified in more than 5000 superfamilies and families by the two most popular databases CATH (Sillitoe et al., 2021) and SCOP (Andreeva et al., 2020). Each protein three-dimensional structure is represented by the coordinates of its atoms. For representing a protein composed by n atoms, 3n numbers are needed. An alternative protein structure representation is based on the distance matrix which is composed by n^2 elements that for the symmetry are reduced to $n(n-1)/2$. Although the distance matrix has more elements than standard representations based on the atom coordinate, it can be advantageous when only low resolution data from NMR are available (Bartoli et al., 2008). A simplified version of the distance matrix is the contact map which considers only the distance between specific atoms of each residue either α or β carbons and a cut-off distance to represent the presence of absence of a contact with a binary number. The possibility of reconstructing the protein structure starting from a reduced representation is an essential aspect for its application to the study of the protein structure. Previous studies have proved that contact maps provide a good representation of the protein backbone (Porto et al., 2004; Vassura et al., 2008). Thus, the contact map, which retains the main information about protein structure, can be used as a proxy for the characterization of the protein folding mechanism.

5. Contact maps and graph Laplacian

Every protein can be represented as a network of interacting particles, where nodes can correspond to single atoms, residues, or even larger motifs, and links to their interactions. Of course, in this approach, a key point consists of how nodes and links are defined. For sake of simplicity, we consider α-Carbons (Cα) as nodes and distance-dependent interactions as links, by imposing that two nodes are connected if their distance d is lower than a specific threshold t. In this way, given the number N of Cα, we could associate to each protein a contact map A, that is a binary NxN symmetric matrix defined as:

$$A_{ij} = \begin{cases} 1 \; if \; d_{ij} \leq t \\ 0 \; if \; d_{ij} > t \end{cases}$$

In network theory, A corresponds to the adjacency matrix of a graph, and represents the starting point for studying the importance of nodes within the network, and the topological structure of the interactions between them.

From A, the Laplacian operator of a network, is derived as:

$$L = D - A$$

where D is the degree matrix, defined as $D_{ij} = k_i \bullet \delta_{ij}$, and k_i represents the degree of node i. In particular, the action of L on a N-dimensional lattice corresponds to the discretization of a N-dimensional elastic membrane, where L's eigenvalues represent the frequencies of the normal modes and L's eigenvectors represent the normal mode solutions or eigenfunctions (Biyikoglu et al., 2007). With this analogy in mind, the eigenvalue decomposition of the Laplacian operator corresponds to searching for extremal values of the Rayleigh functional, vectors x that maximize or minimize the mutual distance between nodes in the network, expressed by the following semi-positive quadratic form:

$$\vec{x}^T L \vec{x} = \sum_{i \sim j} (x_i - x_j)^2$$

The trivial solution corresponds to the 0 eigenvalue, in which all nodes have the same spatial coordinates and thus $x_i = x_j$ for every i,j. The non-trivial solutions seek for a minimal distance by imposing the orthogonality with the constant vector. If we hypothesize that the elastic potential schematized by the Laplacian operator is an approximation around the minimum of the Lennard-Jones potential-like function, modeling the interaction between protein residues, the 3D coordinates of Cα can be estimated by the components of the 3 eigenvectors associated with the 3 smallest positive eigenvalues of the Laplacian operator, thus providing a reconstruction of the 3D protein structure up to a linear transformation.

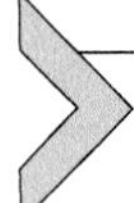

6. Protein folding state discrimination and Laplacian spectrum

In (Menichetti et al., 2016) the properties of the Laplacian spectrum are leveraged to predict protein folding kinetics as two-state, an "all-or-none" type of transition, or as multi-state, in presence of one or more intermediates. The training database for Fisher discriminant analysis consists of 63 manually annotated proteins by Ivankov and Finkelstein (2004)

(25 multi-state, 38 two-state), all proteins with structure available on PDB (https://www.rcsb.org).

PCMs are derived by choosing an upper threshold of 8 Å. Interestingly, among the 5 network observables defined in the paper, whose performances in combination or alone are extremely predictive of folding classes, we find 3 Laplacian-based variables. However, the best accuracy and Matthews correlation coefficient are not achieved by deriving the Laplacian from the original PCM, but by focusing on a modified version that keeps only long-range contacts, while preserving the network connectivity.

With the hypothesis that the most relevant information on folding kinetics is determined by the long-range contacts of the native folded state, a partial removal of the protein backbone, up to the breaking point of the protein into fragments, enhances the role of long-range connections with respect to the protein backbone, while keeping the PCM still connected. The number of diagonals removed varies from protein to protein.

Once the Laplacian spectrum of the modified PCM is computed for each protein, the 3 largest eigenvalues are collected and rescaled by the number of residues N_C. The rescaling corrects for the dependence of the largest eigenvalues λ_N, λ_{N-1}, λ_{N-2} on N_C. According to the vibrational interpretation of the Laplacian, the selected eigenvalues represent the highest vibrational frequencies associated with the small-range structure of the protein, compared to a more global assessment of the long-range vibrations and algebraic connectivity of the protein structure, offered by the Fiedler number (second smallest eigenvalue).

The percentage of correctly classified proteins, when using λ_N, λ_{N-1}, λ_{N-2} separately, is 76.6% $\pm$ 1.3, 76.7% $\pm$ 1.4, and 77.6% $\pm$ 1.1, representing the average values of 10-fold cross-validation over 10,000 resamplings. Similarly, the Matthews correlation coefficient follows as 0.57 ± 0.02, 0.58 ± 0.02, and 0.59 ± 0.02.

Overall, we observe that two-state proteins tend to have larger values of fast-vibrating frequencies, compared to multi-state proteins, and that the vibrational modes (i.e., the corresponding eigenvectors) associated to high frequencies are in general characterized by a strong localization along the vector, corresponding to specific protein regions. If this feature is observed also in our case, and if this can be associated to specific folding/unfolding dynamics, is still an open issue.

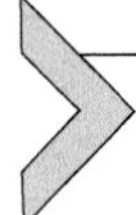

7. A case study: Methods for protein 3D-structure reconstruction

In order to compare and to evaluate two different network-based approaches that we choose to use, we observe that a faithful reconstruction of 3D structure, given the "network" information provided by the contact map only, can be considered a good validation that the network framework adopted can characterize properties associated with protein folding. Therefore, in this section we will show two different network-based methods that allow the reconstruction of the 3D coordinates of Cα, starting from their contact maps.

The first method is based on the first three eigenvectors of the Laplacian operator, as explained in Section 5, that exploits the "vibrational" analogy of the Laplacian operator, as describing a set of unit masses connected by springs with equal stiffness, and which first eigenvectors correspond to the largest-scale vibrational modes. The second method was proposed by Lesne et al. (2014), who devised an algorithm called ShRec3D with the aim of reconstructing the 3D structure of chromosomes starting from Hi-C data (Lieberman-Aiden et al., 2009), which allows the mapping of neighboring DNA fragments, generating an output formally equivalent to a protein contact map, despite the physics underlying chromosome and protein 3D configuration is different (Merlotti et al., 2020), since there are no direct chemical bonds between DNA strands, but rather we can talk about a spatial proximity mediated by other factors (like cohesin, histones and CTCF proteins). Moreover, differently from DNA 3D structure which is still largely unknown particularly at a fine scale of the single nucleotides, for our protein dataset we have the ground truth provided by the protein 3D configuration obtained through X-ray crystallography to be compared with our reconstructions.

The ShRec3D algorithm can be divided into two steps: (1) the computation of Cα distances starting from the contact map and (2) the estimation of Cα spatial coordinates starting from their mutual distances. The first step is performed by measuring the distance between two Cα as the length of the shortest path connecting them in the network provided by the contact map. In fact, it is known that the shortest paths s_{ij} between nodes i and j in a symmetric network satisfy the conditions to be considered a metrics: (1) be $=0$ if $i=j$; (2) be symmetric $s_{ij}=s_{ji}$; (3) satisfy triangular inequality. Thus, the idea behind this approach is that given that the contact map is only an approximation to the real distance matrix between protein residues (identifying only the shortest distances below a threshold), the best approximation to the full distance matrix (that satisfies the conditions for which the theorems of distance geometry hold) is guessed by the distance matrix computed from the shortest paths of the contact map.

The second step is based on the results of distance geometry (Sippl & Scheraga, 1985; Havel et al., 1983) and multidimensional scaling (Torgerson, 1952), which concern the reconstruction of the original spatial structure of a 3-dimensional object (in our case, the proteins) given the full distance matrix between its elements. This requires the spectral decomposition of the Gram matrix, defined according to the following formula:

$$G_{ij} = \frac{1}{2}\left[d_{0i}^2 + d_{0j}^2 - D_{ij}^2\right]$$

where

$$d_{0i}^2 = \frac{1}{N}\sum_{j=1}^{N} D_{ij}^2 - \frac{1}{N^2}\sum_{j=1}^{N}\sum_{k>j}^{N} D_{jk}^2$$

represents the distance between the barycenter O and the point P_i of the 3D object. Cα spatial coordinates are then estimated through the 3 eigenvectors E_l ($l=1, 2, 3$) associated with the 3 largest eigenvalues λ_l ($l=1, 2, 3$) of the Gram matrix, as follows:

$$V_{li} = E_l(i) \times \sqrt{\lambda_l} \text{ with } \sum_{i=1}^{N} E_l^2(i) = 1.$$

where $E_l(i)$ is the i-th component of the eigenvector E_l. In this way, we can obtain a 3D reconstruction of Cα spatial coordinates up to an arbitrary rotation, dilation and possibly mirror symmetry (Lesne et al., 2014).

The two approaches start from different assumptions, but they rely on a similar algebraic structure, since also the Laplacian operator can be seen as a Gram matrix $L=I^T I$, where I is a rectangular incidence matrix that has one row for each link of the network, containing −1 and 1 values in each row in correspondence to the connected nodes (the direction of the link can be arbitrarily chosen, since this direction information is lost in the Laplacian operator).

We reconstructed the 3D structure of 63 proteins from (Menichetti et al., 2016) and characterized by different sizes (from 20 to 8015 Cα) and different folding kinetics. The results were evaluated by computing the Pearson's correlation between Cα-pairwise distances in the reconstructed and real structure: the higher the correlation, the better the reconstruction.

We tested whether and how the following parameters could affect the results: (1) the folding kinetics; (2) the number of Cα composing the proteins. Moreover, the threshold value used to compute the contact map was varied, considering in particular 8 and 12 Å, to see how the variation in the resulting contact map could affect the reconstruction performance.

Looking at the histograms of correlation values represented in Fig. 1, we can notice that the reconstruction through ShRec3D achieved higher performances than the Laplacian-based one. In particular, the former is characterized by similar performances independently on the threshold value used to calculate the contact map, while the latter is characterized by an overall worsening for contact maps calculated using 12 Å as threshold (see Table 1). This result justifies the hypothesis that the shortest path distance matrix provides a more reliable estimation of the original distance matrix than the simple contact map, showing an increasing performance as the correlation value between shortest path distances and true distances increases (see Fig. 2). In fact, if we represent in a scatterplot the former as a function of the latter for each pair of Cα composing a protein, we can see that the structures that are well reconstructed by ShRec3D, show a linear

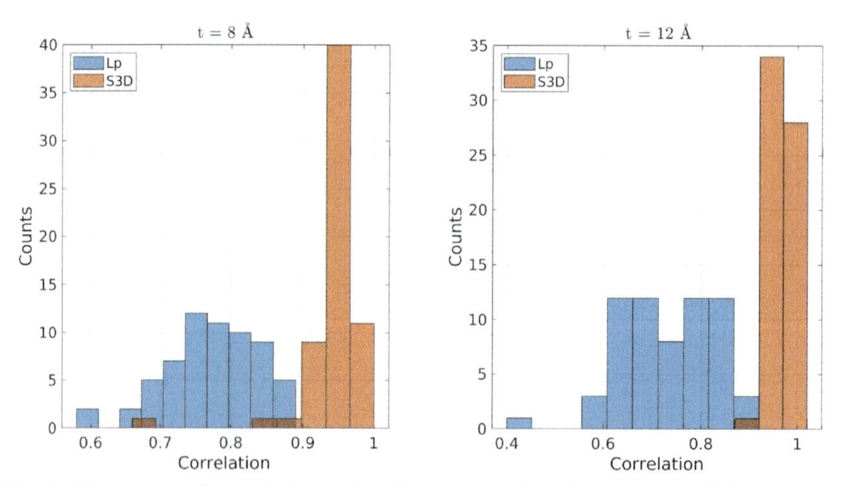

Fig. 1 Histograms of correlation values between real and reconstructed Cα-pairwise distances via Laplacian and ShRec3D embedding, starting from contact maps obtained using different threshold values: 8 and 12 Å.

Table 1 Mean correlation values between real and reconstructed Cα distances via Laplacian and ShRec3D method, starting from different contact maps, obtained using as threshold values 8 and 12 Å.

	$t = 8$ Å	$t = 12$ Å
Laplacian	0.77	0.73
ShRec3D	0.94	0.97

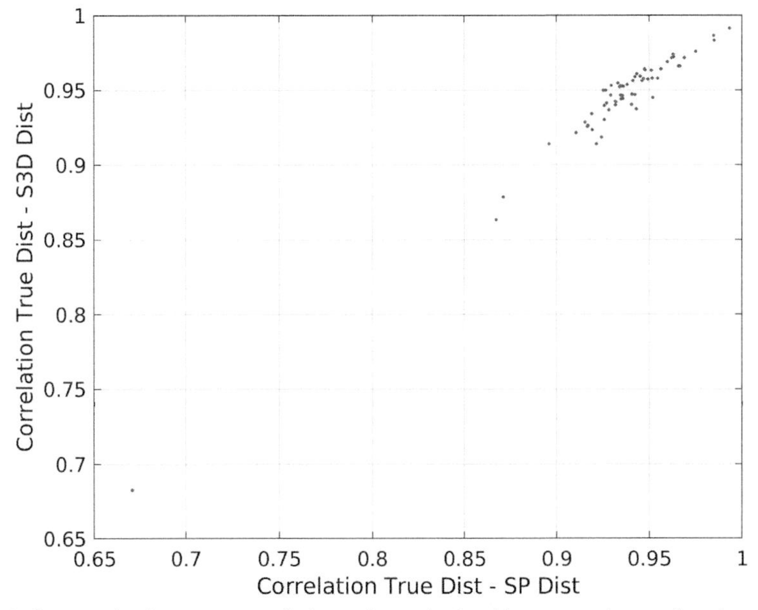

Fig. 2 Scatter plot between correlation values obtained by comparing real and reconstructed distances with Shrec3D (S3D Dist) and correlation values obtained by comparing real distances and shortest-path lengths (SP Dist), starting from an 8 Å contact map.

increasing trend characterized by a lower dispersion (see Fig. 3); whereas the structures that are not well reconstructed are characterized by a higher dispersion (see Fig. 4).

If we consider the 10 proteins with the highest and the lowest correlation values between real and reconstructed distances obtained from an 8 Å contact map, we can notice that Laplacian embedding provides the best results on proteins that do not show a modular structure (see Table 2, Figs. 5 and 6), which corresponds to a contact map characterized by blocks

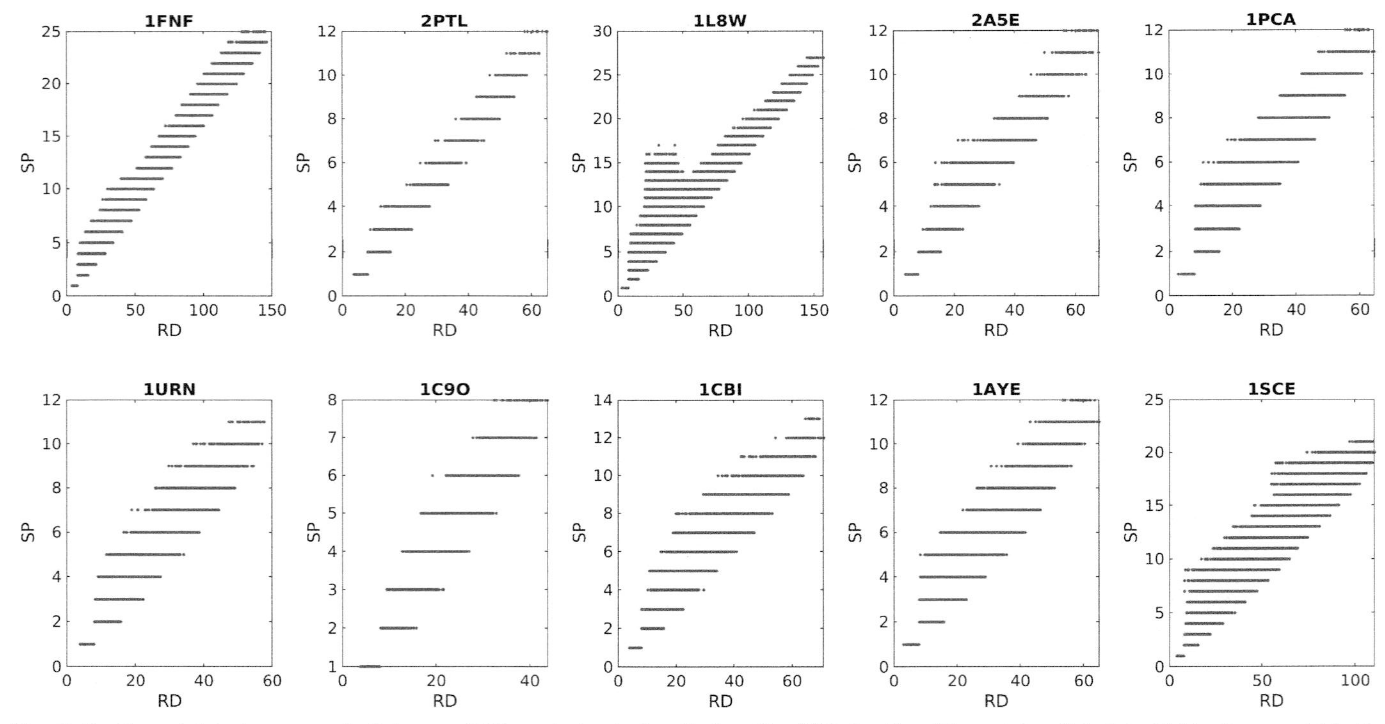

Fig. 3 Scatter plot between real distances (RD) and shortest-path lengths (SP) for the 10 proteins listed in Table 3, on which the *ShRec3D*-based method obtained the *best results*.

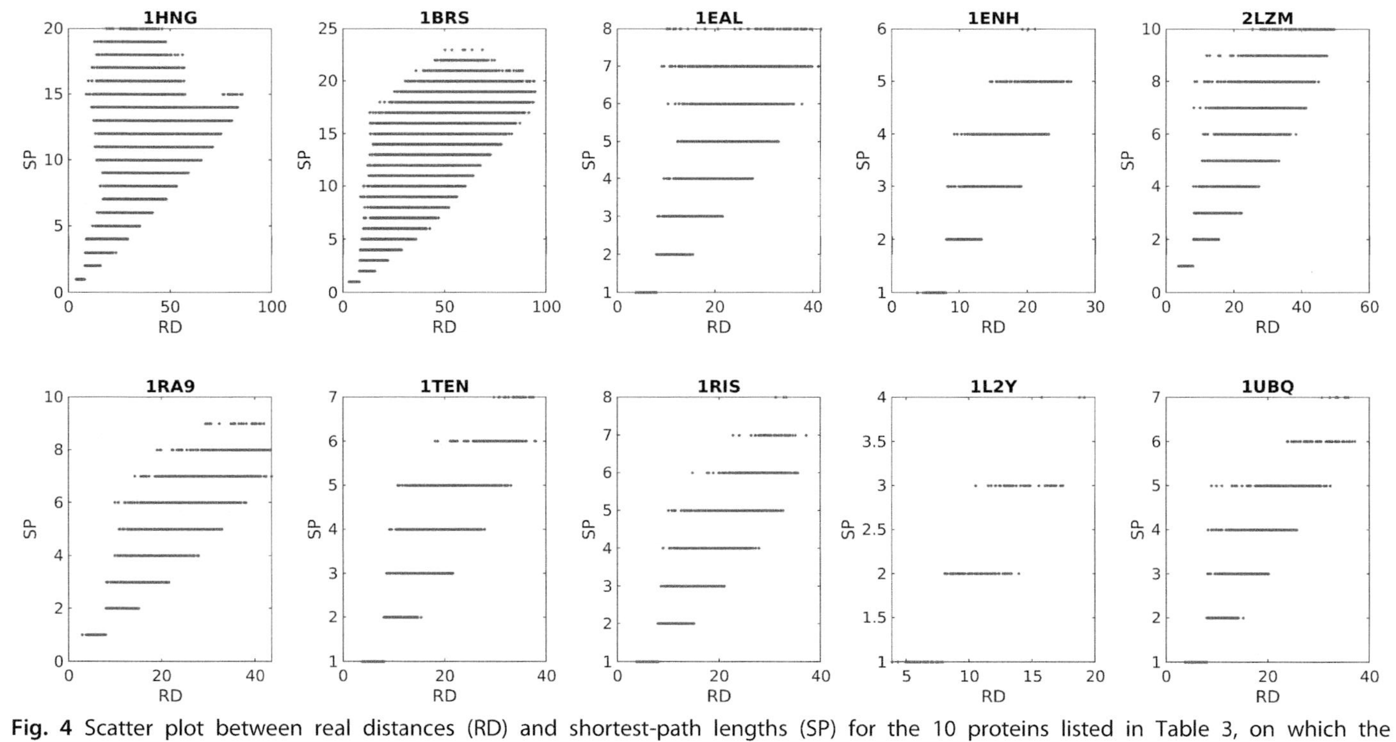

Fig. 4 Scatter plot between real distances (RD) and shortest-path lengths (SP) for the 10 proteins listed in Table 3, on which the *ShRec3D*-based method obtained the *worst results*.

Table 2 *Left*: 10 proteins with the *highest correlation values* between real distances and reconstructed ones via *Laplacian eigenvectors*, starting from an 8 Å contact map. All the proteins are characterized by the absence of a modular structure (MS, shown in Fig. 5 and even more clearly in Fig. 6) and 8 out of 10 belong to the two-state class (FK). *Right*: 10 proteins with the *lowest correlation values* between real distances and reconstructed ones via *Laplacian eigenvectors*, starting from an 8 Å contact map. 8 out of 10 proteins are characterized by a modular structure (MS, shown in Fig. 7 and even more clearly in Fig. 8) and 6 out of 10 belong to the multi-state class (FK).

Protein ID	MS	FK	Protein ID	MS	FK
2CRO	No	Multi-state	1SCE	Yes	Multi-state
1ARR	No	Two-state	1CBI	Yes	Multi-state
1BNZ	No	Two-state	1PBA	No	Two-state
1SRL	No	Two-state	1PHP	Yes	Multi-state
2PTL	No	Two-state	1FNF	Yes	Multi-state
1HRC	No	Two-state	1VII	No	Two-state
1BTA	No	Multi-state	1OPA	Yes	Multi-state
1POH	No	Two-state	1PIN	Yes	Two-state
1YCC	No	Two-state	2LZM	Yes	Multi-state
2ACY	No	Two-state	1C9O	Yes	Two-state

along the diagonal; on the contrary, ShRec3D embedding provides the worst results on proteins characterized by the absence of a modular structure (see Table 3, Figs. 11 and 12). In particular, the 10 proteins with the lowest correlation values between real and Laplacian-reconstructed distances and the 10 proteins with the highest correlation values between real and ShRec3D-reconstructed distances, have four elements in common: 1SCE, 1CBI, 1FNF, and 1C9O, which are all characterized by a contact map with blocks along the diagonal.

If we stratify the dataset according to protein two-state or multi-state folding kinetics, we can see that ShRec3D reaches the best performance when using 12 Å as threshold, independently on the two-state or multi-state class (see Table 4), whereas the Laplacian-based method reaches the best performance on two-state and multi-state proteins at different threshold values, which are respectively 8 and 12 Å (see Table 5).

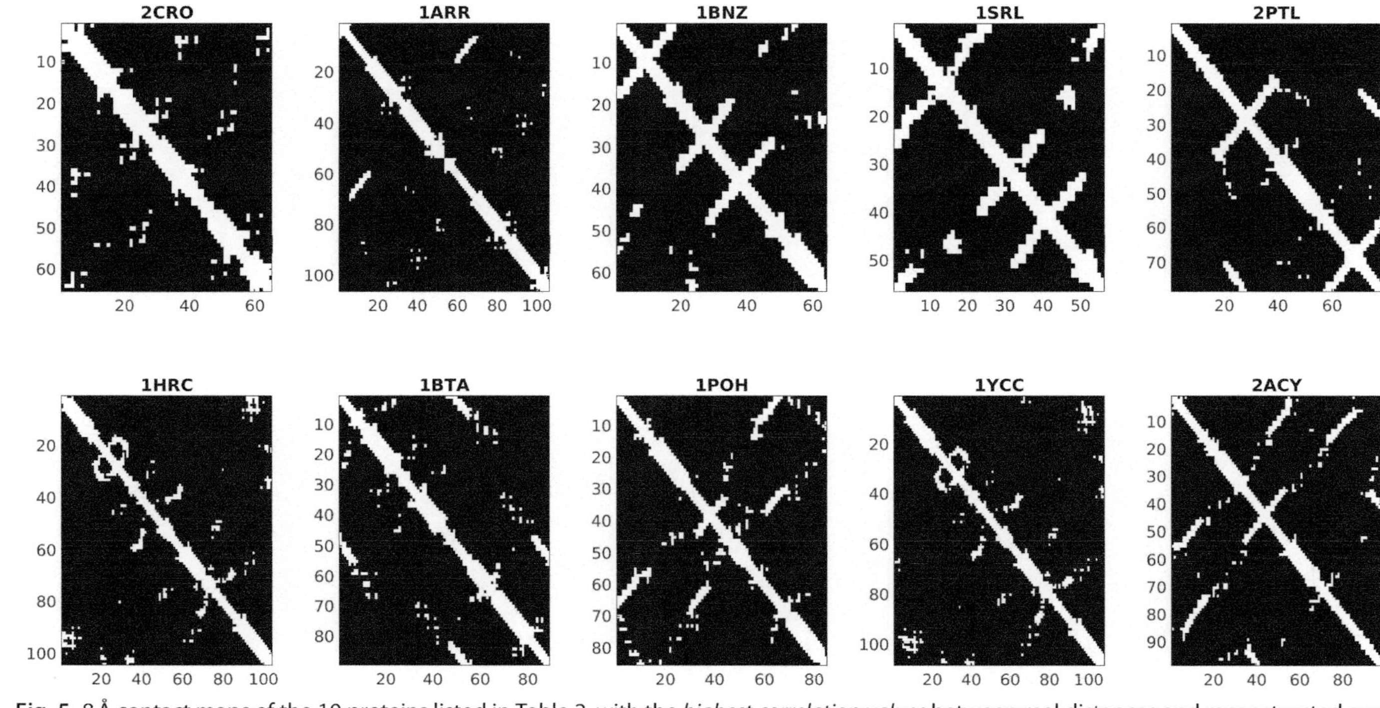

Fig. 5 8 Å contact maps of the 10 proteins listed in Table 2, with the *highest correlation values* between real distances and reconstructed ones via *Laplacian eigenvectors*.

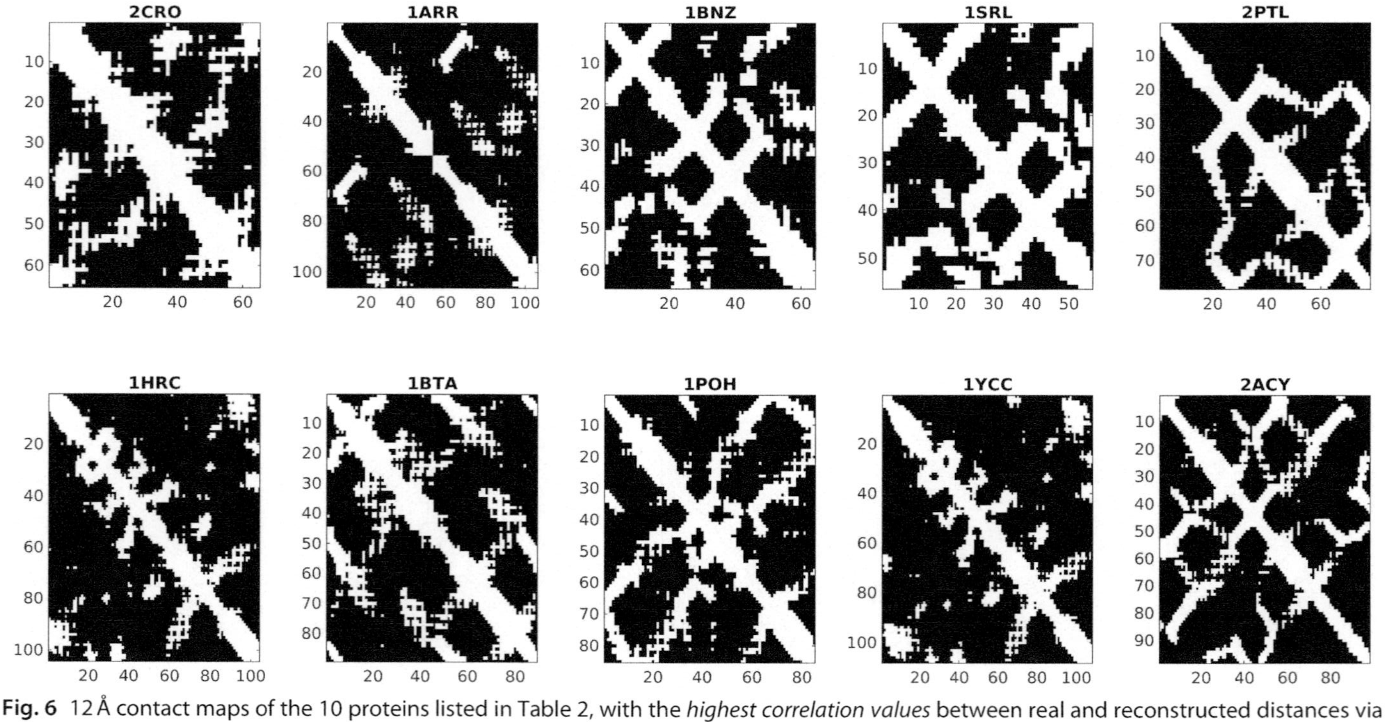

Fig. 6 12 Å contact maps of the 10 proteins listed in Table 2, with the *highest correlation values* between real and reconstructed distances via *Laplacian eigenvectors*.

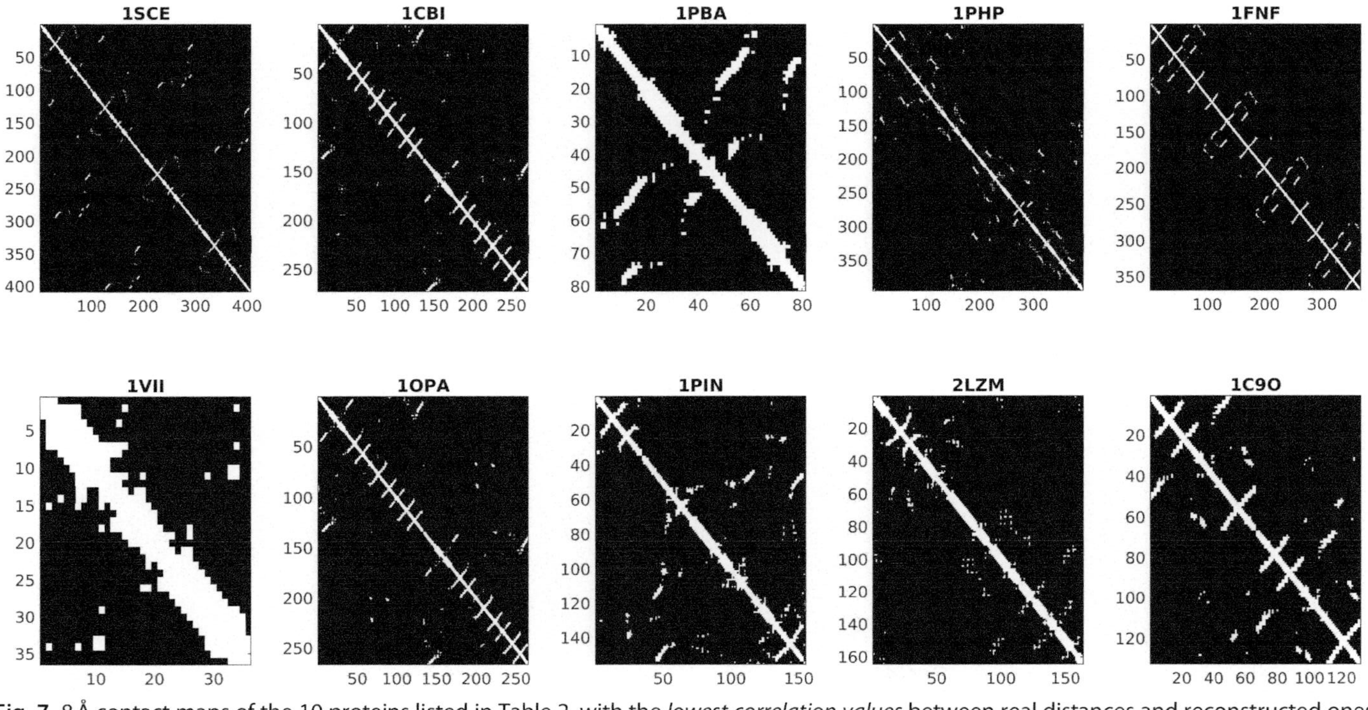

Fig. 7 8 Å contact maps of the 10 proteins listed in Table 2, with the *lowest correlation values* between real distances and reconstructed ones via *Laplacian eigenvectors*.

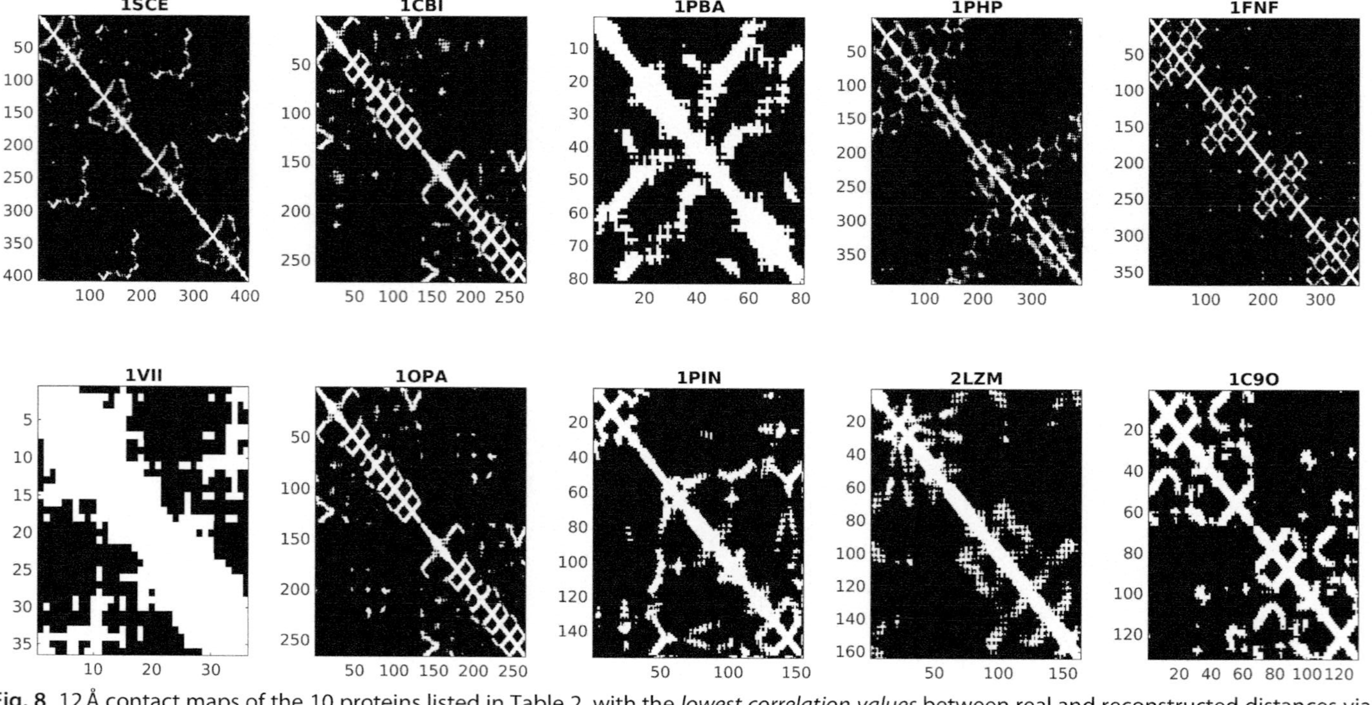

Fig. 8 12 Å contact maps of the 10 proteins listed in Table 2, with the *lowest correlation values* between real and reconstructed distances via *Laplacian eigenvectors*.

Table 3 *Left*: 10 proteins with the *highest correlation values* between real distances and reconstructed ones via *ShRec3D*, starting from an *8 Å contact map*. 6 out of 10 proteins are characterized by a modular structure (MS, shown in Fig. 9 and even more clearly in Fig. 10) and 7 out of 10 belong to the two-state class (FK). *Right*: 10 proteins with the *lowest correlation values* between real distances and reconstructed ones via *ShRec3D*, starting from an 8 Å contact map. 8 out of 10 proteins are characterized by the absence of a modular structure (MS, shown in Fig. 11 and even more clearly in Fig. 12) and 5 out of 10 belong to the multi-state class (FK).

Protein ID	MS	FK	Protein ID	MS	FK
1FNF	Yes	Multi-state	1HNG	Yes	Multi-state
2PTL	No	Two-state	1BRS	Yes	Multi-state
1L8W	Yes	Two-state	1EAL	No	Multi-state
2A5E	No	Multi-state	1ENH	No	Two-state
1PCA	No	Two-state	2LZM	No	Multi-state
1URN	Yes	Two-state	1RA9	No	Multi-state
1C9O	Yes	Two-state	1TEN	No	Two-state
1CBI	Yes	Multi-state	1RIS	No	Two-state
1AYE	No	Two-state	1L2Y	No	Two-state
1SCE	Yes	Multi-state	1UBQ	No	Two-state

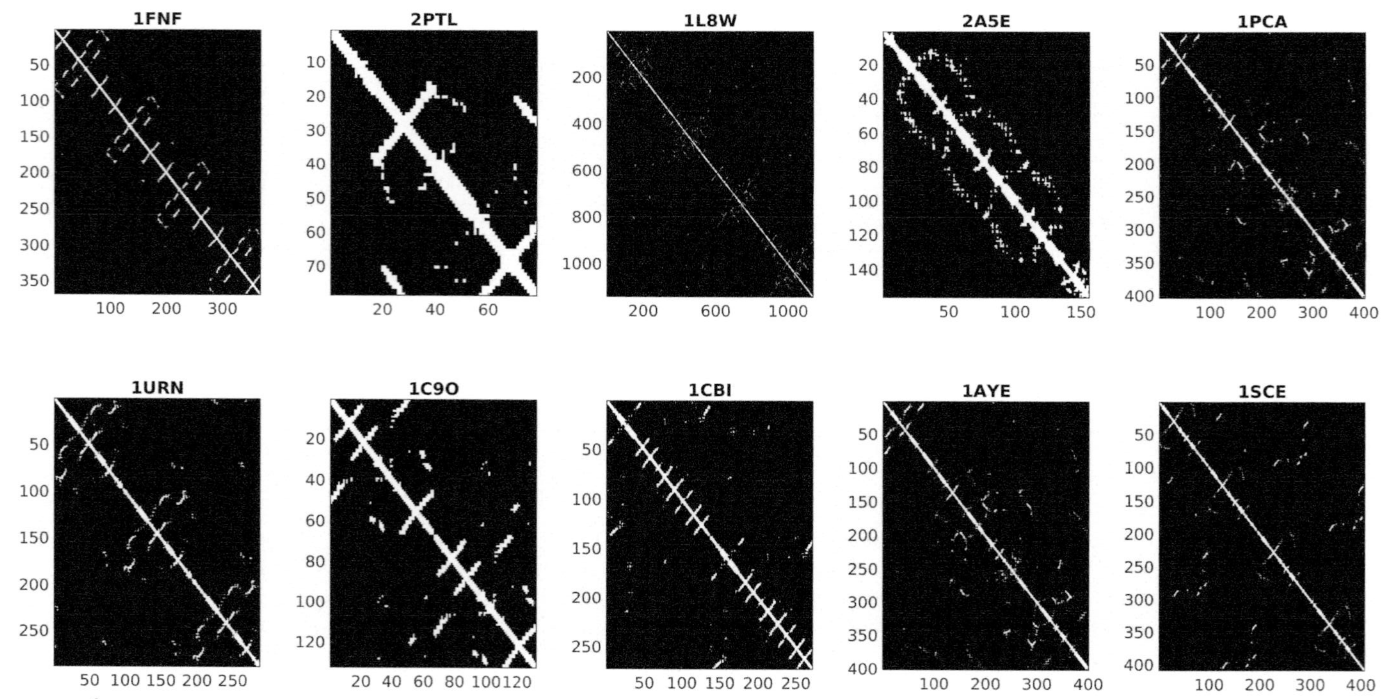

Fig. 9 8 Å contact maps of the 10 proteins listed in Table 3, with the *highest correlation values* between real distances and reconstructed ones via *ShRec3D*.

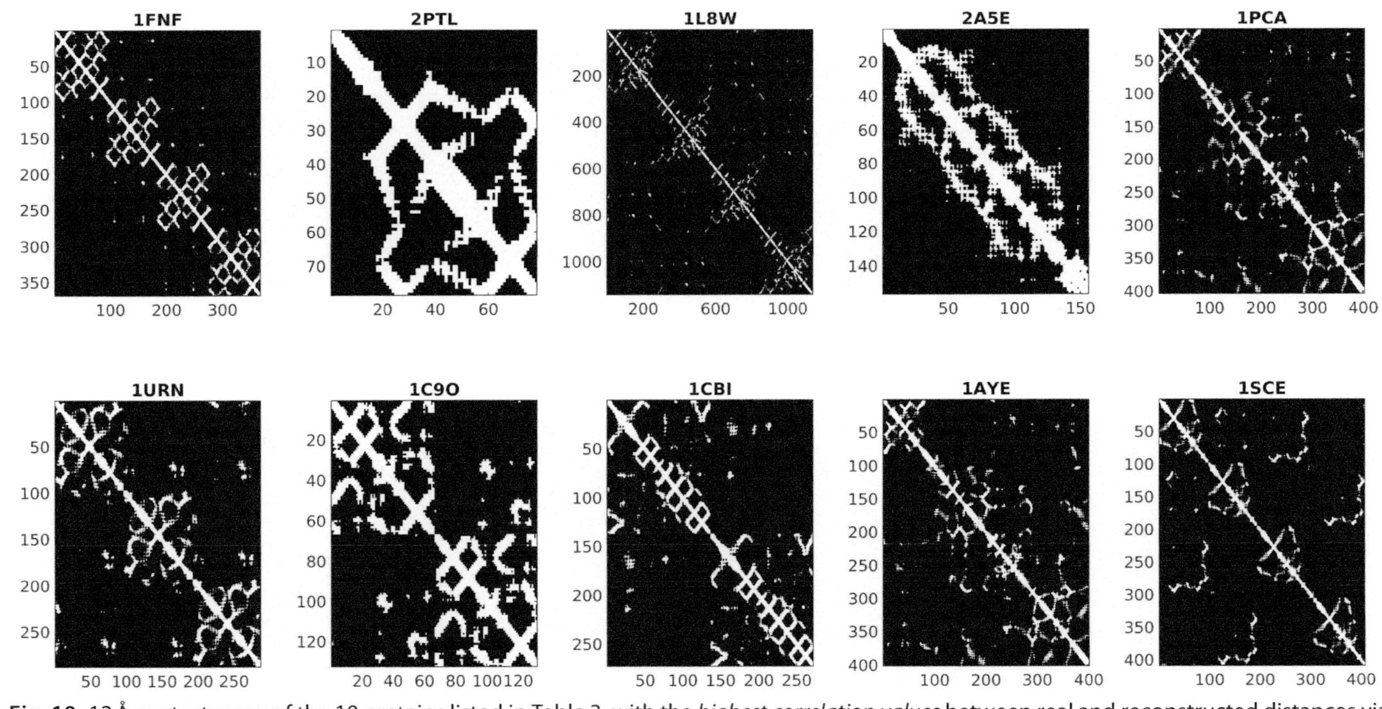

Fig. 10 12 Å contact maps of the 10 proteins listed in Table 3, with the *highest correlation values* between real and reconstructed distances via *ShRec3D*.

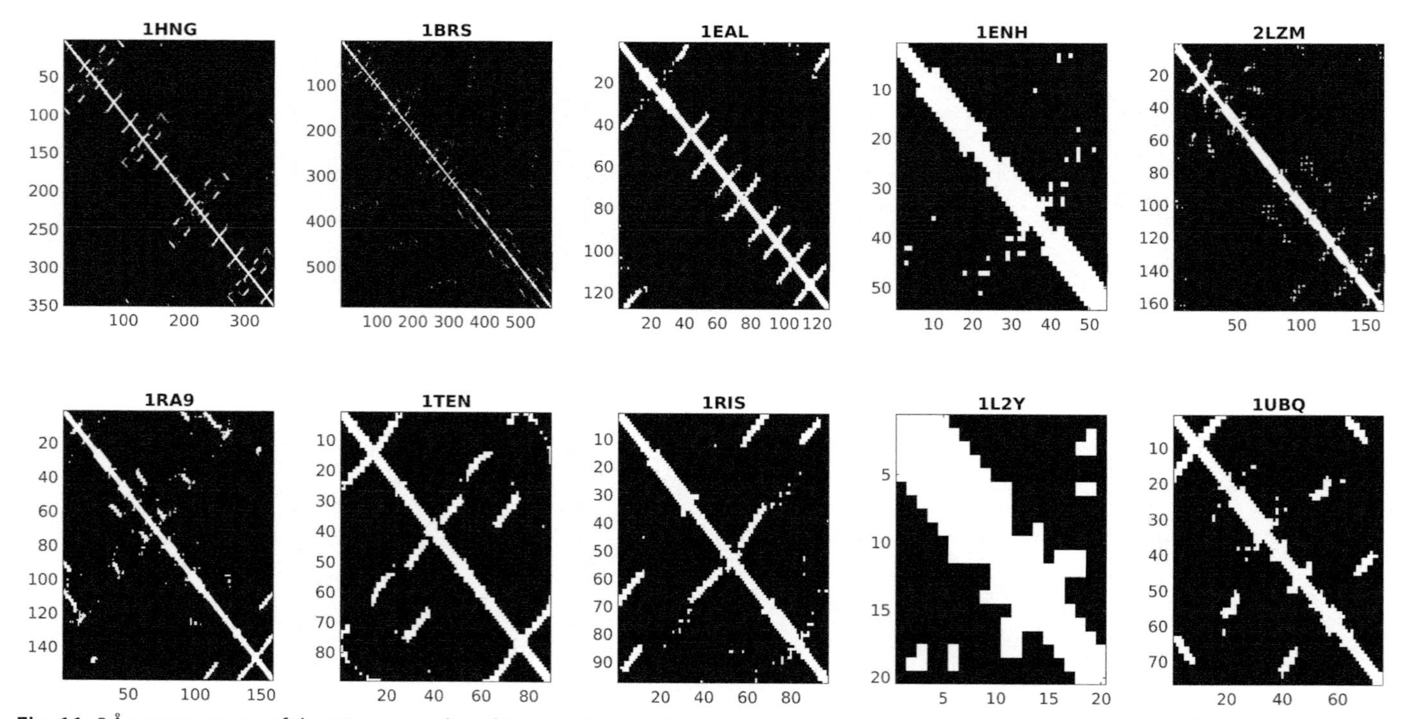

Fig. 11 8 Å contact maps of the 10 proteins listed in Table 3, with the *lowest correlation values* between real distances and reconstructed ones via *ShRec3D*.

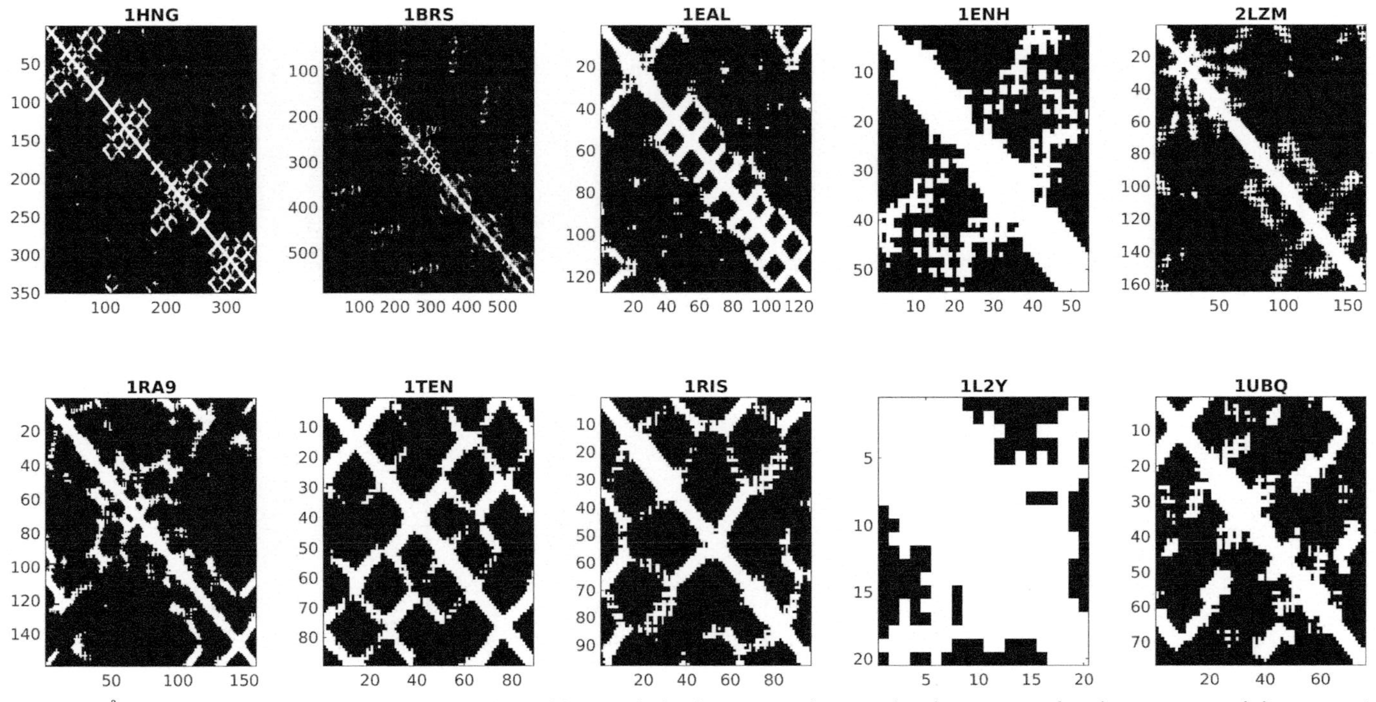

Fig. 12 12 Å contact maps of the 10 proteins listed in Table 3, with the *lowest correlation values* between real and reconstructed distances via *ShRec3D*.

Table 4 Mean correlation values between real and reconstructed Cα distances via *ShRec3D* method, for proteins divided into two-state or multi-state and represented by different contact maps, obtained using as threshold values 8 and 12 Å.

	Two-state	**Multi-state**
t = 8 Å	0.95	0.93
t = 12 Å	0.96	0.97

Table 5 Mean correlation values between real and reconstructed Cα distances via *Laplacian* method, for proteins divided into two-state or multi-state and represented by different contact maps, obtained using as threshold values 8 and 12 Å.

	Two-state	**Multi-state**
t = 8 Å	0.79	0.75
t = 12 Å	0.71	0.78

Moreover, if we represent the correlation values between real and reconstructed Cα distances as a function of the number of residues (see Figs. 13–16), we can notice two different behaviors between the two-state/multi-state classes, depending on the threshold used to calculate the contact map and on the reconstruction method. In fact, if we start from contact maps produced using 8 Å as threshold, we can see that the performance of both methods is almost constant as the length of the number of residues increases, for two-state proteins (see Fig. 15); whereas, in the case of multi-state proteins, this is still true only for ShRec3D reconstruction, while the Laplacian-based one shows a decreasing performance as the number of residues increases (see Fig. 15). If we start from contact maps produced using 12 Å as threshold, we can see that the scenario changes only for two-state proteins, whose performance increases as the number of residues increases, for both methods (see Fig. 16).

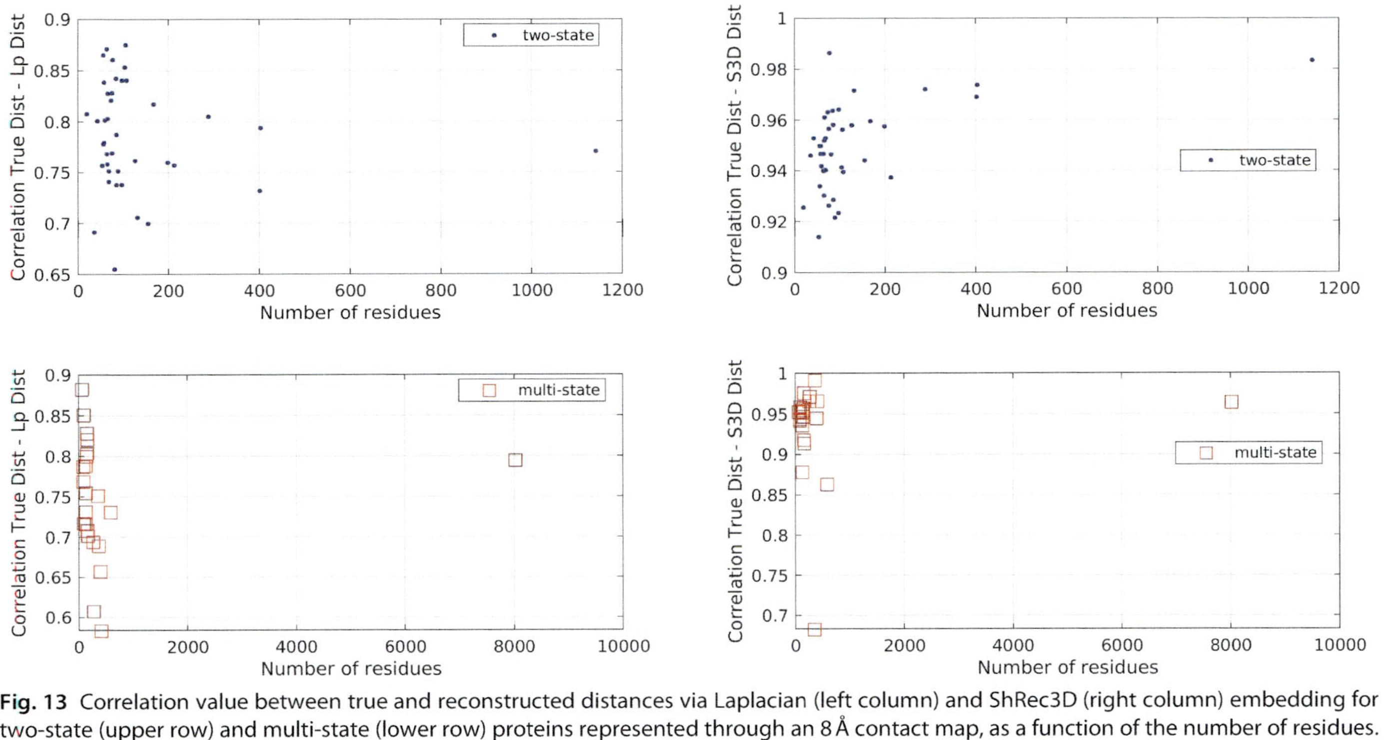

Fig. 13 Correlation value between true and reconstructed distances via Laplacian (left column) and ShRec3D (right column) embedding for two-state (upper row) and multi-state (lower row) proteins represented through an 8 Å contact map, as a function of the number of residues.

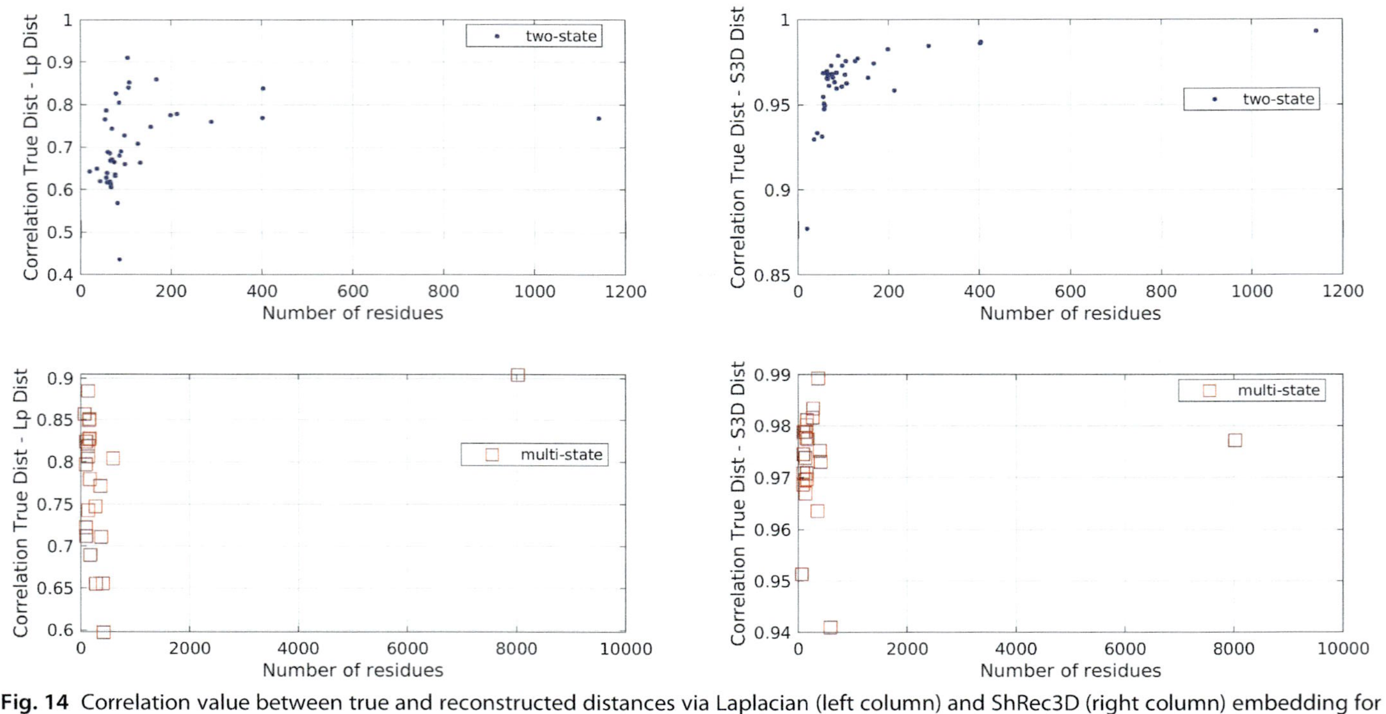

Fig. 14 Correlation value between true and reconstructed distances via Laplacian (left column) and ShRec3D (right column) embedding for two-state (upper row) and multi-state (lower row) proteins represented through a 12 Å contact map, as a function of the number of residues.

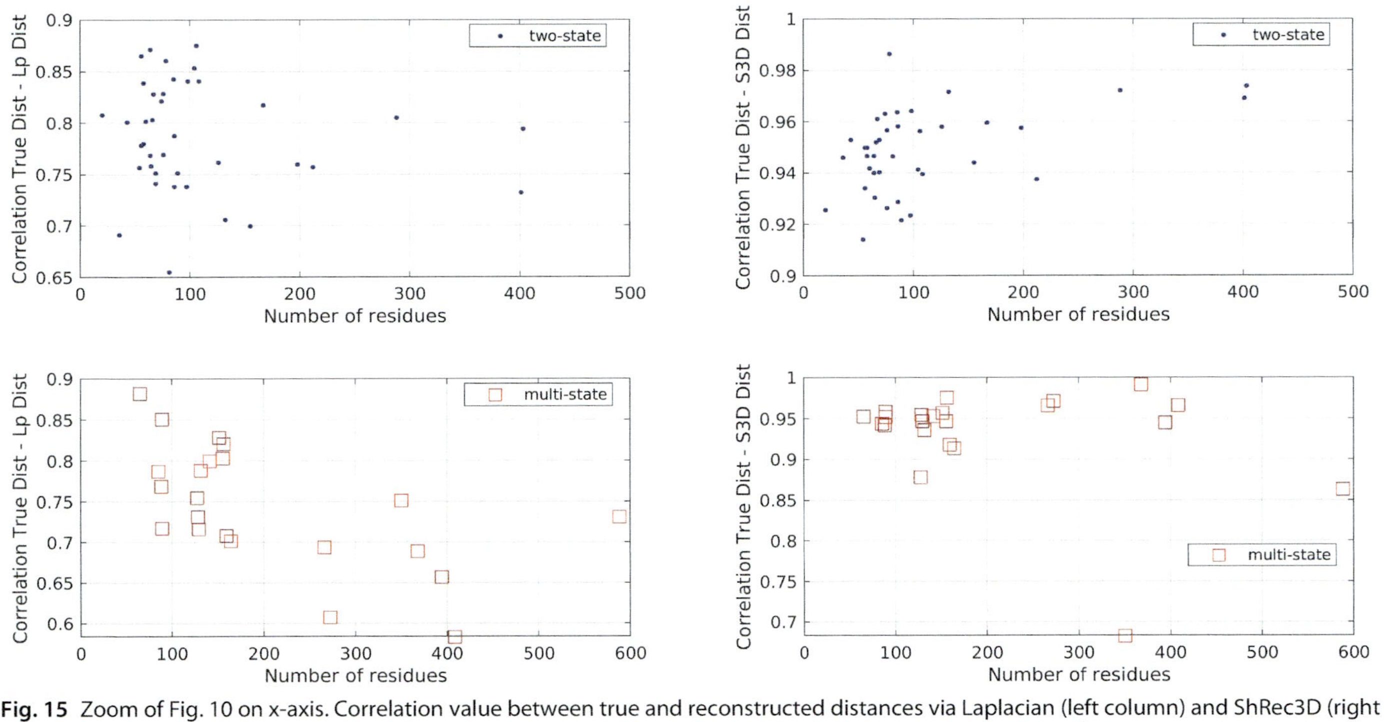

Fig. 15 Zoom of Fig. 10 on x-axis. Correlation value between true and reconstructed distances via Laplacian (left column) and ShRec3D (right column) embedding for two-state (upper row) and multi-state (lower row) proteins represented through an 8 Å contact map, as a function of the number of residues.

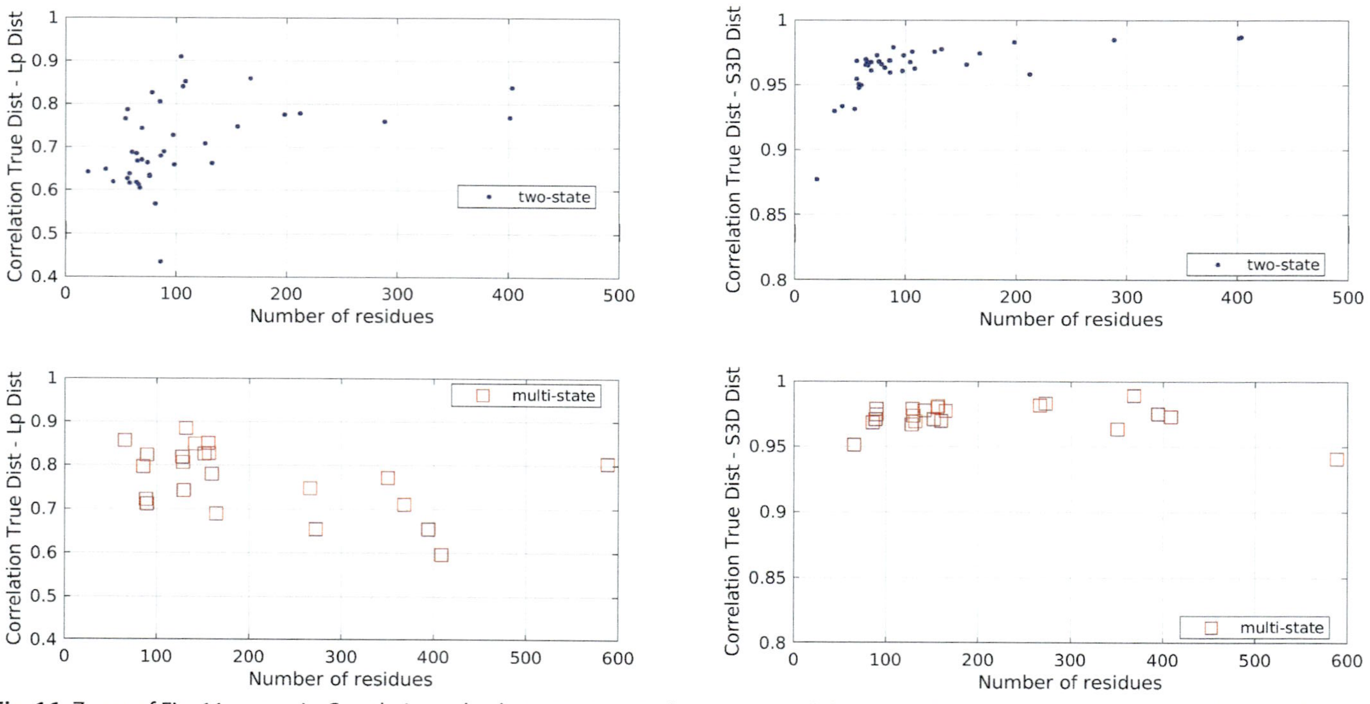

Fig. 16 Zoom of Fig. 11 on x-axis. Correlation value between true and reconstructed distances via Laplacian (left column) and ShRec3D (right column) embedding for two-state (upper row) and multi-state (lower row) proteins represented through a 12 Å contact map, as a function of the number of residues.

8. Discussion

Network-based methods can provide useful tools to characterize properties associated to protein folding, in particular regarding the 3D structure reconstruction. This scope can be interpreted as the identification of an optimal embedding manifold for the network through spectral approaches, related to the recently developing topic of network geometry (Boguñá et al., 2021), that deals with networks characterized by an intrinsic geometric space, in our case the 3D Euclidean space in which the 1D residue chain folds. At difference with less "physical" networks (like social networks, the world wide web, or protein interaction networks, in which there are no physical constraints on the links related to a maximum distance allowed between nodes) the properties of the chain structure of the protein, and the fact that it folds in a physical space, appear to be reflected in the properties of the contact map. In particular, the guess of using the shortest path distance as a proxy for the real residue distance, as proposed within the ShRec3D approach, seems satisfying in many cases, achieving better results than the Laplacian-based approach. As shown in a previous paper nonetheless (Menichetti et al., 2016), observables associated to the Laplacian operator allowed to discriminate between two-state and multi-state proteins, thus for other applications the informative content provided by this network formalism can be relevant as well. Even if the two proposed approaches rely on a common theoretical ground (i.e., the algebra of Gram matrices) the performances of the two methods are independent from each other (see Fig. 17) and seem to be much more influenced by the threshold value chosen for the computation of the contact map rather than the folding kinetics class (two-state/multi-state) or the number of residues. The database we used allowed us to evaluate and compare the two methods for the specific task of protein fold structure reconstruction, but in general the resulting spectral embeddings can be used on larger protein datasets as a pre-processing for unsupervised (i.e., clustering) or supervised (classification or mapping of specific protein chemical/physical properties) studies, and providing an optimal metrics for novel approaches like semi-supervised methods (van Engelen & Hoos, 2020).

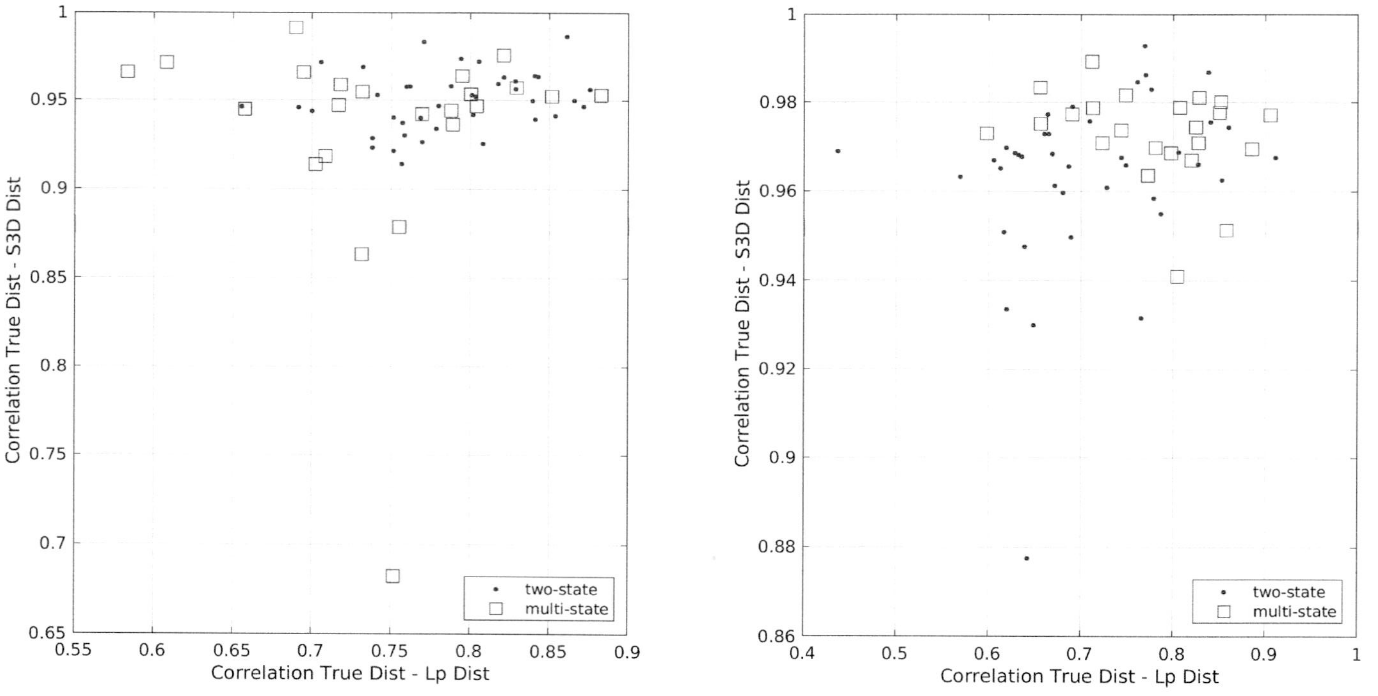

Fig. 17 Correlation value between true distances and reconstructed distances via ShR3c3D method as a function of correlation value between true distances and reconstructed distances via Laplacian-based method, obtained starting from an 8 Å contact map (left) and a 12 Å contact map (right).

References

Amitai, G., et al. (2004). Network analysis of protein structures identifies functional residues. *Journal of Molecular Biology*, *344*, 1135–1146.

Andreeva, A., et al. (2020). The SCOP database in 2020: Expanded classification of representative family and superfamily domains of known protein structures. *Nucleic Acids Research*, *48*, D376–D382.

Anfinsen, C. B. (1973). Principles that govern the folding of protein chains. *Science*, *181*, 223–230.

Bagler, G., & Sinha, S. (2007). Assortative mixing in Protein Contact Networks and protein folding kinetics. *Bioinformatics*, *23*, 1760–1767.

Baker, D., & Sali, A. (2001). Protein structure prediction and structural genomics. *Science*, *294*, 93–96.

Bartoli, L., et al. (2008). The pros and cons of predicting protein contact maps. *Methods in Molecular Biology*, *413*, 199–217.

Biyikoglu, T., et al. (2007). *Laplacian eigenvectors of graphs: Perron-Frobenius and Faber-Krahn type theorems*. Berlin Heidelberg: Springer-Verlag.

Böde, C., et al. (2007). Network analysis of protein dynamics. *FEBS Letters*, *581*, 2776–2782.

Boguñá, M., et al. (2021). Network geometry. *Nature Reviews Physics*, *3*, 114–135.

Bollobas, B. (1998). *Modern graph theory*. New York: Springer-Verlag.

Brinda, K. V., & Vishveshwara, S. (2005). A network representation of protein structures: Implications for protein stability. *Biophysical Journal*, *89*, 4159–4170.

Capriotti, E., & Casadio, R. (2007). K-Fold: A tool for the prediction of the protein folding kinetic order and rate. *Bioinformatics*, *23*, 385–386.

Chang, C. C. H., et al. (2015). Towards more accurate prediction of protein folding rates: A review of the existing Web-based bioinformatics approaches. *Briefings in Bioinformatics*, *16*, 314–324.

Chothia, C., & Lesk, A. M. (1986). The relation between the divergence of sequence and structure in proteins. *The EMBO Journal*, *5*, 823–826.

Compiani, M., & Capriotti, E. (2013). Computational and theoretical methods for protein folding. *Biochemistry*, *52*, 8601–8624.

Dill, K. A. (1990). Dominant forces in protein folding. *Biochemistry*, *29*, 7133–7155.

Dill, K. A., & MacCallum, J. L. (2012). The protein-folding problem, 50 years on. *Science*, *338*, 1042–1046.

Dill, K. A., et al. (2007). The protein folding problem: When will it be solved? *Current Opinion in Structural Biology*, *17*, 342–346.

Dishman, A. F., & Volkman, B. F. (2018). Unfolding the mysteries of protein metamorphosis. *ACS Chemical Biology*, *13*, 1438–1446.

Fariselli, P., et al. (2007). The WWWH of remote homolog detection: The state of the art. *Briefings in Bioinformatics*, *8*, 78–87.

Grabowski, M., et al. (2016). The impact of structural genomics: The first quindecennial. *Journal of Structural and Functional Genomics*, *17*, 1–16.

Greene, L. H. (2012). Protein structure networks. *Briefings in Functional Genomics*, *11*, 469–478.

Grewal, R. K., & Roy, S. (2015). Modeling proteins as residue interaction networks. *Protein and Peptide Letters*, *22*, 923–933.

Gromiha, M. M., & Selvaraj, S. (2001). Comparison between long-range interactions and contact order in determining the folding rate of two-state proteins: Application of long-range order to folding rate prediction. *Journal of Molecular Biology*, *310*, 27–32.

Havel, T. F., et al. (1983). The theory and practice of distance geometry. *Bulletin of Mathematical Biology*, *45*, 665–720.

Hrmova, M., & Fincher, G. B. (2009). Functional genomics and structural biology in the definition of gene function. *Methods in Molecular Biology*, *513*, 199–227.

Huang, L.-T., & Gromiha, M. M. (2010). First insight into the prediction of protein folding rate change upon point mutation. *Bioinformatics*, *26*, 2121–2127.

Ivankov, D. N., & Finkelstein, A. V. (2004). Prediction of protein folding rates from the amino acid sequence-predicted secondary structure. *Proceedings of the National Academy of Sciences of the United States of America, 101*, 8942–8944.

Karplus, M., & Weaver, D. L. (1976). Protein-folding dynamics. *Nature, 260*, 404–406.

Kryshtafovych, A., et al. (2019). Critical assessment of methods of protein structure prediction (CASP)-Round XIII. *Proteins, 87*, 1011–1020.

Lesne, A., et al. (2014). 3D genome reconstruction from chromosomal contacts. *Nature Methods, 11*, 1141–1143.

Lieberman-Aiden, E., et al. (2009). Comprehensive mapping of long-range interactions reveals folding principles of the human genome. *Science, 326*, 289–293.

Magliery, T. J. (2015). Protein stability: Computation, sequence statistics, and new experimental methods. *Current Opinion in Structural Biology, 33*, 161–168.

Menichetti, G., et al. (2016). Network measures for protein folding state discrimination. *Scientific Reports, 6*, 30367.

Merlotti, A., et al. (2020). Merging 1D and 3D genomic information: Challenges in modelling and validation. *Biochimica et Biophysica Acta, Gene Regulatory Mechanisms, 1863*, 194415.

Plaxco, K. W., et al. (1998). Contact order, transition state placement and the refolding rates of single domain proteins. *Journal of Molecular Biology, 277*, 985–994.

Porto, M., et al. (2004). Reconstruction of protein structures from a vectorial representation. *Physical Review Letters, 92*, 218101.

Punta, M., & Rost, B. (2005). Protein folding rates estimated from contact predictions. *Journal of Molecular Biology, 348*, 507–512.

Sanavia, T., et al. (2020). Limitations and challenges in protein stability prediction upon genome variations: Towards future applications in precision medicine. *Computational and Structural Biotechnology Journal, 18*, 1968–1979.

Senior, A. W., et al. (2020). Improved protein structure prediction using potentials from deep learning. *Nature, 577*, 706–710.

Sillitoe, I., et al. (2021). CATH: Increased structural coverage of functional space. *Nucleic Acids Research, 49*, D266–D273.

Sippl, M. J., & Scheraga, H. A. (1985). Solution of the embedding problem and decomposition of symmetric matrices. *Proceedings of the National Academy of Sciences of the United States of America, 82*, 2197–2201.

Taylor, N. R. (2013). Small world network strategies for studying protein structures and binding. *Computational and Structural Biotechnology Journal, 5*, e201302006.

Torgerson, W. S. (1952). Multidimensional scaling: I. Theory and method. *Psychometrika, 17*, 401–419.

van Engelen, J. E., & Hoos, H. H. (2020). A survey on semi-supervised learning. *Machine Learning, 109*, 373–440.

Vassura, M., et al. (2008). Reconstruction of 3D structures from protein contact maps. *IEEE/ACM Transactions on Computational Biology and Bioinformatics, 5*, 357–367.

wwPDB consortium. (2019). Protein Data Bank: The single global archive for 3D macromolecular structure data. *Nucleic Acids Research, 47*, D520–D528.

Yan, W., et al. (2014). The construction of an amino acid network for understanding protein structure and function. *Amino Acids, 46*, 1419–1439.

Zhou, H., & Zhou, Y. (2002). Folding rate prediction using total contact distance. *Biophysical Journal, 82*, 458–463.

CHAPTER SEVEN

Proteomics and systems biology in optic nerve regeneration

Sean D. Meehan[b,c], Leila Abdelrahman[a,d], Jennifer Arcuri[a,b,c], Kevin K. Park[a,c,e], Mohammad Samarah[f], and Sanjoy K. Bhattacharya[a,b,c,]*

[a]Bascom Palmer Eye Institute, University of Miami Miller School of Medicine, Miami, FL, United States
[b]Molecular and Cellular Pharmacology Graduate Program, University of Miami, Miami, FL, United States
[c]Miami Integrative Metabolomics Research Center, University of Miami, Miami, FL, United States
[d]Department of Electrical and Computer Engineering, University of Miami, Miami, FL, United States
[e]Miami Project to Cure Paralysis, University of Miami, Miami, FL, United States
[f]Carroll University, Waukesha, WI, United States
*Corresponding author: e-mail address: sbhattacharya@med.miami.edu

Contents

Abstract

We present an overview of current state of proteomic approaches as applied to optic nerve regeneration in the historical context of nerve regeneration particularly central nervous system neuronal regeneration. We present outlook pertaining to the optic nerve regeneration proteomics that the latter can extrapolate information from multi-systems level investigations. We present an account of the current need of systems level standardization for comparison of proteome from various models and across different pharmacological or biophysical treatments that promote adult neuron regeneration. We briefly overview the need for deriving knowledge from proteomics and integrating with other omics to obtain greater biological insight into process of adult neuron regeneration in the optic nerve and its potential applicability to other central nervous system neuron regeneration.

Advances in Protein Chemistry and Structural Biology, Volume 127
ISSN 1876-1623
https://doi.org/10.1016/bs.apcsb.2021.03.002

1. Introduction to nerve regeneration

The earliest references of severed peripheral nerves' integration finds mention Chirurgia Magna and are attributed to Guy de Chauliac (Holmes, 1951). Scientists have studied peripheral and central nerve regeneration since 1757 (Arnemann, 1786, 1787). The actual evidence of possible regeneration of the nerves was experimentally obtained during 1785–1797 (Cruikshank, 1797). By 1883, the misdirection and looping of the regenerating nerve, particularly in wound healing, were realized.

The involvement of cells in the repair and reconnection finds mention in the earliest literature in the 1700s (Arnemann, 1786, 1787). Concerning proteins' involvement in the repair and regeneration of nerves, the word protein was coined in 1838 (Hartley, 1951; Mulder, 1838). Gerardus Johannes Mulder and Jöns Jacob Berzelius performed elemental analysis, as gravimetric analyses became quite common in that period, and found that nearly all common proteins had the same empirical formula, $C_{400}H_{620}N_{100}O_{120}P_1S_1$. By 1902, proteins composing of polypeptides became well known. Exactly when the proteins were considered to play a role in nerve regeneration is not clear, but during 1880–1892, they were implicated to play some role in regeneration (Holmes, 1951; Holmes & Young, 1942; Howell & Huber, 1892). In 1985, a 37 kDa protein specific for regenerating PNS and CNS nerves was found (Muller, Gebicke-Harter, Hangen, & Shooter, 1985). The sequence or identity of this protein was unknown at the time of discovery. Subsequently, it became identified as brain derived neurotrophic factor (BDNF).

It is well appreciated that PNS neurons regenerate relatively quickly compared to CNS neurons (David & Aguayo, 1981, 1985). It has been subsequently recognized that where the predominant myelination is carried out by the Schwann cells, regeneration is favored, such as most peripheral nerves (Bunge, 2016; Wood et al., 1990). In contrast, where the oligodendroglia are predominant myelinating entities such as most CNS neurons the regenerative capacities are rather limited (Barateiro, Brites, & Fernandes, 2016; Domingues et al., 2018; Miyata, 2019). The command neurons (CNs) differ tremendously (Gamkrelidze, Laurienti, & Blankenship, 1995) with respect to their regenerating potential. The CN1 consists of bundles (fascicles) of central processes of bipolar neurons in the olfactory neuroepithelium (Yamamoto et al., 2004). These bundles travel through the ethmoid bone's cribriform plate and are broken when the brain is removed. When unbroken in the living individual, Schwann cells are myelinated and travel to reach the olfactory bulb where they synapse on secondary special sensory neurons called mitral cells. There are some fundamental differences in the olfactory neurons of CN1 and the ganglion cells of the retina of CN2 that appears to determine regeneration potential. The neuroepithelium containing the bipolar olfactory neurons is pseudostratified. It has basal cells that can replenish the olfactory neuroepithelium, including the olfactory neurons. Unlike olfactory neurons, retinal ganglion cells (RGCs) do not have a basal cell population to replenish them. Also, the central processes of olfactory neurons are myelinated by Schwann cells, which form endoneurial tubes for regenerating olfactory neuron central processes growth cones to grow by guiding them to the location of the correct mitral cells they innervate. Oligodendroglia, not Schwann cells, myelinate retinal ganglion cell central processes that form the optic nerve (Barateiro et al., 2016; Simons & Nave, 2015).

Our understanding of regeneration of CNS neurons have undergone a significant change since the 1990s (Aguayo et al., 1990; Aguayo et al., 1991). Around 1989 the experimental evidence became recognized that the injured or damaged axons in the adult mammalian CNS can be stimulated to regenerate (Cho & So, 1989), subsequent evidence that reducing or eliminating growth inhibitory (Schwab, 1996) and increasing growth promoting (Cui, So, & Yip, 1998) substances could indeed be regenerating for adult CNS neurons/retinal ganglion neurons. The importance of transgene on the regeneration of RGCs became apparent (Hu et al., 2007). The intrinsic and extrinsic nature of neuron growth promoting, and inhibitory substances became increasingly clear and a number of them were identified as proteins. Long-distance regeneration of RGC neurons in the optic nerve using a combination of proteins and metabolites could be demonstrated (Kurimoto et al., 2010). Modulation of proteins to promote long-distance regeneration of neurons in other CNS system was also successfully demonstrated (Fry, Chagnon, Lopez-Vales, Tremblay, & David, 2010). Proteome thus came to occupy a central place toward long-distance axon regeneration. Toward the membrane expansion as an integral part of long-distance axon regeneration, growth cones are thought to be the main drivers (Pfenninger, 2009; Pfenninger & Friedman, 1993; Pfenninger et al., 2003) and proteome is thought to occupy a central position. Growth cone proteome has been the subject of a few investigations (Chauhan et al., 2020; Estrada-Bernal et al., 2012; Igarashi, 2014; Nozumi et al., 2009). The posttranslational state of proteins toward regulating axon regeneration has received due consideration (Igarashi et al., 2020). As is evident from early studies, metabolites such as cyclic adenosine monophosphate (cAMP) have strong influence on long-distance regeneration (Kurimoto et al., 2010). Recognition of intrinsic and extrinsic factors with the majority of them being proteins (part of the functional proteome) juxtaposed with the discovery of facilitating analytical instruments, led to the profiling of regenerating optic nerve proteome from different animal models of optic nerve regeneration. The different animal models show vast differences with respect to regenerating capacity. The different model systems such as zebrafish, frog, axolotl also pose the problem of comparing different regeneration proteome and critical changes that underlie regeneration (Belrose, Prasad, Sammons, Gibbs, & Szaro, 2020; Diekmann, Kalbhen, & Fischer, 2015; Harvey, Baxter, & Granato, 2019; Lee-Liu, Sun, Dovichi, & Larrain, 2018; Rabinowitz et al., 2017; Stelzner, Bohn, & Strauss, 1986; Whitworth et al., 2017; Zammit, Clarke, Golding, Goodbrand, & Tonge, 1993; Zhao et al., 2011). At the same time the problem remains to dissect the tissue level proteome with cellular and their in-situ secreted proteome to delineate which proteome and which specific players contribute to long-distance axon regeneration in the adult mammalian optic nerve and how they relate to parallel regeneration in high regenerative capacity lower vertebrates.

2. Proteomics levels and their integration

In 1983, goldfish regenerating optic nerve studies with [^{35}S] labeled methionine revealed axonal proteins' existence in the regenerating optic nerve that was thought to be of retinal origin (Quitschke & Schechter, 1983a, 1983b). The regenerating goldfish optic nerve after crush was found to undergo significant protein changes (Perry, Burmeister, & Grafstein, 1985). As noted above, around the same time (1985), a 37 kDa protein specific for regenerating mammalian PNS and CNS nerves (Muller et al., 1985) was also identified. Protein modification or phosphorylation was also identified in the regenerating goldfish optic nerve around 1987 (Larrivee & Grafstein, 1987). With the advent of high throughput mass spectrometric qualitative and quantitative methods (Li, Steen, & Gygi, 2003; Pandey & Mann, 2000; Ross et al., 2004), labeled proteomics has become common in many biological fields, including those in neuroregeneration (Belin et al., 2015; Magharious et al., 2011). The high throughput approaches generated "big data." For example, general dataset search engines, Google's dataset search engine (https://datasetsearch.research.google.com/) do not yet integrate the proteomic data with the studies. The proteomic data is usually deposited in specialized databases such as PRIDE and different query methods, and programs are being developed. Whether general dataset engines connect and provide the ability to integrate data from similar or related studies or not, specialized databases are likely to come up with tools that allow better query and integration of data deposited within them. Current thinking regarding optic nerve regeneration or CNS neuro regeneration fields conclude that fractionation is the key to obtaining better or high-resolution molecular definitions (Chauhan & Bhattacharya, 2021). In the current state, this brings the following questions:

(a) *Concerning fractionated proteomics in optic nerve regeneration, what is the completeness of the data.* Currently, the total tissue proteomics, be it retina or optic nerve, and the optic nerve using crush vs regeneration is available for mouse models. Much more in-depth coverage for these tissues is available for genomics or transcriptomics in mice and some in rats. However, this review focuses on proteomics. The proteomics data from optic nerve regeneration or spinal cord regeneration after crush or mechanical injury using multiple pharmacological agents or modalities are not available. Thus, concerning the availability of proteomics data from multiple regeneration studies initiated by different approaches is lacking. There is almost a complete lack of fractionated proteomics data for different cells (Chauhan & Bhattacharya, 2021) except for isolated retinal ganglion cells (Belin et al., 2015).

(b) *How to integrate cell-type-specific data with whole tissue level data.* Not only lack of data exists for fractionated cells, but methodological limitations exist to have different sets of data that will enable better integration of cell-specific data with whole tissue level data (Chauhan & Bhattacharya, 2021). The isolated cells fractionated with fluorescent activated cell sorting, laser capture, or column chromatography approaches may change during fractionation and provide limited material resulting in incomplete proteome coverage. Isolated cells will provide adequate coverage. However, these cells may have a minor or significantly altered proteome than those in-situ. Isolated cells will also identify secretome, but they may not represent in-situ secretome by the same cells. Genetic ablation of cell types followed by fractionation and whole tissue proteome are other approaches, which will help build complete data in the long-run.

(c) *What are the prioritizing orders for focus on pathways. Are their priority targets that help better understand biology and also enable the bench to bedside translation?* Different studies, be it at tissue level or isolated cell level, focus on different pathways (Belin et al., 2015; Kohen & Giger, 2020; Yang et al., 2020). The focus on the pathway may be due to the original study hypothesis or very rarely driven by the unbiased prioritization. For example, the proteomic study of isolated fluorescent cell sorted RGCs (Belin et al., 2015) focused on the cytoskeletal pathway, partly influenced by the hypothesis. Similarly, the focus on the lysolipid conversion pathway was also influenced by the hypothesis of the studies (Kohen & Giger, 2020; Yang et al., 2020). Availability of large datasets at the tissue and the fractionated level may enable prioritizing multiple pathways simultaneously as a cluster and enabling devising ways to influence multiple pathways using pharmacological means simultaneously. Such prioritizing may result in a clustering of the pathways and treat or influence the clusters as groups (Fig. 1).

(d) *How to make analogies across different systems.* Which of the biological processes are more important to prioritize for regeneration? The several processes or pathways simultaneously identified across different systems using high throughput approaches that are important for the system where they are identified. Different systems may provide a picture that may be incomplete from the standpoint of another system even for a single process or a single pathway. However, for extrapolation of a process or pathway information from one system to the other some basic, understanding of why a singular process or pathway is important. For example, draw analogies or prioritization from the proteome from high regenerative organisms such as goldfish (Quitschke & Schechter, 1983b), zebrafish (Veldman, Bemben, & Goldman, 2010; Veldman, Bemben, Thompson, & Goldman, 2007) and *Xenopus laevis* (Lee-Liu et al., 2018), for mouse, large mammals such as cats, monkeys (Rose et al., 2008) and ultimately for humans (Fig. 1).

3. Standardization approaches

Currently, there is no standardization for the degree of optic nerve regeneration in a single animal model or in multiple animal models. The regeneration after optic nerve crush is promoted by targeted disruption of genes (knockout) (Park, Liu, Hu, Kanter, & He, 2010; Park et al., 2008), pharmacological agents (such as Zymosan + CPT-cAMP (Kurimoto et al., 2010), Wnt3a (Patel, Park, & Hackam, 2017) or oncomodulin (Yin et al., 2003)) or biophysical factors (optogenetic stimulation (Park et al., 2015; Sun et al., 2014), electrical stimulation (Corredor & Goldberg, 2009; Goldberg et al., 2002; Gordon, Udina, Verge, & de Chaves, 2009;

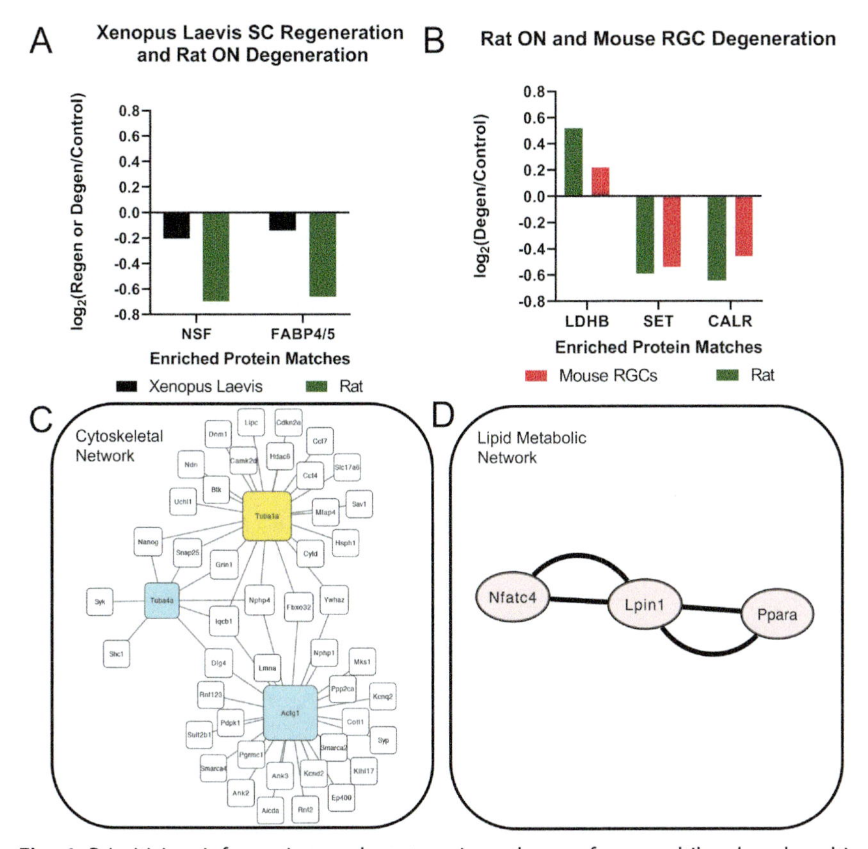

Fig. 1 Prioritizing information and proteomic pathways from multilevel and multisystem proteomics comparisons. (A) Multisystem comparison. The high regeneration organisms (such as frogs) have limited but useful information to offer low regeneration organisms such as rat. Common statistically significant proteins found in *Xenopus laevis* spinal cord regeneration and rat optic nerve degeneration. (B) Multilevel comparisons, for example, the tissue level compared with isolated cells may provide incomplete yet useful information for promoting various aspects of regeneration. Common statistically significant proteins found in degeneration of rat optic nerve and isolated mouse retinal ganglion cell post-optic nerve crush. Expression levels are shown as log2 of expression ratio (regeneration or degeneration to control) in (A) and (B). (C) Prioritizing pathways. Different pathways have emerged from different studies, cytoskeletal pathway (Belin et al., 2015) and glycerolipid catalyzing protein pathway (Yang et al., 2020). Proteomics may help to prioritize multiple pathways as clusters. Clusters can be further prioritized. Protein-protein interaction networks produced based on identified cytoskeletal proteins in mouse RGC degeneration. (D) The lipid metabolic enzymes in optic nerve degeneration. For cytoskeletal network, node size represents the degree of centrality while increased or decreased expressions in degeneration are shown in blue and yellow, respectively. Protein expression levels for Yang et al. (2020) were not available. Networks generated using OmicsNet and Cytoscape. Panel C: *Based on data from Belin, S., Nawabi, H., Wang, C., Tang, S., Latremoliere, A., Warren, P., Schorle, H., Uncu, C., Woolf, C.J., He, Z., et al. (2015). Injury-induced decline of intrinsic regenerative ability revealed by quantitative proteomics.* Neuron, 86*(4), 1000–1014;* Panel D: *Based on data from Yang, C., Wang, X., Wang, J., Wang, X., Chen, W., Lu, N., Siniossoglou, S., Yao, Z., & Liu, K. (2020). Rewiring neuronal glycerolipid metabolism determines the extent of axon regeneration.* Neuron, 105*(2), 276–292.*

Miyake, Yoshida, Inoue, & Hata, 2007)). For most of such regeneration and relevant for future translation, the pharmacological or biophysical factor-induced regeneration experiments (mostly in mice or in rats) report just the one-dimensional length, after 7 days of stimulation, 15 days of stimulation, or 30 days of stimulation (Belin et al., 2015; Park et al., 2010; Park et al., 2008). The length of the sectioned optic nerve is measured and compared to untreated controls under the same or similar conditions. This does not say the optic nerve's total volume and whether they reach optic chiasma and beyond. For functional connectivity, reaching up to the lateral geniculate nucleus (LGN) and innervation of new neurons with existing neurons is essential. The extent of volumetric regeneration and nerve fibers particularly with different pharmacological or biophysical methods necessitates standardization to determine their synergistic effects. The regenerated nerve fibers' ability to reach to optic chiasma, calibrated deflection of a part to ipsi vs contralateral eye, finally reaching to LGN and innervation is critical for functionality restoration. Volumetric standardization and ensuring that it is due to fibers are essential for these reagents so that their potential and synergistic effects are harnessed for eventual translation. We propose here a standardization method fulfilling this gap in the field. We propose that the optic nerve cylinder's measurement is taken from the optic nerve head (ONH) to the optic chiasma for the mice's species and age. It is ensured that at least one fiber traverses all the way (latest on the 30th day after crush with treatment with an agent) from crush to optic chiasma. As depicted in Fig. 2, the CTB-stained cylinder volume is compared to the volume of the ON cylinder for a given age and species of mice (Fig. 3). The CTB-stained volume (Fig. 3A) is compared with the total ON volume (see Fig. 3B) at 65% (maximum volume for optic nerve fibers within the cylinder). In this approach, regeneration can be compared for the day on measurement (7–15 days after pharmacological treatment). For standardization of volume in this method, full-length regeneration (from crush to optic chiasm) must be taken for at least one fiber on the 30th day after crush and treatment with regeneration agent. The 30th-day standard is based on the maximum regeneration observed in PTEN KO mice (Belin et al., 2015) on the 30th day (after crush). Any shortfall in reaching full length should be noted compared to the day of regeneration (7th or 15th day as reported). An algorithm is being developed to automate calculations based on volume data in mice (Fig. 3B) and the approach outlined in Fig. 2 (Box 1).

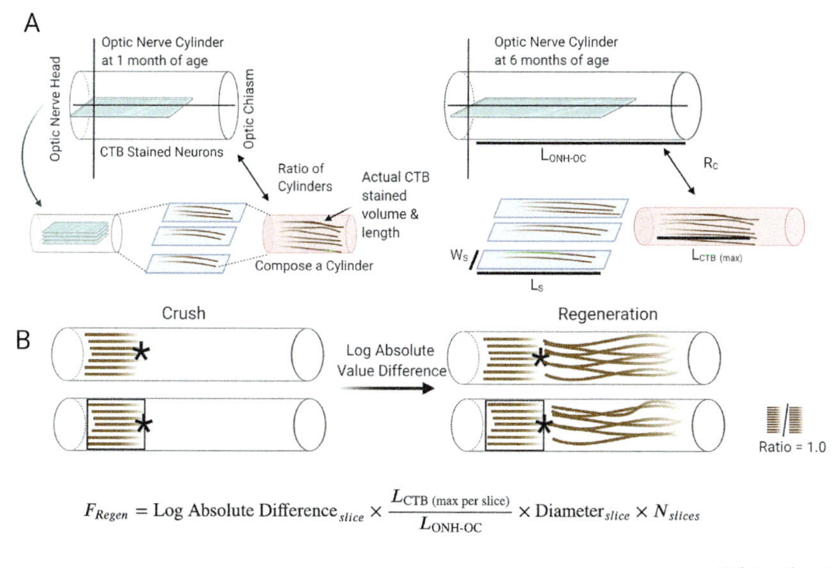

$$F_{Regen} = \text{Log Absolute Difference}_{slice} \times \frac{L_{CTB\ (\text{max per slice})}}{L_{ONH\text{-}OC}} \times \text{Diameter}_{slice} \times N_{slices}$$

Fig. 2 Method for standardization of optic nerve (ON) regeneration in mouse. (A) The cylinder length of the ON from the optic nerve head (ONH) to the optic chiasma (OC) needs to be recorded for age and species of the mouse. This is termed an optic nerve cylinder (ONC). The current approach is to draw an approximate plane passing through the middle of the ONH as depicted. The serial sections are then stained with cholera toxin B (CTB)*. A cylinder drawn from image boundary (Ls, Ws) measurements is an image cylinder (IC). The CTB-stained volume is the nerve fiber or axon volume (AV), reaching 65% maximum (consistent with most estimates). The length of the ONC cylinder is $L_{ONH\text{-}OC}$, whereas the maximum length of the CTB-stained axon is termed $L_{CTB\ (max)}$. At least one fiber should reach maximum length at or before 30th days post after crush (with pharmacological treatment) and should be recorded [based on PTEN KO mice on the 30th day post crush (Belin et al., 2015 cited in the text)]. If the maximum $L_{CTB\ (max)}$ less than $L_{ONH\text{-}OC}$, then the standardization method needs to determine a correction factor for a short length. The volume ratio of IC to ONC multiplied by the AV is the standardized regeneration. Left: The concept of extracting slices from the optic nerve, with each slice containing a specific fiber density. When the slices are stacked together, they approximate the total composition of the optic nerve. Right: Fiber length is critical in approximating total regeneration. After 6 months of age, to calculate regeneration, we also need to consider the difference per slice of the fiber length compared to

(Continued)

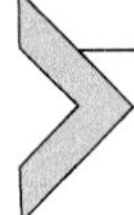

4. Insight about biological knowledge from proteomics and connectivity with other omics

Comparison of proteomics of optic nerve regeneration of tissues or fractionated cells from different systems and induced by different pharmacological or biophysical means may expand our understanding of systems biology and similarities across the treatment with different pharmacological means (Ebrahimi et al., 2018; Glanzer et al., 2007; Koh et al., 2020; Lasseck, Schroer, Koenig, & Thanos, 2007; Lauzi et al., 2019; Ma et al., 2010; Magharious et al., 2011; Perry et al., 1985; Prokosch, Chiwitt, Rose, & Thanos, 2010; Rose et al., 2008; Schroer, Volk, Liedtke, & Thanos, 2007; Smedowski et al., 2016; Tan et al., 2012; Tezel, 2016; Thanos, Bohm, Schallenberg, & Oellers, 2012; Tokuda et al., 2015; Wei, Jiang, Gao, & Wang, 2015; Zupanc, 2006).

In the optic nerve context, knowledge gained on neuronal regeneration is likely to fulfill the goal of restoring peripheral vision in glaucoma (Calkins et al., 2017; Goldberg, Guido, & Agi Workshop, 2016). The knowledge of regeneration is likely to restore the lost vision due to traumatic optic nerve injuries as well (Benowitz, He, & Goldberg, 2017; Goldberg et al., 2002; Goldberg et al., 2016). The traumatic injuries could be due to military relevant injuries, industrial accidents, or simply untoward accidents in civilian life. With respect to vision restoration the goals are to enable straight line growth of neurons up to optic nerve chiasm. The next desirable outcome

Fig. 2—Cont'd the total approximated optic nerve length. (B) The AV determination must include the deduction of crush from regeneration volume and ensure the staining intensity is 1.0. To calculate regeneration, we can take the log absolute value difference of the crush (control) to another nerve with regeneration. The equation highlights how combining the log absolute difference with the length ratio and the approximate diameter of the slice can be multiplied by the number of slices (N) to yield a regeneration factor (F_{Regen}). (C) Published images of the optic nerve (Yang et al., 2020 cited in the text) were used for demonstration of a computer program that enables difference matrix calculation and image generation (see Box 1). We processed two images, the control and another after regeneration, resizing and superimposing them to calculate the absolute difference matrix. The computer logarithmic scale was used to generate the image provided here. *Note here we have alluded to simplistic staining of the nerve section using an antibody against CTB with prior intravitreal injection of animals with unconjugated CTB. The approach described here will also work with the animals that received fluorophore conjugated CTB injection intravitreally and they will not require staining.

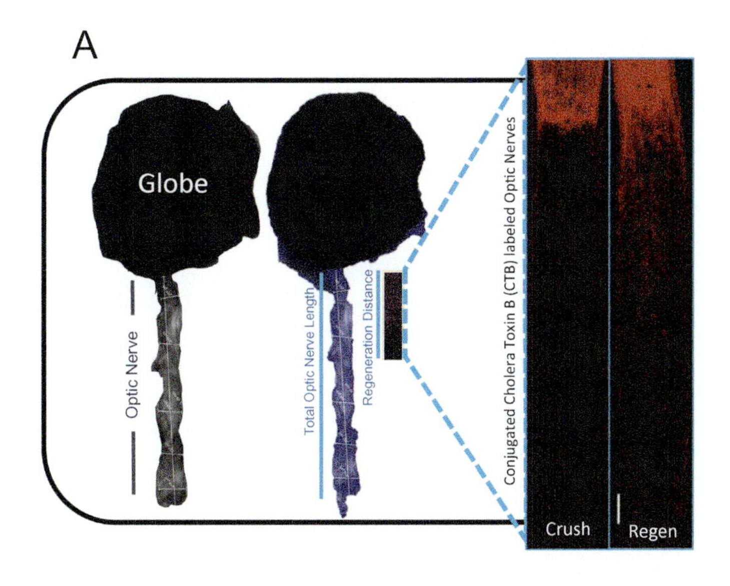

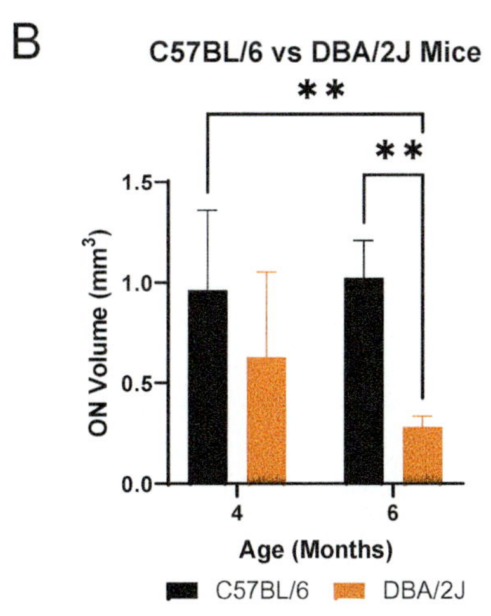

Fig. 3 Optic nerve volume and imaging regeneration. (A) Representative mouse optic nerve brightfield and merged with DAPI staining images. The small yellow boxed figure depicts the distance of a regenerating optic nerve at 2 weeks post injury. This regeneration figure serves as a comparison to the total optic nerve length in the DAPI stained image. The zoomed in figure (larger yellow box) shows a crush control (nonregenerative) and a regenerative image stained using FITC-conjugated cholera toxin B. (B) Optic nerve volumes for C57BL6 ($n = 12$, 4 m; $n = 7$, 6 m) and DBA2J ($n = 6$, 4 m; $n = 4$, 6 m) mice aged 4 and 6 months. Volume equation assumes cylindrical shape. Statistical analysis performed with 2-way ANOVA (*$\leq$0.05, **$\leq$0.01, ***$\leq$0.001). DBA/2J has been presented for comparison with a different mouse strain, this strain is not normally utilized for regeneration experiments, however, they are a good model for studying RGC degeneration due to spontaneous intraocular elevation and induced degeneration.

BOX 1 Standardized regeneration index

The method to quantify and standardize regeneration index F_{Regen} is based on pixel-wise differences in the slice-snapshots within the nerve. In ideal case, the ratio of intensity of crush vs regeneration should be 1.0, (see Fig. 2B). To calculate the slice-wise difference, the following approach was taken to calculate F_{Regen} and generate an image:

1. The two slices from the same anatomical location in the optic nerve need to be selected. One slice is the image from crush only and the other represents X time points after regeneration.
2. Next, the images should be converted into grayscale and resized to render them identical in size.
3. The pixel-wise absolute value difference for the two images is calculated, and,
4. Logarithmic difference matrix is computed.

An interactive implementation for this method is found at the following GitHub link: https:/github.com/leilaabdel/optic-nerve-regen

Alternatively, summing the absolute difference, that is, computing the Manhattan Norm allows to enter it into the equation shown in Fig. 2B. The F_{Regen} value multiplied by W_s and L_s enables approximate estimation of regenerated fiber volume. Measurement of these parameters with R_c will provide regeneration volume from optic nerve head to the optic chiasm and compare approximate regeneration across the models.

is to enable calibrated deflection at the chiasm so that ipsi- and contra-lateral each can receive input from them (Fig. 4). Several biological processes are associated with regeneration and restoration of function. Not all biological functions and their nuances are even well understood currently. However, insight for some important biological processes, as noted below, can be derived:

(a) *Axon regeneration of truncated neurons involves membrane expansion.* It is the specialized plasma membrane of neurons that enables them to generate and propagate ion currents. Plasma membrane of the neurons can reach a surface area of millions of square micrometers for retinal ganglion cells with long axons (Pfenninger, 2009; Pfenninger et al., 2003). Distal tips of severed neurites continue to grow for several hours suggesting the availability of membrane expansion machinery for these tips (Hughes, 1953). Local protein synthesis in dendrites (Antar, Li, Zhang, Carroll, & Bassell, 2006; Eom, Antar, Singer, & Bassell, 2003; Zhang, Singer, & Bassell, 1999) and crosstalk with mitochondria (Rossoll &

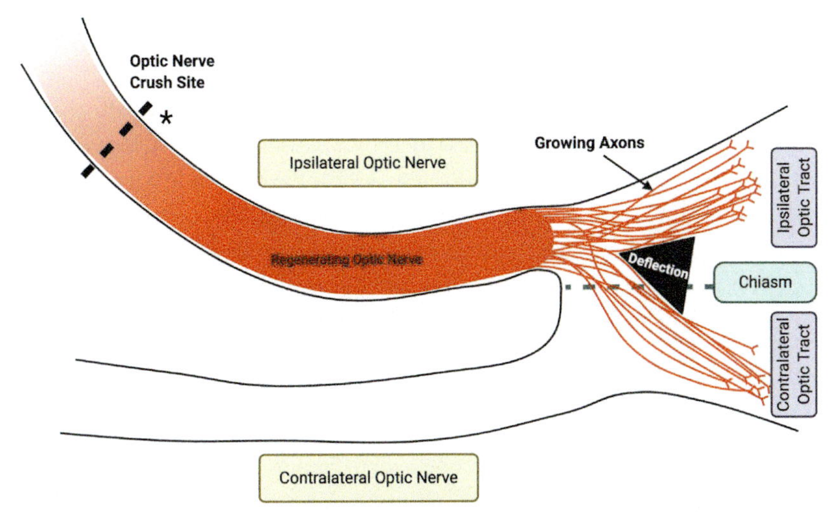

Fig. 4 Schematic diagram of the desired regenerative growth in the optic nerve. An optic nerve subjected to crush is shown in red. An asterisk indicates the site of the crush. A small but significant percentage of the regenerated neurons are needed to be deflected to go toward the contralateral tract as indicated. The red fibers represent regenerating neurons. This schematic is based on actual growth demonstrated by light sheet microscopy in Belin et al. (2015).

Bassell, 2019) have been demonstrated. However, the generation of lipids and membrane expansion for neuronal axons for regeneration is still unclear. Thus far perikaryon of the neuron is thought to be the primary site of synthesis of large molecules. Perikaryon is rich in the rough endoplasmic reticulum and Golgi complexes. These are the sites of synthesis of membrane proteins and lipid synthesis. Neuronal perikaryon synthesizes precursors of multiple vesicular components needed for membrane biogenesis. These generated components are exported to axons and dendrites for membrane expansion. These are therefore often referred to as plasmalemmal precursor vesicles or PPVs (Bunge, 1973; Lockerbie, Miller, & Pfenninger, 1991; Pfenninger & Friedman, 1993). Different types of PPVs have been reported. At least three different types of PPVs carrying membrane proteins and lipids are transported in the axon by different microtubule vesicle motors of the kinesin family directed to specific targets (Pfenninger et al., 2003). Available current evidence suggests that besides perikaryon protein synthesis, lipid remodeling and lipid synthesis occur both in the axons and dendrites. As noted above, transport of mRNA and local protein synthesis occurs at dendrites

(Eom et al., 2003; Rossoll & Bassell, 2019). The membrane extension in neurons involves regulated exocytosis. The specific PPVs cluster at specific growth cone or membrane insertion sites consistent with that the axons' distal sites have regulated membrane insertion involving regulated exocytosis (Chieregatti & Meldolesi, 2005). Exocytosis involves two distinct steps: (1) docking of vesicles or PPVs and (2) vesicle fusion. Current proteomic approaches combined with new tools such as PROTAC (PROteolysis TArgeting Chimera) (Cromm & Crews, 2017; Cyrus et al., 2011) are poised to provide new insights into these processes. Axonal microtubules are oriented uniformly such that their plus end is pointed toward growth cones. In contrast, the dendrites may have a random orientation (Baas, Black, & Banker, 1989; Deitch & Banker, 1993). Several proteins have been identified as axonal markers, such as GAP43, l1, IGF1R, TAU1 (Burack, Silverman, & Banker, 2000; Goslin & Banker, 1990; Tucker & Matus, 1988). Overall proteome analysis of regenerating neurons from different systems (mice, rats, zebrafish, frog) will help identify an expanded list of axon membrane expansion proteins. The incorporation of specialized fractionation approaches will enable identifying proteins involved with different biological processes of axon regeneration. Other outcomes include an expansion in regeneration promoted by different pharmacological or biophysical treatments such as optogenetic light and/or electrical stimulation (Goldberg et al., 2002).

(b) *The processes leading to the formation of growth cones in the injured neurons after crush.* Axons like dendrites also carry localized mRNA and are capable of performing local protein synthesis. Regulation of protein synthesis from localized mRNA in axons by external stimuli is vital for facilitating neuronal regeneration (Willis & Twiss, 2006). Using dorsal root ganglion cell neurons and retinal explants, it has been shown that protein synthesis and degradation underlie growth cone formation after axotomy (Verma et al., 2005). However, a direct demonstration of sealing and growth cone on truncated neurons in-situ remains to be done (Fig. 5). In addition to protein synthesis, both caspase-dependent and independent pathways are involved in growth cone formation and axon regeneration after axotomy (Verma et al., 2005; Willis & Twiss, 2006). The axons' ability to regenerate their growth cones depends on axonal type and age, and several independent signaling pathways and factors (e.g., calcium, cyclic AMP) (Chierzi, Ratto, Verma, & Fawcett, 2005) influence regeneration. Axon transport pathways have been

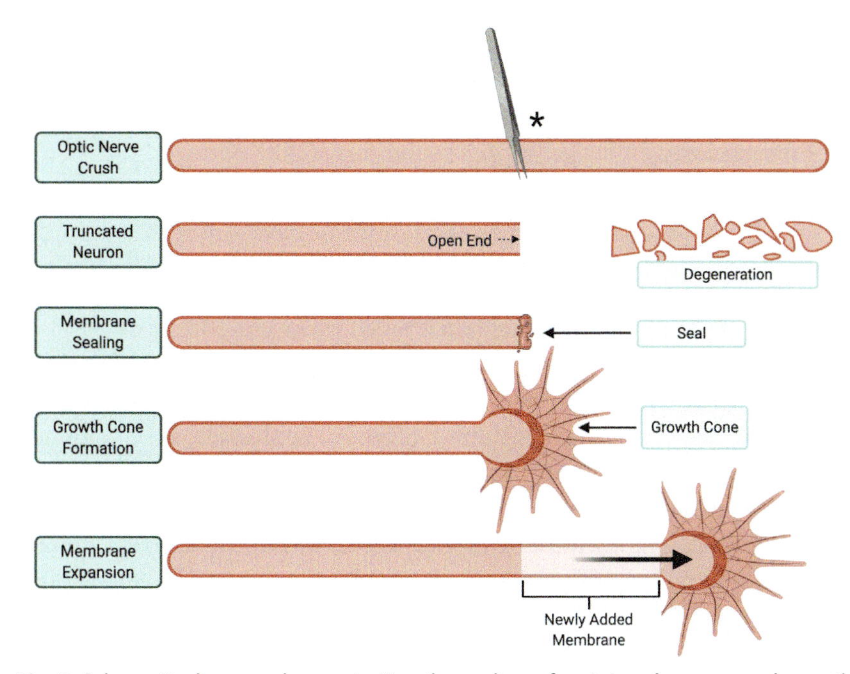

Fig. 5 Schematic diagram demonstrating the sealing of an injured neuron and growth cone formation for expansion. The injury to neurons whether natural or experimentally induced is represented by a pair of forceps. The site of injury is indicated by an asterisk. The injury follows the sealing of the injured end of the neuron and subsequent growth cone formation. The growth cone promotes membrane expansion. These steps and protein players currently lack details within the literature.

shown to converge on local protein synthesis during development (Li et al., 2004), and it is likely to occur in regenerating axons for long-distance growth. Despite substantial advancement, the processes that lead to the formation of growth cone in injured neurons remains poorly understood (Blanquie & Bradke, 2018; Fawcett & Verhaagen, 2018). Proteomic analysis with other omics analyses may provide insight into the pathways involved in growth cone formation and the processes that lead up to that point.

(c) *Process of developing myelination, myelination patterns.* The retinal ganglion cells are usually myelinated after entering the optic nerve, at the lamina cribrosa region. The normal myelination after regeneration of mammalian RGCs has recently received renewed interest (Franklin, Frisen, & Lyons, 2020).

(d) *Connectivity with non-neuronal cells*. Often this connectivity involves lateral connectivity with oligodendrocytes and other astroglia cells. These connectivities are essential in addition to more widely studies synaptic connectivity.

5. Conclusion

Major advancement in neuronal regeneration particularly regeneration of neurons in the optic nerve has taken place but a lot remains to be done. Proteomic analyses of regeneration promoted by different pharmacological or biophysical methods will provide greater insights. Comparison of proteomics with other omics (transcriptomics, lipidomics, metabolomics) will help identify convergence of different pathways important for regeneration. Comparison of proteomics of high (e.g., zebra fish) and low regenerative organisms (e.g., mouse) will enable comparison of different protein pathways and understand comparative system differences. The standardization of structural regeneration remains a critical hurdle for comparison of different regeneration across a single model system or across different model systems. These advances will provide greater insights and help comparison across different treatments within a single model system or regeneration across different model systems.

Acknowledgments

This work was supported by NIH grants: U01EY027257, P30EY014801, US Department of Defense grant W81XWH1910845 and an unrestricted grant from Research to Prevent Blindness. We thank all authors in this field whose work was cited and those we were unable to acknowledge in this review. We have refrained from compartment level proteomics in this area due to its very emergent and incomplete nature.

References

Aguayo, A. J., Bray, G. M., Rasminsky, M., Zwimpfer, T., Carter, D., & Vidal-Sanz, M. (1990). Synaptic connections made by axons regenerating in the central nervous system of adult mammals. *The Journal of Experimental Biology*, *153*, 199–224.

Aguayo, A. J., Rasminsky, M., Bray, G. M., Carbonetto, S., McKerracher, L., Villegas-Perez, M. P., et al. (1991). Degenerative and regenerative responses of injured neurons in the central nervous system of adult mammals. *Philosophical Transactions of the Royal Society of London. Series B, Biological Sciences*, *331*(1261), 337–343.

Antar, L. N., Li, C., Zhang, H., Carroll, R. C., & Bassell, G. J. (2006). Local functions for FMRP in axon growth cone motility and activity-dependent regulation of filopodia and spine synapses. *Molecular and Cellular Neurosciences*, *32*(1–2), 37–48.

Arnemann, J. (1786). *Ueber die Reproduktion der Nerven*. Göttingen: Göttingen, J.C. Dieterich.

Arnemann, J. (1787). *Versuche über die Regeneration an lebenden Thieren/1 Über die Regeneration der Nerven: Mit IV Kupfertafeln*. Göttingen: Göttingen: Dieterich.

Baas, P. W., Black, M. M., & Banker, G. A. (1989). Changes in microtubule polarity orientation during the development of hippocampal neurons in culture. *The Journal of Cell Biology*, *109*(6 Pt. 1), 3085–3094.

Barateiro, A., Brites, D., & Fernandes, A. (2016). Oligodendrocyte development and myelination in neurodevelopment: Molecular mechanisms in health and disease. *Current Pharmaceutical Design*, *22*(6), 656–679.

Belin, S., Nawabi, H., Wang, C., Tang, S., Latremoliere, A., Warren, P., et al. (2015). Injury-induced decline of intrinsic regenerative ability revealed by quantitative proteomics. *Neuron*, *86*(4), 1000–1014.

Belrose, J. L., Prasad, A., Sammons, M. A., Gibbs, K. M., & Szaro, B. G. (2020). Comparative gene expression profiling between optic nerve and spinal cord injury in *Xenopus laevis* reveals a core set of genes inherent in successful regeneration of vertebrate central nervous system axons. *BMC Genomics*, *21*(1), 540.

Benowitz, L. I., He, Z., & Goldberg, J. L. (2017). Reaching the brain: Advances in optic nerve regeneration. *Experimental Neurology*, *287*(Pt. 3), 365–373.

Blanquie, O., & Bradke, F. (2018). Cytoskeleton dynamics in axon regeneration. *Current Opinion in Neurobiology*, *51*, 60–69.

Bunge, M. B. (1973). Fine structure of nerve fibers and growth cones of isolated sympathetic neurons in culture. *The Journal of Cell Biology*, *56*(3), 713–735.

Bunge, M. B. (2016). Efficacy of Schwann cell transplantation for spinal cord repair is improved with combinatorial strategies. *The Journal of Physiology*, *594*(13), 3533–3538.

Burack, M. A., Silverman, M. A., & Banker, G. (2000). The role of selective transport in neuronal protein sorting. *Neuron*, *26*(2), 465–472.

Calkins, D. J., Pekny, M., Cooper, M. L., Benowitz, L., Initiative, L. I., & Participants, G. N. (2017). The challenge of regenerative therapies for the optic nerve in glaucoma. *Experimental Eye Research*, *157*, 28–33.

Chauhan, M. Z., Arcuri, J., Park, K. K., Zafar, M. K., Fatmi, R., Hackam, A. S., et al. (2020). Multi-omic analyses of growth cones at different developmental stages provides insight into pathways in adult neuroregeneration. *iScience*, *23*(2), 100836.

Chauhan, M. Z., & Bhattacharya, S. K. (2021). Multi-omics insights into neuronal regeneration and re-innervation. *Neural Regeneration Research*, *16*(2), 296–297.

Chieregatti, E., & Meldolesi, J. (2005). Regulated exocytosis: New organelles for non-secretory purposes. *Nature Reviews. Molecular Cell Biology*, *6*(2), 181–187.

Chierzi, S., Ratto, G. M., Verma, P., & Fawcett, J. W. (2005). The ability of axons to regenerate their growth cones depends on axonal type and age, and is regulated by calcium, cAMP and ERK. *The European Journal of Neuroscience*, *21*(8), 2051–2062.

Cho, E. Y., & So, K. F. (1989). De novo formation of axon-like processes from axotomized retinal ganglion cells which exhibit long distance growth in a peripheral nerve graft in adult hamsters. *Brain Research*, *484*(1–2), 371–377.

Corredor, R. G., & Goldberg, J. L. (2009). Electrical activity enhances neuronal survival and regeneration. *Journal of Neural Engineering*, *6*(5), 055001.

Cromm, P. M., & Crews, C. M. (2017). Targeted protein degradation: From chemical biology to drug discovery. *Cell Chemical Biology*, *24*(9), 1181–1190.

Cruikshank, W. E. (1797). Experiments on the nerves, particularly on their reproduction; and on the spinal marrow of living animals: From the same work. *Medical Facts and Observations*, 7, 136–154.

Cui, Q., So, K. F., & Yip, H. K. (1998). Major biological effects of neurotrophic factors on retinal ganglion cells in mammals. *Biological Signals and Receptors*, 7(4), 220–226.

Cyrus, K., Wehenkel, M., Choi, E. Y., Han, H. J., Lee, H., Swanson, H., et al. (2011). Impact of linker length on the activity of PROTACs. *Molecular BioSystems*, 7(2), 359–364.

David, S., & Aguayo, A. J. (1981). Axonal elongation into peripheral nervous system "bridges" after central nervous system injury in adult rats. *Science*, *214*(4523), 931–933.

David, S., & Aguayo, A. J. (1985). Axonal regeneration after crush injury of rat central nervous system fibres innervating peripheral nerve grafts. *Journal of Neurocytology*, *14*(1), 1–12.

Deitch, J. S., & Banker, G. A. (1993). An electron microscopic analysis of hippocampal neurons developing in culture: Early stages in the emergence of polarity. *The Journal of Neuroscience*, *13*(10), 4301–4315.

Diekmann, H., Kalbhen, P., & Fischer, D. (2015). Characterization of optic nerve regeneration using transgenic zebrafish. *Frontiers in Cellular Neuroscience*, *9*, 118.

Domingues, H. S., Cruz, A., Chan, J. R., Relvas, J. B., Rubinstein, B., & Pinto, I. M. (2018). Mechanical plasticity during oligodendrocyte differentiation and myelination. *Glia*, *66*(1), 5–14.

Ebrahimi, M., Ai, J., Biazar, E., Ebrahimi-Barough, S., Khojasteh, A., Yazdankhah, M., et al. (2018). In vivo assessment of a nanofibrous silk tube as nerve guide for sciatic nerve regeneration. *Artificial Cells, Nanomedicine, and Biotechnology*, *46*(Suppl. 1), 394–401.

Eom, T., Antar, L. N., Singer, R. H., & Bassell, G. J. (2003). Localization of a beta-actin messenger ribonucleoprotein complex with zipcode-binding protein modulates the density of dendritic filopodia and filopodial synapses. *The Journal of Neuroscience*, *23*(32), 10433–10444.

Estrada-Bernal, A., Sanford, S. D., Sosa, L. J., Simon, G. C., Hansen, K. C., & Pfenninger, K. H. (2012). Functional complexity of the axonal growth cone: A proteomic analysis. *PLoS One*, *7*(2), e31858.

Fawcett, J. W., & Verhaagen, J. (2018). Intrinsic determinants of axon regeneration. *Developmental Neurobiology*, *78*(10), 890–897.

Franklin, R. J. M., Frisen, J., & Lyons, D. A. (2020). Revisiting remyelination: Towards a consensus on the regeneration of CNS myelin. *Seminars in Cell & Developmental Biology*, *S1084-9521*, 30157–30159.

Fry, E. J., Chagnon, M. J., Lopez-Vales, R., Tremblay, M. L., & David, S. (2010). Corticospinal tract regeneration after spinal cord injury in receptor protein tyrosine phosphatase sigma deficient mice. *Glia*, *58*(4), 423–433.

Gamkrelidze, G. N., Laurienti, P. J., & Blankenship, J. E. (1995). Identification and characterization of cerebral ganglion neurons that induce swimming and modulate swim-related pedal ganglion neurons in *Aplysia brasiliana*. *Journal of Neurophysiology*, *74*(4), 1444–1462.

Glanzer, J. G., Enose, Y., Wang, T., Kadiu, I., Gong, N., Rozek, W., et al. (2007). Genomic and proteomic microglial profiling: Pathways for neuroprotective inflammatory responses following nerve fragment clearance and activation. *Journal of Neurochemistry*, *102*(3), 627–645.

Goldberg, J. L., Espinosa, J. S., Xu, Y., Davidson, N., Kovacs, G. T., & Barres, B. A. (2002). Retinal ganglion cells do not extend axons by default: Promotion by neurotrophic signaling and electrical activity. *Neuron*, *33*(5), 689–702.

Goldberg, J. L., Guido, W., & Agi Workshop, P. (2016). Report on the national eye institute audacious goals initiative: Regenerating the optic nerve. *Investigative Ophthalmology & Visual Science*, *57*(3), 1271–1275.

Gordon, T., Udina, E., Verge, V. M., & de Chaves, E. I. (2009). Brief electrical stimulation accelerates axon regeneration in the peripheral nervous system and promotes sensory axon regeneration in the central nervous system. *Motor Control*, *13*(4), 412–441.

Goslin, K., & Banker, G. (1990). Rapid changes in the distribution of GAP-43 correlate with the expression of neuronal polarity during normal development and under experimental conditions. *The Journal of Cell Biology*, *110*(4), 1319–1331.

Hartley, H. (1951). Origin of the word 'protein'. *Nature*, *168*(4267), 244.

Harvey, B. M., Baxter, M., & Granato, M. (2019). Optic nerve regeneration in larval zebrafish exhibits spontaneous capacity for retinotopic but not tectum specific axon targeting. *PLoS One*, *14*(6), e0218667.

Holmes, W. (1951). The repair of nerves by suture. *Journal of the History of Medicine and Allied Sciences*, *6*(1), 44–63.

Holmes, W., & Young, J. Z. (1942). Nerve regeneration after immediate and delayed suture. *Journal of Anatomy*, 77(Pt. 1), 63–96.10.

Howell, W. H., & Huber, G. C. (1892). A physiological, histological and clinical study of the degeneration and regeneration in peripheral nerve fibres after severance of their connections with the nerve centres. *The Journal of Physiology*, *13*(5), 335–406.311.

Hu, Y., Arulpragasam, A., Plant, G. W., Hendriks, W. T., Cui, Q., & Harvey, A. R. (2007). The importance of transgene and cell type on the regeneration of adult retinal ganglion cell axons within reconstituted bridging grafts. *Experimental Neurology*, *207*(2), 314–328.

Hughes, A. (1953). The growth of embryonic neurites; a study of cultures of chick neural tissues. *Journal of Anatomy*, *87*(2), 150–162.

Igarashi, M. (2014). Proteomic identification of the molecular basis of mammalian CNS growth cones. *Neuroscience Research*, *88*, 1–15.

Igarashi, M., Kawasaki, A., Ishikawa, Y., Honda, A., Okada, M., & Okuda, S. (2020). Phosphoproteomic and bioinformatic methods for analyzing signaling in vertebrate axon growth and regeneration. *Journal of Neuroscience Methods*, *339*, 108723.

Koh, K., Park, M., Bae, E. S., Duong, V. A., Park, J. M., Lee, H., et al. (2020). UBA2 activates Wnt/beta-catenin signaling pathway during protection of R28 retinal precursor cells from hypoxia by extracellular vesicles derived from placental mesenchymal stem cells. *Stem Cell Research & Therapy*, *11*(1), 428.

Kohen, R., & Giger, R. J. (2020). Greasing the wheels of regeneration. *Neuron*, *105*(2), 207–209.

Kurimoto, T., Yin, Y., Omura, K., Gilbert, H. Y., Kim, D., Cen, L. P., et al. (2010). Long-distance axon regeneration in the mature optic nerve: Contributions of oncomodulin, cAMP, and pten gene deletion. *The Journal of Neuroscience*, *30*(46), 15654–15663.

Larrivee, D. C., & Grafstein, B. (1987). Phosphorylation of proteins in normal and regenerating goldfish optic nerve. *Journal of Neurochemistry*, *49*(6), 1747–1757.

Lasseck, J., Schroer, U., Koenig, S., & Thanos, S. (2007). Regeneration of retinal ganglion cell axons in organ culture is increased in rats with hereditary buphthalmos. *Experimental Eye Research*, *85*(1), 90–104.

Lauzi, J., Anders, F., Liu, H., Pfeiffer, N., Grus, F., Thanos, S., et al. (2019). Neuroprotective and neuroregenerative effects of CRMP-5 on retinal ganglion cells in an experimental in vivo and in vitro model of glaucoma. *PLoS One*, *14*(1), e0207190.

Lee-Liu, D., Sun, L., Dovichi, N. J., & Larrain, J. (2018). Quantitative proteomics after spinal cord injury (SCI) in a regenerative and a nonregenerative stage in the frog *Xenopus laevis*. *Molecular & Cellular Proteomics*, *17*(4), 592–606.

Li, C., Sasaki, Y., Takei, K., Yamamoto, H., Shouji, M., Sugiyama, Y., et al. (2004). Correlation between semaphorin3A-induced facilitation of axonal transport and local activation of a translation initiation factor eukaryotic translation initiation factor 4E. *The Journal of Neuroscience*, *24*(27), 6161–6170.

Li, J., Steen, H., & Gygi, S. P. (2003). Protein profiling with cleavable isotope-coded affinity tag (cICAT) reagents: The yeast salinity stress response. *Molecular & Cellular Proteomics*, *2*(11), 1198–1204.

Lockerbie, R. O., Miller, V. E., & Pfenninger, K. H. (1991). Regulated plasmalemmal expansion in nerve growth cones. *The Journal of Cell Biology*, *112*(6), 1215–1227.

Ma, C., Gao, Y., Chai, G., Su, H., Wang, N., Yang, Y., et al. (2010). Djrho2 is involved in regeneration of visual nerves in Dugesia japonica. *Journal of Genetics and Genomics*, *37*(11), 713–723.

Magharious, M., D'Onofrio, P. M., Hollander, A., Zhu, P., Chen, J., & Koeberle, P. D. (2011). Quantitative iTRAQ analysis of retinal ganglion cell degeneration after optic nerve crush. *Journal of Proteome Research*, *10*(8), 3344–3362.

Miyake, K., Yoshida, M., Inoue, Y., & Hata, Y. (2007). Neuroprotective effect of transcorneal electrical stimulation on the acute phase of optic nerve injury. *Investigative Ophthalmology & Visual Science, 48*(5), 2356–2361.

Miyata, S. (2019). Cytoskeletal signal-regulated oligodendrocyte myelination and remyelination. *Advances in Experimental Medicine and Biology, 1190*, 33–42.

Mulder, G. J. (1838). Sur la composition de quelques substances animales. *Bulletin des Sciences Physiques et Naturelles en Néerlande, 104*, 9.

Muller, H. W., Gebicke-Harter, P. J., Hangen, D. H., & Shooter, E. M. (1985). A specific 37,000-dalton protein that accumulates in regenerating but not in nonregenerating mammalian nerves. *Science, 228*(4698), 499–501.

Nozumi, M., Togano, T., Takahashi-Niki, K., Lu, J., Honda, A., Taoka, M., et al. (2009). Identification of functional marker proteins in the mammalian growth cone. *Proceedings of the National Academy of Sciences of the United States of America, 106*(40), 17211–17216.

Pandey, A., & Mann, M. (2000). Proteomics to study genes and genomes. *Nature, 405*(6788), 837–846.

Park, S., Koppes, R. A., Froriep, U. P., Jia, X., Achyuta, A. K., McLaughlin, B. L., et al. (2015). Optogenetic control of nerve growth. *Scientific Reports, 5*, 9669.

Park, K. K., Liu, K., Hu, Y., Kanter, J. L., & He, Z. (2010). PTEN/mTOR and axon regeneration. *Experimental Neurology, 223*(1), 45–50.

Park, K. K., Liu, K., Hu, Y., Smith, P. D., Wang, C., Cai, B., et al. (2008). Promoting axon regeneration in the adult CNS by modulation of the PTEN/mTOR pathway. *Science, 322*(5903), 963–966.

Patel, A. K., Park, K. K., & Hackam, A. S. (2017). Wnt signaling promotes axonal regeneration following optic nerve injury in the mouse. *Neuroscience, 343*, 372–383.

Perry, G. W., Burmeister, D. W., & Grafstein, B. (1985). Changes in protein content of goldfish optic nerve during degeneration and regeneration following nerve crush. *Journal of Neurochemistry, 44*(4), 1142–1151.

Pfenninger, K. H. (2009). Plasma membrane expansion: A neuron's Herculean task. *Nature Reviews. Neuroscience, 10*(4), 251–261.

Pfenninger, K. H., & Friedman, L. B. (1993). Sites of plasmalemmal expansion in growth cones. *Brain Research. Developmental Brain Research, 71*(2), 181–192.

Pfenninger, K. H., Laurino, L., Peretti, D., Wang, X., Rosso, S., Morfini, G., et al. (2003). Regulation of membrane expansion at the nerve growth cone. *Journal of Cell Science, 116*(Pt. 7), 1209–1217.

Prokosch, V., Chiwitt, C., Rose, K., & Thanos, S. (2010). Deciphering proteins and their functions in the regenerating retina. *Expert Review of Proteomics, 7*(5), 775–795.

Quitschke, W., & Schechter, N. (1983a). In vitro protein synthesis in the goldfish retinotectal pathway during regeneration: Evidence for specific axonal proteins of retinal origin in the optic nerve. *Journal of Neurochemistry, 41*(4), 1137–1142.

Quitschke, W., & Schechter, N. (1983b). Specific optic nerve proteins during regeneration of the goldfish retinotectal pathway. *Brain Research, 258*(1), 69–78.

Rabinowitz, J. S., Robitaille, A. M., Wang, Y., Ray, C. A., Thummel, R., Gu, H., et al. (2017). Transcriptomic, proteomic, and metabolomic landscape of positional memory in the caudal fin of zebrafish. *Proceedings of the National Academy of Sciences of the United States of America, 114*(5), E717–E726.

Rose, K., Schroer, U., Volk, G. F., Schlatt, S., Konig, S., Feigenspan, A., et al. (2008). Axonal regeneration in the organotypically cultured monkey retina: Biological aspects, dependence on substrates and age-related proteomic profiling. *Restorative Neurology and Neuroscience, 26*(4–5), 249–266.

Ross, P. L., Huang, Y. N., Marchese, J. N., Williamson, B., Parker, K., Hattan, S., et al. (2004). Multiplexed protein quantitation in *Saccharomyces cerevisiae* using amine-reactive isobaric tagging reagents. *Molecular & Cellular Proteomics, 3*(12), 1154–1169.

Rossoll, W., & Bassell, G. J. (2019). Crosstalk of local translation and mitochondria: Powering plasticity in axons and dendrites. *Neuron, 101*(2), 204–206.

Schroer, U., Volk, G. F., Liedtke, T., & Thanos, S. (2007). Translin-associated factor-X (Trax) is a molecular switch of growth-associated protein (GAP)-43 that controls axonal regeneration. *The European Journal of Neuroscience, 26*(8), 2169–2178.

Schwab, M. E. (1996). Structural plasticity of the adult CNS. Negative control by neurite growth inhibitory signals. *International Journal of Developmental Neuroscience, 14*(4), 379–385.

Simons, M., & Nave, K. A. (2015). Oligodendrocytes: Myelination and axonal support. *Cold Spring Harbor Perspectives in Biology, 8*(1), a020479.

Smedowski, A., Liu, X., Pietrucha-Dutczak, M., Matuszek, I., Varjosalo, M., & Lewin-Kowalik, J. (2016). Predegenerated Schwann cells—A novel prospect for cell therapy for glaucoma: Neuroprotection, neuroregeneration and neuroplasticity. *Scientific Reports, 6*, 23187.

Stelzner, D. J., Bohn, R. C., & Strauss, J. A. (1986). Regeneration of the frog optic nerve. Comparisons with development. *Neurochemical Pathology, 5*(3), 255–288.

Sun, L., Shay, J., McLoed, M., Roodhouse, K., Chung, S. H., Clark, C. M., et al. (2014). Neuronal regeneration in *C. elegans* requires subcellular calcium release by ryanodine receptor channels and can be enhanced by optogenetic stimulation. *The Journal of Neuroscience, 34*(48), 15947–15956.

Tan, H., Kang, X., Zhong, Y., Shen, X., Cheng, Y., Jiao, Q., et al. (2012). Erythropoietin upregulates growth associated protein-43 expression and promotes retinal ganglion cell axonal regeneration in vivo after optic nerve crush. *Neural Regeneration Research, 7*(4), 295–301.

Tezel, G. (2016). Applying proteomics to research for optic nerve regeneration in glaucoma: What's on the horizon? *Expert Review of Proteomics, 13*(11), 979–981.

Thanos, S., Bohm, M. R., Schallenberg, M., & Oellers, P. (2012). Traumatology of the optic nerve and contribution of crystallins to axonal regeneration. *Cell and Tissue Research, 349*(1), 49–69.

Tokuda, K., Kuramitsu, Y., Byron, B., Kitagawa, T., Tokuda, N., Kobayashi, D., et al. (2015). Up-regulation of DRP-3 long isoform during the induction of neural progenitor cells by glutamate treatment in the ex vivo rat retina. *Biochemical and Biophysical Research Communications, 463*(4), 593–599.

Tucker, R. P., & Matus, A. I. (1988). Microtubule-associated proteins characteristic of embryonic brain are found in the adult mammalian retina. *Developmental Biology, 130*(2), 423–434.

Veldman, M. B., Bemben, M. A., & Goldman, D. (2010). Tuba1a gene expression is regulated by KLF6/7 and is necessary for CNS development and regeneration in zebrafish. *Molecular and Cellular Neurosciences, 43*(4), 370–383.

Veldman, M. B., Bemben, M. A., Thompson, R. C., & Goldman, D. (2007). Gene expression analysis of zebrafish retinal ganglion cells during optic nerve regeneration identifies KLF6a and KLF7a as important regulators of axon regeneration. *Developmental Biology, 312*(2), 596–612.

Verma, P., Chierzi, S., Codd, A. M., Campbell, D. S., Meyer, R. L., Holt, C. E., et al. (2005). Axonal protein synthesis and degradation are necessary for efficient growth cone regeneration. *The Journal of Neuroscience, 25*(2), 331–342.

Wei, J., Jiang, H., Gao, H., & Wang, G. (2015). Raf-1 kinase inhibitory protein (RKIP) promotes retinal ganglion cell survival and axonal regeneration following optic nerve crush. *Journal of Molecular Neuroscience, 57*(2), 243–248.

Whitworth, G. B., Misaghi, B. C., Rosenthal, D. M., Mills, E. A., Heinen, D. J., Watson, A. H., et al. (2017). Translational profiling of retinal ganglion cell optic nerve regeneration in *Xenopus laevis*. *Developmental Biology, 426*(2), 360–373.

Willis, D. E., & Twiss, J. L. (2006). The evolving roles of axonally synthesized proteins in regeneration. *Current Opinion in Neurobiology*, *16*(1), 111–118.

Wood, P., Moya, F., Eldridge, C., Owens, G., Ranscht, B., Schachner, M., et al. (1990). Studies of the initiation of myelination by Schwann cells. *Annals of the New York Academy of Sciences*, *605*, 1–14.

Yamamoto, T., Nishimura, Y., Matsuura, T., Shibuya, H., Lin, M., & Asahara, T. (2004). Cerebellar activation of cortical motor regions: Comparisons across mammals. *Progress in Brain Research*, *143*, 309–317.

Yang, C., Wang, X., Wang, J., Wang, X., Chen, W., Lu, N., et al. (2020). Rewiring neuronal glycerolipid metabolism determines the extent of axon regeneration. *Neuron*, *105*(2), 276–292.

Yin, Y., Cui, Q., Li, Y., Irwin, N., Fischer, D., Harvey, A. R., et al. (2003). Macrophage-derived factors stimulate optic nerve regeneration. *The Journal of Neuroscience*, *23*(6), 2284–2293.

Zammit, P. S., Clarke, J. D., Golding, J. P., Goodbrand, I. A., & Tonge, D. A. (1993). Macrophage response during axonal regeneration in the axolotl central and peripheral nervous system. *Neuroscience*, *54*(3), 781–789.

Zhang, H. L., Singer, R. H., & Bassell, G. J. (1999). Neurotrophin regulation of beta-actin mRNA and protein localization within growth cones. *The Journal of Cell Biology*, *147*(1), 59–70.

Zhao, Y., Ju, F., Zhao, Y., Wang, L., Sun, Z., Liu, M., et al. (2011). The expression of alphaA- and betaB1-crystallin during normal development and regeneration, and proteomic analysis for the regenerating lens in *Xenopus laevis*. *Molecular Vision*, *17*, 768–778.

Zupanc, G. K. (2006). Neurogenesis and neuronal regeneration in the adult fish brain. *Journal of Comparative Physiology. A, Neuroethology, Sensory, Neural, and Behavioral Physiology*, *192*(6), 649–670.

CHAPTER EIGHT

Proteomics of pseudoexfoliation materials in the anterior eye segment

Jada Morris[a,b,c], Ciara Myer[a,c], Tara Cornet[a,c], Anna K. Junk[a,c,d], Richard K. Lee[a,b,c], and Sanjoy K. Bhattacharya[a,b,c,*]

[a]Bascom Palmer Eye Institute, University of Miami Miller School of Medicine, Miami, FL, United States
[b]Vision Science and Investigative Ophthalmology Graduate Program, University of Miami, Miami, FL, United States
[c]Miami Integrative Metabolomics Research Center, University of Miami, Miami, FL, United States
[d]Miami Veterans Affairs Health Care System, Miami, FL, United States
*Corresponding author: e-mail address: sbhattacharya@med.miami.edu

Contents

Abstract

Pseudoexfoliation syndrome (PEX) is characterized by the production of white extracellular fluffy clumps of microfibrillar material that aggregates in various organs throughout the body but is known to cause disease in the eye. The accumulation of PEX material (PEXM) in the anterior segment ocular structures is believed to cause an increase in intraocular pressure (IOP) resulting in pseudoexfoliation glaucoma (PEXG). The onset of PEXG is often bilateral but asymmetric—one eye often presents with glaucoma prior to the other eye. Proteomics has been used to identify key proteins involved in PEXM formation with the end goal of developing effective treatments for PEX and PEXG which may act through inhibiting the formation of the PEX aggregates. To date, a variety of proteins with various molecular functions have been identified from extracted anterior segment structures and fluids, such as aqueous humor (AH) and blood serum of patients affected by PEX. From past studies, some proteins identified in AH, lens capsule epithelium, iris tissue, and blood serum samples include vitamin D binding protein (GC), apolipoprotein A4 (APOA4), lysyl oxidase like-1 (LOXL1), complement C3, beta-crystalline

Advances in Protein Chemistry and Structural Biology, Volume 127
ISSN 1876-1623
https://doi.org/10.1016/bs.apcsb.2021.03.004

B1, and B2, and antithrombin-III (SERPINC1). Each of these proteins have been observed in eyes with PEX at varying levels within the different eye structures. In this review, we further examine the anterior segment ocular proteomics of PEXM from past studies to better understand the mechanism of PEX and PEXG development. Both genetic and environmental risk factors have been implicated to be involved in the development of PEX and PEXG. This field is at an early stage of investigation identifying how these factors modify proteins both at the expression and functional level to cause changes leading to the pathophysiology of PEX glaucoma.

1. Introduction

Pseudoexfoliation syndrome (PEX) is a systemic, age-related disease most prominently manifested as a sight threatening disease associated with the anterior segment of the eye. PEX is characterized by an accumulation of clumps of white extracellular deposits. PEX material (PEXM) presents as white, fluffy proteinaceous deposits on the crystallin lens, zonules, pupillary border of the iris, and other ocular tissues (Myer et al., 2020; Vazquez & Lee, 2014). Aberrant extracellular matrix metabolism is believed to be an underlying mechanism of PEX pathobiology associated with the aggregation of elastic fiber components (Dewundara & Pasquale, 2015). Tissues such as blood vessel walls, the trabecular meshwork (TM), and lamina cribrosa are composed of elastic fibers and are therefore affected by PEX (Vazquez & Lee, 2014). PEX is categorized as a type of elastosis characterized by the production and degeneration of elastic microfibrils and abnormal aggregation of these fibers. The material is believed to originate from zonular fibers and other tissues which consist of elastic microfibrils (Naumann, 1998; Netland, Ye, Streeten, & Hernandez, 1995; Vazquez & Lee, 2014). Additional cell types that contribute to the active production of PEX material include smooth muscle cells, vascular endothelial cells, and melanocytes (Naumann, 1998). PEX was first described by John Gustaf Lindberg over 100 years ago in 1917 (Grzybowski, Kanclerz, & Ritch, 2019). Since then, although there have been many new discoveries concerning the components of PEXM, the exact mechanism and biochemical composition of PEXM remains unknown (Schlötzer-Schrehardt, 2018). Since PEX is an age-related disorder, the prevalence of this condition increases with age (Schlötzer-Schrehardt, 2009). The PEX glaucoma represent only a subset of overall total glaucoma patients. Individuals with PEX may have up to a 15% risk of developing pseudoexfoliation glaucoma (PEXG). In Scandinavian countries, PEXG is the most common form of glaucoma.

In some estimates, in Scandinavian countries, it may occur with an incidence of 25% of all open-angle secondary glaucoma's, making PEXG one of the most common identifiable cause of open-angle secondary glaucoma in those regions (Ritch & Schlötzer-Schrehardt, 2001; Vazquez & Lee, 2014). PEX can present unilaterally or bilaterally. Unilateral involvement is a precursor to bilateral involvement, and patients with bilateral PEX are at risk of developing glaucoma that may initially only affect one eye before developing it in the fellow eye (Ritch & Schlötzer-Schrehardt, 2001).

PEXG is a form of secondary open-angle glaucoma with a causative molecular etiology hypothesized to be PEXM deposit formation that leads to blockage of the aqueous TM anterior chamber drainage system which regulates the intraocular pressure (IOP) (Fig. 1), thereby causing fluctuations in IOP (Lee, 2008). PEXG is diagnosed by slit-lamp examination, whereby PEXM is observed on the lens capsule or pupillary border (Lee, 2008; Ritch & Schlötzer-Schrehardt, 2001). PEXG is caused by PEXM inhibiting

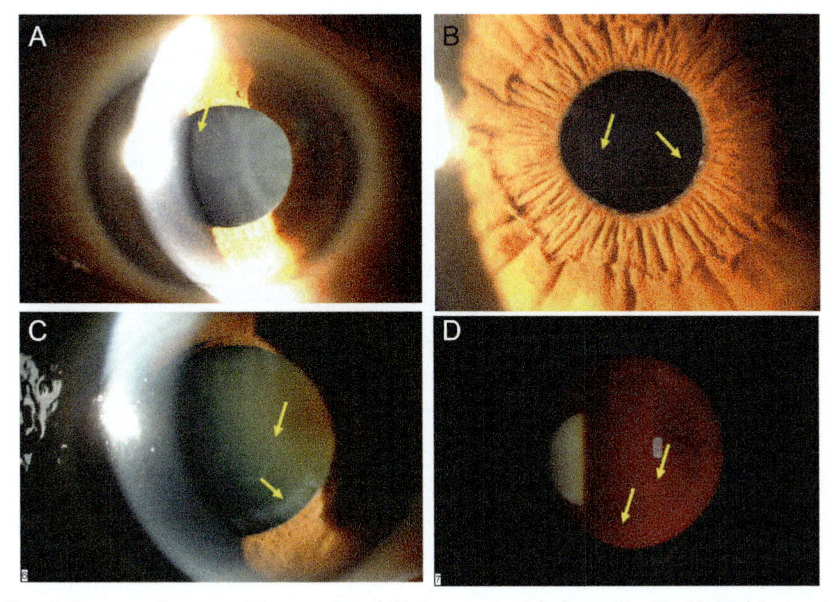

Fig. 1 Image of eyes with pseudoexfoliation material deposits. (A, B) Right eye in 82-year-old White male with PEX in bulls' eye and moth-eaten pupillary border with PEX material on lens capsule and pupillary border. (C) Dense cataract with PXF in bulls' eye in right eye—82-year-old Hispanic woman. (D) Bulls eye of PEXM also seen with slit lamp retro-illumination of the crystalline lens. The visible fluffy PEXM deposit is indicated with an arrow.

the normal outflow of AH, consequently raising the IOP which leads to optic nerve head damage. Maintaining IOP homeostasis is critical to prevent glaucoma development, as IOP is the primary modifiable risk factor affecting optic nerve damage (Acott et al., 2014). The IOP does not have to be high because at any IOP level in eyes with PEX there can be glaucomatous damage (Ritch & Schlötzer-Schrehardt, 2001). For this reason, more studies are needed to pinpoint more reliable risk factors that can be modified to preemptively treat PEXG. Patients with PEX have about a 40% chance of being diagnosed with PEXG initially or developing glaucoma within 10 years; however, some patients may never develop glaucoma (Ritch & Schlötzer-Schrehardt, 2001). Among the types of glaucoma, PEXG is more severe with poorer prognosis, worse visual field damage, a more rapid clinical course, poorer response to glaucoma medications, and higher rates of cataract surgical complications (Palko, Qi, & Sheybani, 2017; Vazquez & Lee, 2014).

PEXM is associated with extracellular degenerative changes in the iris, lens, cornea, and other structures in the eye (Lee, 2008). The proteome, lipidome, and metabolome are altered in the disease state and it is important to identify these differences. Proteomics enables identification of proteins associated with PEX compared to control eyes. Thus, proteomic analyses may help identify differentially expressed proteins that may be potentially associated with the pathogenesis of PEXG. Clinical diagnosis of PEXM via slit-lamp examination on anterior segment ocular structures is sufficient for diagnosing this syndrome. However, uncovering the exact composition of PEXM and determining a definitive molecular pathobiology of the protein interactions may lead to molecular diagnosis and treatment of PEXG.

In this review, we focus on the total protein content of the tissues and fluids in the anterior eye segment and their importance in PEXM formation and PEXG. We categorized proteomic analyses of PEXG in the literature as quantitative, qualitative, or mixed methods based on the methods utilized in reported studies (Fig. 2). Identification of key proteins in the anterior eye segment that are either upregulated or downregulated in this ocular disease is expected to help understand the complex mechanisms involved in the onset and possibly progression of PEXG. Uncovering the proteins that contribute to the formation of PEXM will provide insight into the role that the anterior eye segment structures and their constituent proteins play in the pathogenesis of PEXG. Expanding our current understanding of proteome changes associated with PEXG could lead to the development of molecular therapeutic strategies to prevent progression of this sight threatening disease.

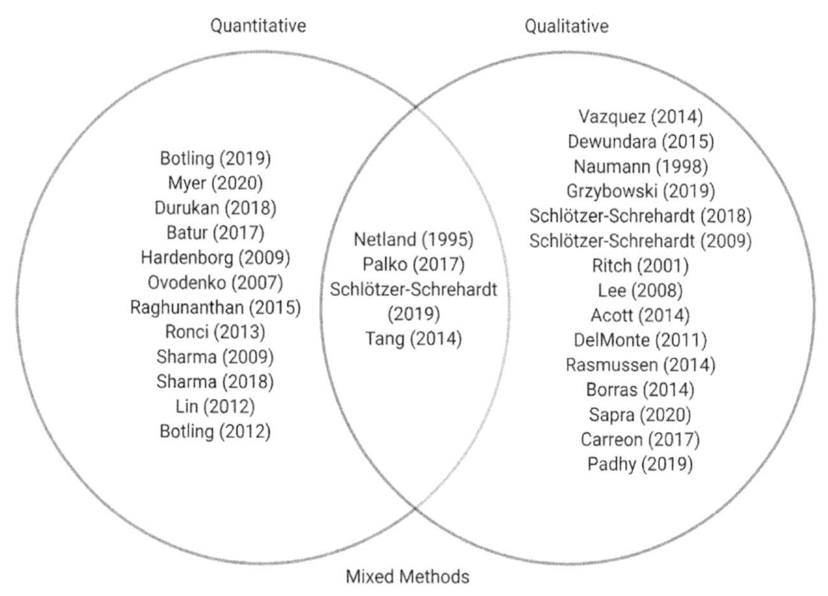

Fig. 2 Venn diagram of the quantitative, qualitative, and mixed methods studies. Each study referenced in the present review is categorized as a quantitative, qualitative, or mixed methods study.

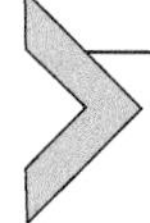

2. Anterior segment anatomy and aqueous outflow in pseudoexfoliation glaucoma

The anterior eye segment consists of the cornea, lens, TM, and iris. A schematic of the eye anterior segment includes the specialized structures of the eye (i.e., cornea, iris, lens, etc.) and the conventional AH outflow pathway (Fig. 3). The cornea is made up of five layers: the epithelium, Bowman's layer, stroma, Descemet's membrane, and the endothelium (DelMonte & Kim, 2011). When afflicted with PEX or PEXG, the corneal endothelium is one of the ocular structures where PEX fibrillar material accumulates, specifically on the corneal endothelium's cellular surface (Palko et al., 2017). Eyes with PEX demonstrate lower corneal endothelium cell density and morphological alterations in cell size and shape (Durukan, 2018). PEXM has also been found along the corneal epithelial basement membrane and corneal stroma (Palko et al., 2017). In a literature review comparing how various corneal structures are altered by PEX and PEXG, lower endothelial cell density was a common finding (Palko et al., 2017), possibly due to damage inflicted by elevated IOP levels and the severity of PEX (Palko et al., 2017).

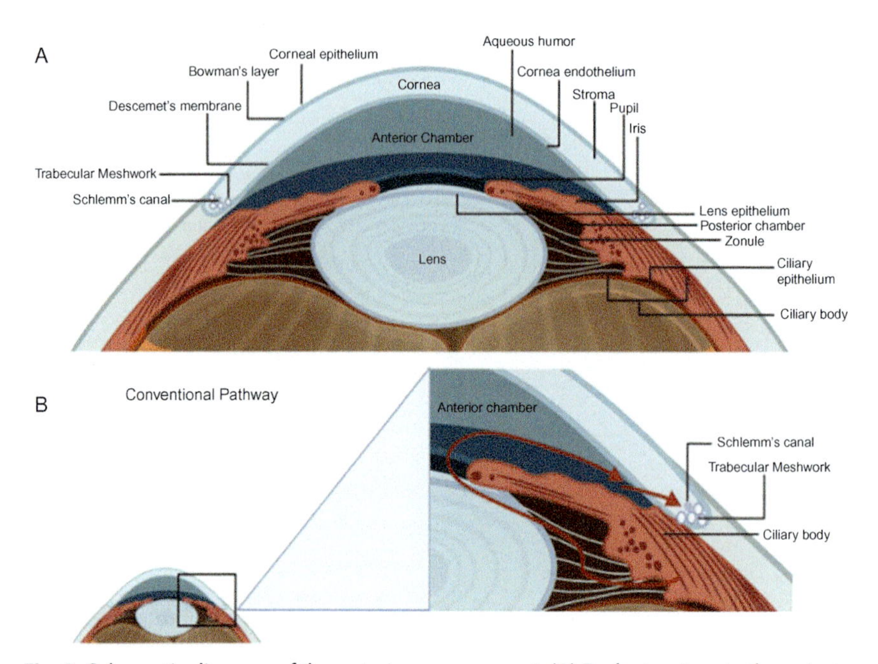

Fig. 3 Schematic diagram of the anterior eye segment. (A) Each structure in the anterior eye segment is identified. (B) The conventional pathway of aqueous humor (AH) flow begins in the ciliary body and travels through the posterior chamber into the anterior chamber. The AH exits through the trabecular meshwork and into Schlemm's canal.

Moving from the cornea deeper into the anterior chamber, the next structure involved in PEXG is the iris. The iris is composed of the iris pigmented epithelium, dilator and sphincter muscles, stroma connective tissue containing melanocytes, and an anterior layer of cells (Borras, 2014). The iris has had the most and earliest interest with hypothesized involvement in the pathogenesis of PEXM (Schlötzer-Schrehardt, 2009). PEXM can accumulate in the iris vessels leading to a reduction in iris-capillary flow in patients affected by PEX and secondary open-angle glaucoma (Naumann, 1998). The anatomic properties of the iris have been visualized by anterior segment optical coherence tomography (OCT) in patients with PEX (Batur, Seven, Tekin, & Yasar, 2017). The mid-peripheral iris thickness is significantly thinner in the PEX group compared to the control non-PEX group (Batur et al., 2017). Thinning in the iris occurs predominantly in the dilator muscles and the pupil dilates poorly in PEX eyes, which could be due to atrophy or degeneration of iris muscle cells and increased iris rigidity (Batur et al., 2017).

The lens is located behind the pupil and iris. The PEX extracellular microfibrillar material has been hypothesized by many to be actively produced by the lens epithelium near the zonular apparatus attachment to the anterior lens capsule (Naumann, 1998). In a lens densitometry study in patients with clinically unilateral PEX (Durukan, 2018), patients with PEXM had higher average lens densitometry measurements compared to controls (Durukan, 2018). Although the sample size of the study was small, their findings provide enough reason to conduct future studies that focus on the effects of PEX on the crystalline lens.

As the AH transits from the anterior chamber, it is filtered through the TM. The TM is composed of three structures: the inner uveal meshwork, the corneoscleral meshwork, and the juxtacanalicular (JCT) or cribriform region (Rasmussen & Kaufman, 2014). The JCT is next to the inner wall of Schlemm's canal (SC) and both regions have the highest AH outflow resistance (Rasmussen & Kaufman, 2014). Aging alters the integrity of the TM thereby leading to its degradation and dysfunction. The extracellular PEXM aggregates beneath the inner wall endothelium of SC, the uveal, and JCT regions, and is hypothesized to interfere with aqueous outflow (Rasmussen & Kaufman, 2014). In the uveal meshwork, PEXM washes into the meshwork from the aqueous but may also be produced by TM structures and accumulate to form a physical resistance (Rasmussen & Kaufman, 2014).

The AH is a clear, viscous fluid that bathes the anterior eye tissues. AH is produced and secreted by the ciliary body into the posterior chamber, then travels to the anterior chamber via the pupil, where it drains through the TM and SC (Fig. 3) (Sunderland & Sapra, 2021). Most AH outflow occurs through the conventional pathway passing through the three TM structures and the SC inner wall endothelium (Acott et al., 2014; Carreon, van der Merwe, Fellman, Johnstone, & Bhattacharya, 2017). Proteinaceous PEXM in the AH is hypothesized to accumulate in the TM as AH exits through its filter-like structure, although TM resistance can also originate from dysfunction of the TM itself (Vazquez & Lee, 2014). The TM and its channels are believed to be a space of elevated AH outflow resistance due to the buildup of PEXM in the juxtacanalicular portion of the meshwork next to SC (Naumann, 1998). Excessive resistance can lead to elevated intraocular pressures, optic nerve damage, and development of optic neuropathies such as glaucoma. Increased aqueous protein levels stemming from a blood-aqueous barrier defect and melanin granules from the iris can also contribute to AH outflow resistance (Naumann, 1998). The iris pigment epithelium is a tissue that could also produce components of the PEXM as well and can block the AH outflow pathway (Borras, 2014).

The composition of AH is similar to plasma but has a lower protein content. AH protein content may be altered by pathologic conditions. In eyes with PEX, more functional proteins associated with transport, binding to immune response, and defense have been identified in the AH (Hardenborg et al., 2009). PEX AH and control AH have differences in protein content (Hardenborg et al., 2009). Some proteins found only in PEX AH include apolipoprotein A-1, β-2-glycoprotein 1 (transport and protein binding), α-1-acid glycoprotein 1, Ig γ-3 chain C region, α-1B-glycoprotein (immune response and defense), ribonuclease pancreatic, antithrombin-III (metabolism/enzyme and inhibitors) and β-crystallin S (structural component of the lens) (Hardenborg et al., 2009). Alteration of the normal protein content in PEX is due to upregulation or downregulation of certain genes and their corresponding proteins. For example, complement factors such as complement factor C3 and complement C4-A were identified in controls but not in PEX while vitamin D binding protein and β-crystallins were present in the PEX group but not in the control group (Hardenborg et al., 2009).

3. Genetic risk factors and proteins

Genetically predisposed populations and environmental factors, such as stress, can upregulate or downregulate extracellular matrix genes which potentially lead to PEXM formation subsequent and PEXG. Extracellular matrix and basement membrane proteins like fibronectin, fibrillin-1, fibulin-2, vitronectin, laminin, proteoglycan core proteins such as syndecan-3, versican, and the multifunctional glycoprotein clusterin are commonly associated with PEX (Ovodenko et al., 2007). In addition to these few proteins identified within PEXM, we also want to highlight three genes, LOXL1, TGFβ3, and FBLN5, which are associated with PEX and may influence the expression of the previously mentioned proteins.

A common gene that contributes to PEXM formation is LOXL1, which is the strongest associated genetic risk factor worldwide and stems from single nucleotide polymorphisms in the lysyl oxidase like-1 (LOXL1) gene (Schlötzer-Schrehardt & Zenkel, 2019). LOXL1 is essential to the biogenesis of connective tissue and catalyzes crosslinking with collagen and elastin (Tang et al., 2014). LOXL1 is located on chromosome 15q24.1 and was discovered in a genome wide association scan (GWAS) of Scandinavian populations in patients with PEX (Schlötzer-Schrehardt & Zenkel, 2019). Allelic variants of this gene can be either protective or risk factors for developing PEX depending on the study population (Schlötzer-Schrehardt, 2009). For example, one allele may be protective for an Asian population but concurrently a risk factor for a Scandinavian population. The expression of LOXL1 in ocular structures is differentially regulated based on the stage of PEX progression. For instance, an increase in LOXL1 mRNA expression has been noted in early PEX while levels are decreased in advanced stages—independent of glaucoma presence compared to controls (Schlötzer-Schrehardt, 2009). The LOXL1 protein also appears to be an important component of intra- and extra-ocular PEX deposits and co-localizes with another protein, clusterin (an extracellular protein chaperone). Clusterin is a fluid phase inhibitor of the complement activation pathway and recognizes and prevents abnormal aggregation of misfolded proteins in extracellular spaces (Ovodenko et al., 2007; Schlötzer-Schrehardt, 2009). In anterior segment tissues, clusterin expression was observed to be significantly decreased in PEX AH (Schlötzer-Schrehardt, 2009). Clusterin and LOXL1 are two proteins expressed by genes that are genetic risk factors in the pathogenesis of PEX (Padhy, Kapuganti, Hayat, Mohanty, & Alone, 2019).

LOXL1 expression in the TM is upregulated by transforming growth factor β (TGF-β), which consists of a family of proteins (TGFβ1, TGFβ2, and TGFβ3) (Schlötzer-Schrehardt & Zenkel, 2019). These proteins are involved with extracellular matrix remodeling and found to be elevated in glaucoma patients' AH with TGFβ3 being significantly elevated in PEX AH (Raghunathan et al., 2015). Human TM cells treated with TGFβ3 developed stiffer cells compared to controls, partially implicating the reduced outflow of AH via TM cells secondary to increased IOP (Raghunathan et al., 2015). The biophysical and biochemical properties of matrix proteins deposited by TM cells are significantly altered (Table 1) (Raghunathan et al., 2015). TGFβ1 stimulates clusterin synthesis at the mRNA and protein levels in various cell types. TGFβ1 was also significantly elevated in PEX AH, as it is a key component in fibrillar deposit formation (Ovodenko et al., 2007). The TGFβ1 gene is upregulated in anterior segment tissues while also regulating other genes expressed in PEX tissues, such as fibrillin-1, latent-transforming growth factor binding proteins (LTBP-1 and -2), tropoelastin, TGase-2, clusterin, and LOXL1 (Schlötzer-Schrehardt, 2009). Fibrillin-1, a glycoprotein capable of interacting with other matrix proteins like LTBP-2, fibronectin, fibulin-2, and versican, is also a major component of PEXM and elastic microfibrils (Schlötzer-Schrehardt, 2009).

Table 1 Assessment of quality of proteomic studies: A QUADOMICS scoring approach.

References	1	2	3	4	5	6	7	8	9	10	11	12	13	14	15	16	Score
																	(Out of 14)
Botling Taube, Konzer, Alm, and Bergquist (2019)	N	Not applicable	Y	Y	Y	Y	Y	Y	Y	Y	Y	Y	N	Not applicable	Y	Y	10
Hardenborg et al. (2009)	N		Y	Y	Y	Y	Y	Y	Y	Y	Y	Y	N		Y	Y	12
Ovodenko et al. (2007)	N		Y	Y	Y	Y	Y	Y	Y	Y	Y	Y	N		Y	Y	12
Padhy et al. (2019)	N		Y	Y	Y	Y	Y	N	Y	Y	Y	Y	N		Y	Y	11
Raghunathan et al. (2015)	Y		Y	N	Y	Y	Y	Y	Y	Y	Y	Y	N		Y	Y	12
Ronci, Sharma, Martin, Craig, and Voelcker (2013)	N		Y	N	Y	Y	Y	Y	Y	Y	Y	Y	N		Y	Y	11
Sharma et al. (2009)	N		Y	Y	Y	Y	Y	Y	Y	Y	Y	Y	Y		Y	Y	13
Taube et al. (2012)	N		Y	Y	Y	Y	Y	Y	Y	Y	Y	Y	N		Y	Y	12
Sharma et al. (2018)	N		Y	Y	Y	Y	Y	Y	Y	Y	Y	Y	N		Y	Y	13

Quality of referenced studies were assessed using published QUADOMICS approach/tool. Each PEX proteomics study was scored out of a total of 14 criteria: *Y*, present; *N*, absent. Criteria 2 and 14 are not applicable to any of the referenced studies. The criterions are: **1**. The selection criteria are clearly described. Criteria 1 requires a flow chart depicting the transition from theoretical population to final study population. **2**. The spectrum of patients are representative of the theoretical practice test patient population. **3**. The type of sample is fully described. **4**. The procedures and timing of biological sample collection are described with detail including clinical characteristics (4.1. Clinical and physiological factors 4.2. Diagnostic and treatment procedures). **5**. The handling and pre-analytical procedures are reported in sufficient detail and congruent for all samples. Conversely, reported differences in procedures and their effect on the results were assessed. **6**. The time period between the reference standard and the index test is reasonably short to ensure that the target condition did not change between the tests. **7**. Is the reference standard likely to correctly classify the target condition? **8**. The whole sample or a random selection of the sample was verified using a reference standard. **9**. Patients received the same reference standard despite the index test results. **10**. The execution of the index test is described in sufficient detail enabling replication. **11**. Was the execution of the reference standard described in sufficient detail to permit its replication? **12**. Were the index test results interpreted blinded to the reference standard? **13**. Were the reference standard results interpreted without knowledge of the results of the index test? **14**. Are the clinical data congruent with respect to timing of the test results interpretation (akin to that in the practice test)? The last criterion could be combined with two additional criteria: (a) were uninterpretable/intermediate test results reported and (b) was the presence of overfitting avoided.

The protein precursor of LOXL1 also targets elastic microfibrils by binding to tropoelastin and fibulin-5 (FBLN5) (Schlötzer-Schrehardt, 2009). Two risk variants for PEX have been identified in the FBLN5 gene (Padhy et al., 2019). FBLN5 is an extracellular scaffold protein that plays a role in elastogenesis and has many roles that affect the extracellular matrix, including assisting the deposition and activation of LOXL1 (Table 1) (Padhy et al., 2019). Two noncoding risk variants of the FBLN5 gene, rs7149187: G>A and rs929608:T>C, were discovered to be genetically associated with PEX (Padhy et al., 2019). The presence of single nucleotide polymorphism (SNP) rs7149187 was significant in individuals with PEX and PEXG, while the SNP rs929608 was only found to be significant in individuals with PEX (Padhy et al., 2019). Overall, FBLN5 expression was significantly decreased in only PEX individuals but showed no difference between control and PEXG groups (Padhy et al., 2019). Expression of this gene was also measured in the lens capsules of PEX patients compared to controls, which revealed a significant decrease in expression of FBLN5 (Padhy et al., 2019). When FBLN5 is downregulated, impaired elastic fiber formation occurs because FBLN5 does not interact with LOXL1 as much. Despite having a diminished presence, the direct effect of FBLN5 expression on the pathogenesis of PEX needs to be evaluated further.

The biochemical composition of the extracellular matrix has been studied using mass spectrometry, which has identified 4935 proteins. Researchers have identified nine specific proteins with altered expression patterns by TGFβ3 treatment (Raghunathan et al., 2015). Of the nine proteins, secreted frizzled-related protein 1 and 4 (SFRP1 and SFRP4), plasminogen activator inhibitor type 1 (SERPINE1), angiopoietin-like 4 (ANGPTL4), periostin (POSTN), gremlin 1 (GREM1), and fibroblast growth factor 5 (FGF5) were overexpressed while interalpha-trypsin inhibitor heavy chain 3 (ITIH3) and melanotransferrin (MF12) were decreased in TGFβ3 treated cells (Table 2) (Raghunathan et al., 2015). Gremlin is an antagonist of bone morphogenetic protein signaling that has been implicated in promoting TGFβ2 signaling in TM cells, which results in increased outflow resistance (Raghunathan et al., 2015). Angiopoietin-like 4 is a glycoprotein and periostin is a matricellular protein whose level of expression correlates with collagen crosslinking, which would result in a stiffer extracellular matrix (Raghunathan et al., 2015). SFRP1 in past studies has demonstrated its ability to elevate IOP and reduce outflow facility and as a result of TGFβ3 treatment, the secretion of both SFRP1 and SFRP4 was increased in TM cells (Raghunathan et al., 2015).

Table 2 Table of significantly altered (downregulated or upregulated) proteins from PEX proteomics studies.

References	Proteomics methodology	Tissue	Significantly altered proteins in PEX	Analysis type
Botling Taube et al. (2019)	HPLC-LTQ-Orbitrap Velos Pro MS	Aqueous humor	**Complement C3**; Kininogen-1 (KNG1); **Antithrombin-III (SERPINC1); Vitamin D-binding protein (GC)**; Retinol binding protein (RBP3); Glutathione peroxidase 3 (GPX3); Carboxypeptidase E (CPE); Calsyntenin-1 (CLSTN1)	Quantitative
Hardenborg et al. (2009)	capLC MALDI-TOF/TOF MS	Aqueous humor	Apolipoprotein A-I (APOA1); **Vitamin D-binding protein (VTDB)**; β-2-Glycoprotein 1 (APOH); α-1-Acid glycoprotein 1 (A1AG1); Ig-γ-3 chain C region (IGHG3); α-1B-glycoprotein (A1BG); Ribonuclease pancreatic (RNAS1); **Antithrombin-III (ANT3)**; β-Crystallin S (CRBS)	Quantitative
Ovodenko et al. (2007)	Q-ToF-MS and LC-MS/MS	Anterior lens capsules	Fibulin-2; Desmocollin-2; Syndecan-3; Versican; **Clusterin**	Quantitative
Padhy et al. (2019)	Bradford protein assay	Lens capsules	Fibulin-5 (FBLN5)	Quantitative
Raghunathan et al. (2015)	LC-MS/MS	Trabecular meshwork cells	Secreted frizzled-related protein 1 and 4 (SFRP1 and SFRP4); Plasminogen activator inhibitor type 1 (SERPINE1); Angiopoietin-like 4 (ANGPTL4); Periostin (POSTN); Gremlin 1 (GREM1); Fibroblast growth factor 5 (FGF5); Interalpha-trypsin inhibitor heavy chain 3 (ITIH3); Melanotransferrin (MF12)	Quantitative

Ronci et al. (2013)	MALDI-LIFT-MS/MS	Lens capsules	**LOXL1; APOE**	Quantitative
Sharma et al. (2009)	LC-MS/MS	Anterior lens capsules	**LOXL1; Apolipoprotein E (APOE); Latent-transforming growth factor beta-binding protein 2 (LTBP-2); Complement C3; Clusterin (CLU)**	Quantitative
Taube et al. (2012)	iTRAQ capLC and CE MALDI MS and MS/MS	Aqueous humor	Beta crystalline B2; Osteopontin; Angiotensinogen	Quantitative
Sharma et al. (2018)	LC-MS/MS	Anterior lens capsules	**Lysyl oxidase homolog 1 (LOXL1); Apolipoprotein E (APOE); Clusterin (CLU); Complement C3**; Emilin-1 (EMILIN1); Fibrillin-1 (FBN1); Fibronectin (FN1); **Latent-transforming growth factor beta-binding protein 2 (LTBP-2)**; Metalloproteinase inhibitor 3 (TIMP3); Vitronectin (VTN)	Quantitative

The quantitative studies highlighted here utilize a variety of protein identification and analysis methods and type of sample used. The level or relative amount of several of the proteins identified were found to be significantly altered in PEX compared to controls. Highlighted in bold are the proteins found in multiple studies.

4. Origin of ocular anterior segment proteins

Within PEXM, hundreds to thousands of proteins can be quantified using fractionation and identification techniques followed by bioinformatic analyses resulting in only a handful of proteins showing significant alterations. In this review, across nine quantitative proteomic studies (Table 1) 41 proteins were classified as being significantly differentially regulated in PEXM—whether they were downregulated or upregulated when compared to controls. Assessing the quality of these studies, we applied QUADOMICS which is an adaptation of QUADAS (Quality Assessment of Diagnostic Accuracy Assessment) and assigned a score to each proteomic study (Table 1) (Lumbreras et al., 2008). The studies that met our criteria underwent a 16-item assessment to determine the quality of each paper and influenced the decision of what identified proteins would be highlighted in this review.

Among the QUADOMICS verified studies, a handful of proteins were commonly identified. Complement C3, antithrombin-III (SERPINC1 or ANT3), vitamin D binding protein (GC or VTDB), clusterin (CLU), apolipoprotein E (APOE), LOXL1, and latent-transforming growth factor beta-binding protein 2 (LTBP-2) were identified in at least two or more of the nine studies (Botling Taube et al., 2019; Hardenborg et al., 2009; Ovodenko et al., 2007; Sharma et al., 2009; Sharma et al., 2018). Interestingly, some of these proteins were found in different tissues and fluids, for example, the AH, lens capsule, and TM cells. Understanding where these proteins originate, including how and where they are deposited as components of PEXM, will help pinpoint which proteins are involved in the initiation of PEXM formation and in the pathway for PEXG development.

Two proteins found on human lens capsules affected by PEX are apolipoprotein E (APOE) and previously discussed LOXL1 (Ronci et al., 2013). LOXL1 was consistently observed only on PEX affected lens capsules, solidifying the role of LOXL1 protein overexpression in PEX (Ronci et al., 2013). However, APOE was found in abundance in both the lens capsule pupillary region of PEX patients and controls (Ronci et al., 2013). APOE is a low-density lipoprotein receptor ligand which has been implicated in other diseases that involve pathologic protein aggregation, such as Alzheimer's disease and age-related macular degeneration (Sharma et al., 2009). In another study, Sharma et al. conducted an analysis of the PEXM deposited specifically on the anterior lens capsule, which revealed peptides that corresponded to hemoglobin (HB) subunits HB alpha, beta, gamma, and delta (Sharma et al., 2018). The study also identified novel proteins such as apolipoprotein A-I (APOA1), apolipoprotein A-IV (APOA4), crystallin alpha A (CRYAA), fibrinogen beta (FBG) and peroxiredoxin-2 (PRDX2) (Sharma et al., 2018). Each of these proteins have different functions and have some stake in protein aggregation, inflammatory diseases, and oxidative stress conditions. Although they were found on lens capsules, these proteins could have been deposited by the circulating AH (Sharma et al., 2018).

Other proteins differentially identified in PEX AH samples include beta-crystalline B1 (CRYBB1), CRYBB2, and gamma-crystalline D (CRYGD) (Botling Taube et al., 2019). Crystallins are lens structural proteins. Their integrity diminishes with age resulting in protein aggregation (Botling Taube et al., 2019). Although not definitively proven, these crystallins may also be elevated as a stress-response in eyes with PEX (Botling Taube et al., 2019). In a later study (Taube et al., 2012), AH samples from 20 patients with PEX and 18 patients with only cataracts were analyzed first by isobaric tagging for relative and absolute protein quantification (iTRAQ), then two separation methods, liquid chromatography (LC) and capillary electrophoresis (CE) followed by MALDI mass spectrometry and MS/MS. A total of 64 proteins showed increased concentrations and 77 proteins showed decreased concentrations in the PEX samples. Higher levels of β-crystalline B2 and lower levels of angiotensinogen and osteopontin were measured in the PEX AH samples only. The role of β-crystalline B2 in eyes with PEX is unknown. However, angiotensinogen is involved with IOP regulation and osteopontin, a matricellular protein, has functions ranging from inflammation, tissue repair, and immunological responses (Taube et al., 2012).

Among several other studies, PEX AH has increased concentrations of oxidative stress markers and decreased antioxidants (Botling Taube et al., 2019). For example, proteomic analysis of the AH identified 71 proteins in PEX compared to control samples. Of these proteins, those identified as significantly upregulated in PEX samples were complement C3, kininogen-1 (KNG1), antithrombin-III (SERPINC1), and vitamin D binding protein (GC). Other proteins were downregulated, such as calsyntenin-1 (CLSTN1), retinol-binding protein 3 (RBP3), carboxypeptidase E (CPE) and glutathione peroxidase 3 (GPX3) (Botling Taube et al., 2019). Complement C3, HNK3, and SERPINC1 all play a role in immune system function in which HNK3 is involved with inflammatory response and SERPINC1, also expressed in the

TM, SC and lens epithelium, has anti-inflammatory properties (Botling Taube et al., 2019). RBP3, which is produced by the ciliary epithelium, is a transport protein in the retinal pigment epithelium; GPX3 is known to reduce reactive oxygen species (ROS) to water by oxidizing glutathione to glutathione disulfide, and CPE has a neuroprotective role in the central nervous system when exposed to oxidative or behavioral stress (Botling Taube et al., 2019). GC, a plasma protein, has numerous roles ranging from immune and inflammatory responses to regulating vitamin D and transportation. GC is particularly interesting because of how little is known about its role in the network of PEXM formation, thus future studies of this protein are warranted.

5. Environmental factors and interplay

PEX is multifactorial, with oxidative stress, genetics, and environmental factors influencing the pathogenesis of PEX (Botling Taube et al., 2019). Oxidative stress is defined as an increase in the intracellular concentration of ROS which affects the protective antioxidant defense system (Schlötzer-Schrehardt, 2018). Oxidative stress plays an important role in the pathogenesis of PEX and glaucoma because the homeostatic balance of oxidant to antioxidant is disrupted. Ultraviolet B (UV-B) is an environmental factor known to contribute to PEX deposit formation. Radiation from sunlight accelerates the production of vitamin D (Lin et al., 2012). When there is excess vitamin D, GC is expressed to maintain homeostasis. If vitamin D metabolism is dysregulated, crystallization may begin to take place because GC has large positive and negative patches which provide nucleation centers that may initiate precipitation of PEXM proteins. Proteins including GC have been analyzed for their charged patches (Fig. 4). A study was conducted to measure vitamin D levels in the AH, vitreous humor, and tear fluid (Lin et al., 2012). This study also investigated if vitamin D is synthesized by the corneal endothelium after UV-B exposure. The specific form of vitamin D assessed was vitamin D3, which is derived from sunlight-mediated UV-B conversion of 7-dehydrocholesterol to vitamin D3 (Lin et al., 2012). GC is known to have a higher binding affinity for vitamin D3 compared to other vitamin D forms (Lin et al., 2012). Their results revealed significantly elevated vitamin D3 along with other vitamin D variants following UV-B exposure. Future studies understanding this mechanism are needed.

Many proteins associated with PEX have been identified, and their role in the pathogenesis of PEX has been determined. The next step is to map out the orderly assembly process of PEXM formation. Several proteins and genes

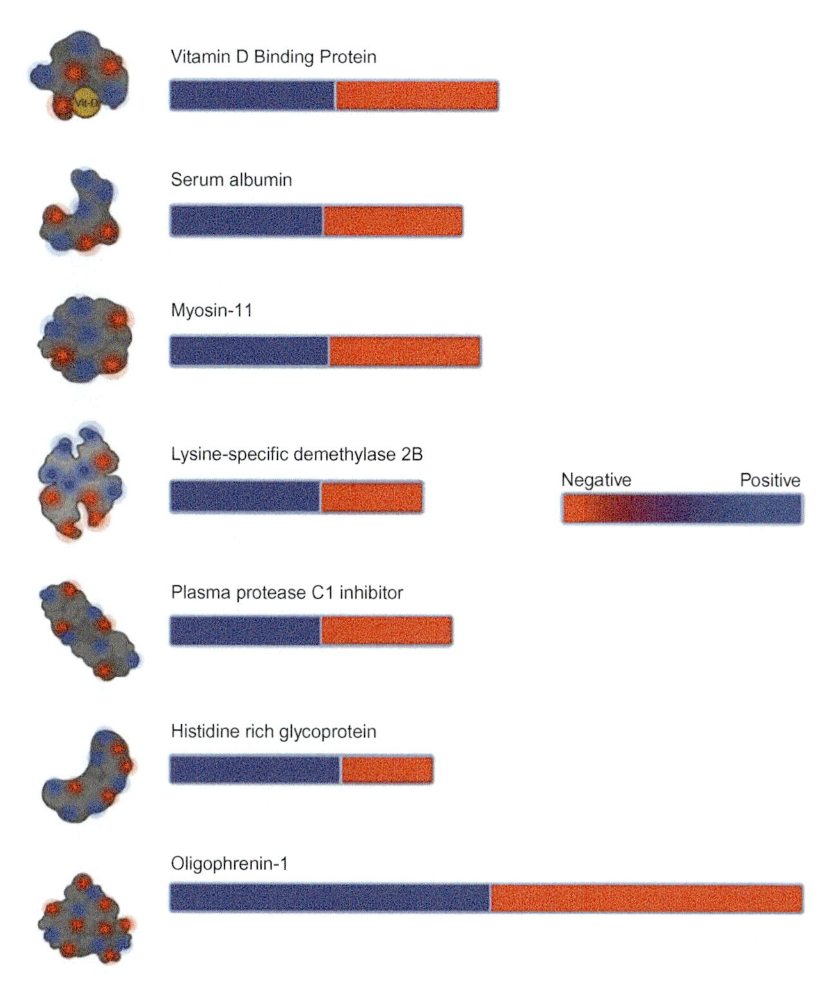

Fig. 4 Structural models of the positive and negative patches found on key proteins involved in pseudoexfoliation material formation. These proteins include vitamin D binding protein, serum albumin, myosin-11, lysine-specific demethylase 2B, plasma protease C1 inhibitor, histidine-rich glycoprotein, and oligophrenin-1. Next to each protein is a bar that also shows the percent positive vs negative patches. Blue represents the percent positive patches and red represents the percent negative patches.

can confidently be accepted as key components of PEX material across all populations, such as the LOXL1 gene. Using the current information about the proteome and recognizing the knowledge gaps in understanding PEX, more analyses are still needed. For instance, extensive research has been conducted with the LOXL1 gene, but there are still genes and their corresponding proteins that need more attention like GC and the specific role of LOXL1 in PEXG pathogenesis is not clear.

6. Conclusion

The significant accumulation of PEXM in the anterior eye segment is a key characteristic in the progression of PEXG. To provide treatments and preventative measures for this disorder, knowing the growth process and where the accumulation begins is essential to understanding the pathological mechanism. Additionally, we only reviewed proteins present in the ocular anterior segment, but there are other studies that have analyzed the blood serum for PEX proteins. Features in common with PEXM and other amyloid disorders, such as Alzheimer's disease, have been suggested by the presence of amyloid-associated proteins, and Alzheimer's peptides and biomarkers in PEX AH (Schlötzer-Schrehardt, 2018). The effect of exfoliative material in PEX on other bodily tissues and systems has not been determined to have similar negative effects as in ocular tissues. Organs such as the skin, heart, lungs, cerebral meninges, and kidney have been shown to contain PEXM most likely secondary to their proximity to elastic and collagen fibers (Streeten, Dark, Wallace, Li, & Hoepner, 1990; Streeten, Li, Wallace, Eagle Jr., & Keshgegian, 1992). Given its systemic nature, PEX is a complex disease and the mechanism of its pathogenesis is still not completely understood. Future molecular and cellular biological studies will enhance our understanding of PEXM formation and PEXG pathophysiology that could lead to the development of new therapeutics and pharmacological agents that treat this sight threatening disease.

Acknowledgments

This research was supported by a grant from The Glaucoma Foundation of New York, an unrestricted fund from Research to Prevent Blindness to University of Miami. R.K. Lee is partially supported by the Walter G. Ross Foundation. We thank all authors in this field whose work was cited and those we were unable to acknowledge in this review.

References

Acott, T. S., Kelley, M. J., Keller, K. E., Vranka, J. A., Abu-Hassan, D. W., Li, X., et al. (2014). Intraocular pressure homeostasis: Maintaining balance in a high-pressure environment. *Journal of Ocular Pharmacology and Therapeutics*, *30*(2–3), 94–101.

Batur, M., Seven, E., Tekin, S., & Yasar, T. (2017). Anterior lens capsule and iris thicknesses in pseudoexfoliation syndrome. *Current Eye Research*, *42*(11), 1445–1449.

Borras, T. (2014). The cellular and molecular biology of the iris, an overlooked tissue: The iris and pseudoexfoliation glaucoma. *Journal of Glaucoma*, *23*(8 Suppl. 1), S39–S42.

Botling Taube, A., Konzer, A., Alm, A., & Bergquist, J. (2019). Proteomic analysis of the aqueous humour in eyes with pseudoexfoliation syndrome. *The British Journal of Ophthalmology*, *103*(8), 1190–1194.

Carreon, T., van der Merwe, E., Fellman, R. L., Johnstone, M., & Bhattacharya, S. K. (2017). Aqueous outflow—A continuum from trabecular meshwork to episcleral veins. *Progress in Retinal and Eye Research*, *57*, 108–133.

DelMonte, D. W., & Kim, T. (2011). Anatomy and physiology of the cornea. *Journal of Cataract and Refractive Surgery*, *37*(3), 588–598.

Dewundara, S., & Pasquale, L. R. (2015). Exfoliation syndrome: A disease with an environmental component. *Current Opinion in Ophthalmology*, *26*(2), 78–81.

Durukan, I. (2018). Evaluation of corneal and lens clarity in unilateral pseudoexfoliation syndrome: A densitometric analysis. *Clinical & Experimental Optometry*, *101*(6), 740–746.

Grzybowski, A., Kanclerz, P., & Ritch, R. (2019). The history of exfoliation syndrome. *Asia-Pacific Journal of Ophthalmology (Philadelphia, Pa.)*, *8*(1), 55–61.

Hardenborg, E., Botling-Taube, A., Hanrieder, J., Andersson, M., Alm, A., & Bergquist, J. (2009). Protein content in aqueous humor from patients with pseudoexfoliation (PEX) investigated by capillary LC MALDI-TOF/TOF MS. *Proteomics. Clinical Applications*, *3*(3), 299–306.

Lee, R. K. (2008). The molecular pathophysiology of pseudoexfoliation glaucoma. *Current Opinion in Ophthalmology*, *19*(2), 95–101.

Lin, Y., Ubels, J. L., Schotanus, M. P., Yin, Z., Pintea, V., Hammock, B. D., et al. (2012). Enhancement of vitamin D metabolites in the eye following vitamin D3 supplementation and UV-B irradiation. *Current Eye Research*, *37*(10), 871–878.

Lumbreras, B., Porta, M., Marquez, S., Pollan, M., Parker, L. A., & Hernandez-Aguado, I. (2008). QUADOMICS: An adaptation of the quality assessment of diagnostic accuracy assessment (QUADAS) for the evaluation of the methodological quality of studies on the diagnostic accuracy of '-omics'-based technologies. *Clinical Biochemistry*, *41*(16–17), 1316–1325.

Myer, C., Abdelrahman, L., Banerjee, S., Khattri, R. B., Merritt, M. E., Junk, A. K., et al. (2020). Aqueous humor metabolite profile of pseudoexfoliation glaucoma is distinctive. *Molecular Omics*, *16*(5), 425–435.

Naumann, G. (1998). Pseudoexfoliation syndrome for the comprehensive ophthalmologist intraocular and systemic manifestations. *Ophthalmology*, *105*(6), 951–968.

Netland, P. A., Ye, H., Streeten, B. W., & Hernandez, M. R. (1995). Elastosis of the lamina cribrosa in pseudoexfoliation syndrome with glaucoma. *Ophthalmology*, *102*(6), 878–886.

Ovodenko, B., Rostagno, A., Neubert, T. A., Shetty, V., Thomas, S., Yang, A., et al. (2007). Proteomic analysis of exfoliation deposits. *Investigative Ophthalmology & Visual Science*, *48*(4), 1447–1457.

Padhy, B., Kapuganti, R. S., Hayat, B., Mohanty, P. P., & Alone, D. P. (2019). De novo variants in an extracellular matrix protein coding gene, fibulin-5 (FBLN5) are associated with pseudoexfoliation. *European Journal of Human Genetics*, *27*(12), 1858–1866.

Palko, J. R., Qi, O., & Sheybani, A. (2017). Corneal alterations associated with pseudoexfoliation syndrome and glaucoma: A literature review. *J. Ophthalmic Vis. Res.*, *12*(3), 312–324.

Raghunathan, V. K., Morgan, J. T., Chang, Y. R., Weber, D., Phinney, B., Murphy, C. J., et al. (2015). Transforming growth factor beta 3 modifies mechanics and composition of extracellular matrix deposited by human trabecular meshwork cells. *ACS Biomaterials Science & Engineering*, *1*(2), 110–118.

Rasmussen, C. A., & Kaufman, P. L. (2014). The trabecular meshwork in normal eyes and in exfoliation glaucoma. *Journal of Glaucoma*, *23*(8 Suppl. 1), S15–S19.

Ritch, R., & Schlötzer-Schrehardt, U. (2001). Exfoliation syndrome. *Survey of Ophthalmology*, *45*(4), 265–315.

Ronci, M., Sharma, S., Martin, S., Craig, J. E., & Voelcker, N. H. (2013). MALDI MS imaging analysis of apolipoprotein E and lysyl oxidase-like 1 in human lens capsules affected by pseudoexfoliation syndrome. *Journal of Proteomics*, *82*, 27–34.

Schlötzer-Schrehardt, U. (2009). Molecular pathology of pseudoexfoliation syndrome/glaucoma—New insights from LOXL1 gene associations. *Experimental Eye Research, 88*(4), 776–785.

Schlötzer-Schrehardt, U. (2018). Molecular biology of exfoliation syndrome. *Journal of Glaucoma, 27*(Suppl. 1), S32–S37.

Schlötzer-Schrehardt, U., & Zenkel, M. (2019). The role of lysyl oxidase-like 1 (LOXL1) in exfoliation syndrome and glaucoma. *Experimental Eye Research, 189*, 107818.

Sharma, S., Chataway, T., Burdon, K. P., Jonavicius, L., Klebe, S., Hewitt, A. W., et al. (2009). Identification of LOXL1 protein and apolipoprotein E as components of surgically isolated pseudoexfoliation material by direct mass spectrometry. *Experimental Eye Research, 89*(4), 479–485.

Sharma, S., Chataway, T., Klebe, S., Griggs, K., Martin, S., Chegeni, N., et al. (2018). Novel protein constituents of pathological ocular pseudoexfoliation syndrome deposits identified with mass spectrometry. *Molecular Vision, 24*, 801–817.

Streeten, B. W., Dark, A. J., Wallace, R. N., Li, Z. Y., & Hoepner, J. A. (1990). Pseudoexfoliative fibrillopathy in the skin of patients with ocular pseudoexfoliation. *American Journal of Ophthalmology, 110*(5), 490–499.

Streeten, B. W., Li, Z. Y., Wallace, R. N., Eagle, R. C., Jr., & Keshgegian, A. A. (1992). Pseudoexfoliative fibrillopathy in visceral organs of a patient with pseudoexfoliation syndrome. *Archives of Ophthalmology (Chicago, Ill: 1960), 110*(12), 1757–1762.

Sunderland, D. K., & Sapra, A. (2021). *Physiology, aqueous humor circulation*. StatPearls. Internet.

Tang, J. Z., Wang, X. Q., Ma, H. F., Wang, B., Wang, P. F., Peng, Z. X., et al. (2014). Association between polymorphisms in lysyl oxidase-like 1 and susceptibility to pseudoexfoliation syndrome and pseudoexfoliation glaucoma. *PLoS One, 9*(3), e90331.

Taube, A. B., Hardenborg, E., Wetterhall, M., Artemenko, K., Hanrieder, J., Andersson, M., et al. (2012). Proteins in aqueous humor from cataract patients with and without pseudoexfoliation syndrome. *European Journal of Mass Spectrometry (Chichester, England), 18*(6), 531–541.

Vazquez, L. E., & Lee, R. K. (2014). Genomic and proteomic pathophysiology of pseudoexfoliation glaucoma. *International Ophthalmology Clinics, 54*(4), 1–13.

CHAPTER NINE

A systems biology approach to COVID-19 progression in population

Magdalena Djordjevic[a], Andjela Rodic[b], Igor Salom[a], Dusan Zigic[a], Ognjen Milicevic[c], Bojana Ilic[a], and Marko Djordjevic[b,*]

[a]Institute of Physics Belgrade, University of Belgrade, Belgrade, Serbia
[b]Computational Systems Biology Group, Faculty of Biology, University of Belgrade, Belgrade, Serbia
[c]Department for Medical Statistics and Informatics, Faculty of Medicine, University of Belgrade, Belgrade, Serbia
*Corresponding author: e-mail address: dmarko@bio.bg.ac.rs

Contents

Abstract

A number of models in mathematical epidemiology have been developed to account for control measures such as vaccination or quarantine. However, COVID-19 has brought unprecedented social distancing measures, with a challenge on how to include these in a manner that can explain the data but avoid overfitting in parameter inference. We here develop a simple time-dependent model, where social distancing effects are introduced analogous to coarse-grained models of gene expression control in systems biology. We apply our approach to understand drastic differences in COVID-19 infection and fatality counts, observed between Hubei (Wuhan) and other Mainland China provinces. We find that these unintuitive data may be explained through an interplay of differences in transmissibility, effective protection, and detection efficiencies between Hubei and other provinces. More generally, our results demonstrate that regional differences may drastically shape infection outbursts. The obtained results demonstrate the applicability of our developed method to extract key infection parameters directly from publically available data so that it can be globally applied to outbreaks of COVID-19

Advances in Protein Chemistry and Structural Biology, Volume 127
ISSN 1876-1623
https://doi.org/10.1016/bs.apcsb.2021.03.003

in a number of countries. Overall, we show that applications of uncommon strategies, such as methods and approaches from molecular systems biology research to mathematical epidemiology, may significantly advance our understanding of COVID-19 and other infectious diseases.

1. Introduction

As the novel COVID-19 disease caused by the SARS-CoV-2 virus took the world by a storm, the new pandemic quickly gained priority in scientific research in a wide range of biological and medical science disciplines. Despite that their prior expertise was in unrelated research fields, many researchers have successfully adapted their approaches and methods to examine various aspects of this viral infection and, thus, contributed to finding the necessary solutions. The systems biology community is not an exception (Alon, Mino, & Yashiv, 2020; Bar-On, Flamholz, Phillips, & Milo, 2020; Djordjevic, Djordjevic, Ilic, Stojku, & Salom, 2021; Eilersen & Sneppen, 2020; Karin et al., 2020; Saad-Roy et al., 2021; Vilar & Saiz, 2020; Wong et al., 2020): those involved in modeling the dynamics of biological systems at the molecular and cellular level can directly apply the similar methodology in epidemiological studying of the virus spread—and this exactly is the central point of the present paper. In particular, dynamic models of biochemical reaction networks, in which the reaction kinetics follow the law of mass action, are analogous to compartmental epidemiological models which, instead of concentrations of chemical species, track the prevalence of individuals in defined population classes over time (Voit, Martens, & Omholt, 2015). Moreover, gene expression dynamics is usually a result of the interplay between the changing rate of cell growth, on which the global physiological rates of molecule synthesis and degradation depend, and complex transcription regulation (Djordjevic, Rodic, & Graovac, 2019). Therefore, modeling dynamics of gene circuits implies combining kinetic models, often relying on the law of mass action, with appropriate non-linear functions describing the regulation part. In the case of the COVID-19 epidemic, one can note that the virus transmission in a population, driven by the biological capacity of the particular virus in the given environment, is coupled with strong, time-dependent regulation, represented by the epidemic mitigation measures imposed by governments. These similarities between the modeled systems may facilitate the application of the systems biology techniques to the epidemiology field of research. In this paper, we will show how such an approach can be used to assess the basic parameters of the COVID-19 epidemic progression in a given population. In particular, we will use the analogy outlined above to study the COVID-19 spread in Mainland China and test the hypothesis about possible reasons for the uneven disease spread in China provinces.

Our interest in Mainland China infection progression comes from Fig. 1. The progression seems highly intriguing, as Hubei (with only 4% of China population) shows an order of magnitude larger number of detected infection cases (Fig. 1A) and two orders of magnitude higher fatalities (Fig. 1B) compared to the *total* sum in all other Mainland China provinces. The epidemic was unfolding well before the Wuhan closure (with the reported symptom onset of the first patient on December 1, 2019) and within the period of huge population movement, which started 2 weeks before January 25 (the Chinese Lunar New Year) (Chen, Yang, Yang, Wang, & Bärnighausen, 2020). As a rough baseline, a modeling study of the infection spread from Wuhan (Wu, Leung, & Leung, 2020) estimated more than 10^5 new cases per day in Chongqing alone—instead, the actual (reported) peak number for *all* Mainland China provinces outside Hubei was just 831.

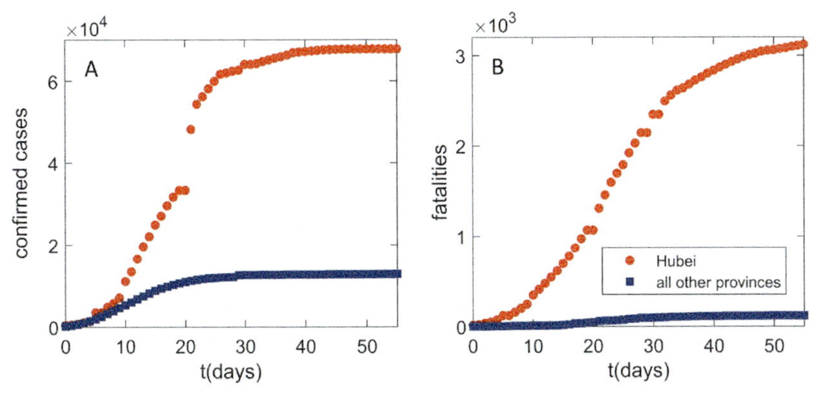

Fig. 1 Infection and fatality counts for Hubei vs all other provinces. The number of (A) detected infections, (B) fatality cases. Zero on the horizontal axis corresponds to the time from which the data (Hu et al., 2020) are taken (January 23), which also coincides with the Wuhan closure. Red circles correspond to the observed Hubei counts. Blue squares correspond to the sum of the number of counts for all other provinces. The figure illustrates a puzzling difference in the number of counts between Hubei alone and the sum of all other Mainland China provinces.

Consequently, it is a notable challenge for computational modeling to understand drastic differences in COVID-19 infection and fatality counts observed between Hubei (Wuhan) and other Mainland China provinces. These drastic differences may be a consequence of an interplay between the virus transmissibility (influenced by environmental and demographic factors) and the effectiveness of the protection measures. Both can significantly change between different provinces (more generally different countries/regions), and the model has to infer this from available data (commonly the number of confirmed cases, publicly available for a large number of countries/regions).

The study presented here will therefore demonstrate the usefulness of the systems biology approach to the analysis of non-trivial COVID-19 data from China. In particular, the developed method will allow us to analyze the puzzling differences in dynamics trajectories in Mainland China provinces, and it will also turn out to be more generally applicable for understanding regional differences in outburst dynamics. The surprising differences in COVID-19 progression in different provinces may put strong constraints on the underlying infection progression parameters and allow us to understand:

i. What interplay between the inherent disease transmissibility and the effects of social distancing is responsible for the large difference in the count numbers between Hubei and the rest of Mainland China? Addressing this question in a proper way would make easier to comprehend how regional differences may shape the infection outbursts, which is important both locally (for explaining this puzzle), and more generally in the context of global COVID-19 pandemics progression.

ii. What is the Infected Fatality Rate (IFR, the number of fatalities per total number of *infected* cases) in China? Case Fatality Rate (CFR, the number of fatalities per *confirmed/detected* cases) can be obtained directly from the data but is highly sensitive to the testing coverage. IFR is a more fundamental mortality parameter, as it does not depend on the testing coverage, but is however much harder to determine, due to the unknown number of infected cases.

Addressing these questions allows understanding both the different response policies, and the inherent risks posed by the pandemics and will enable future cross-country comparisons. The developed methodology (i) demonstrates the usefulness of applying transdisciplinary expertise to efficiently analyze problems of nationwide importance, (ii) allows to readily analyze future outbreaks of COVID-19 and other infectious diseases, as it depends only on inference from straightforward and publically available data.

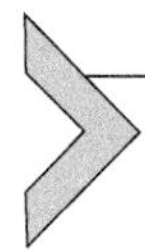

2. An overview of compartmental models of epidemic progression

In epidemiology, for practical and ethical reasons, it is fairly impossible to conduct scientific experiments in controlled conditions in order to investigate the spread of the disease in the human population (Brauer, 2008). Therefore, epidemiologists usually resort to collecting data from clinical reports on the observed situation in the field and, then, using mathematical models to interpret these data, i.e., to infer the principles underlying the process of disease spreading. These principles may point to potentially successful control strategies, as well as to the probable future status of the disease in the population. Epidemiological data can often be incomplete or inaccurate due to poorly controlled or non-standardized collection methods, which significantly complicates modeling. However, even a qualitative agreement of the model with the data can provide useful information of great practical importance. Hence, model predictions are widely used for making various estimates and answering important questions about the seriousness of the epidemic consequences. For example, how many people will be infected, require treatment, or die, or how many patients should the public health facilities expect at any given time? Also, how long will the epidemic last? To what extent could quarantine and self-isolation of the infected contribute to mitigating the effects of the epidemic? Model predictions guide the development of strategies to control the epidemic spread, including vaccination programs.

When the goal is to discover the general principles of epidemic progression, simple mathematical models, which can be solved and analyzed with a "pencil on paper," are a logical choice as they give insight into the properties of the examined process despite failing to reproduce it in detail. In 1927, Kermack and McKendrick formulated a simple model that predicted behavior similar to that observed in numerous epidemics (Kermack & McKendrick, 1927). It was a type of compartmental model describing the infection spread in a population by analogy with a system of vessels connected by pipes through which a fluid flows. Namely, the population is divided into compartments, and assumptions are made about the nature and the rate of the flow between them. The structure of the compartmental model—which sections and how many of them it will contain and how they will be connected—depends on the characteristics of transmission of a given infectious disease and whether the past disease provides immunity to re-infections or not. The model set by these two scientists is known as SIR (from **S**usceptible–**I**nfected–**R**ecovered). It divides the population into three classes which correspond to compartments (Fig. 2): Susceptible (*S*) class includes healthy individuals susceptible to infection, which have never been exposed to the virus; Those who are infected and can infect others belong to the Infected (*I*) class; Recovered (*R*) class encompasses those who are excluded from the population, either by quarantining the infected, or by acquiring immunity through recovery from disease or immunization, or by the death of the infected (Brauer, 2008).

Mathematically, this model is represented by a system of ordinary differential equations. The time derivative of the number of individuals in a compartment, i.e., the rate of their change, is given by the difference between the rates at which the compartment is filled and emptied. Analogous to the processes in which chemical species (e.g., proteins) are degraded or converted into others within a biochemical reaction network (Ingalls, 2013), the rate of transition of individuals from one compartment to another follows the law of mass action. For example, a person moves from compartment S to compartment I at the rate which is proportional to the product of the S and I, as the encounter with an infected person enables virus transmission to the susceptible one (Voit et al., 2015).

By formulating such (or similar) models, one assumes that the epidemic is a deterministic process. Namely, the state of the population at all times is completely determined by its previous state and the rules described by the model. This is a reasonable approximation in cases where the numbers of individuals in the compartments are large, i.e., in a commonly considered

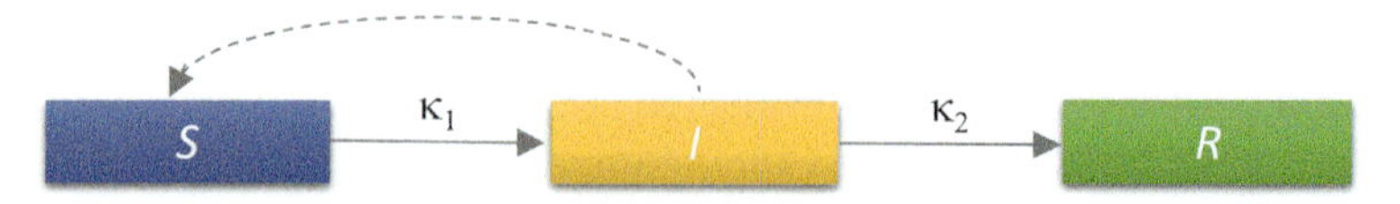

Fig. 2 Schematic representation of the SIR model. Rectangles denote model compartments containing susceptible (*S*), infected (*I*), and recovered (*R*) individuals in the population of size *N*. Permitted directions of flow between compartments are denoted by arrows, with the rates of flow indicated above them. The rates are expressed according to the law of mass action, where κ_1 and κ_2 are the rate constants. The dashed curve corresponds to bimolecular reaction, where newly infected are generated through interactions (contacts) between susceptible and already infected individuals.

deterministic range (>10). Such approximation (i.e., deterministic modeling) is well suited for the spread of COVID-19, which is up to now known for a large number of individuals in all compartments.

3. Systems biology approach to compartmental modeling of the COVID-19 epidemic

The above-introduced SIR model is likely the simplest compartmental model in mathematical epidemiology and many subsequent models are derivatives of this basic form. Among others, these extensions have also been developed toward including control measures such as vaccination or quarantine (Diekmann, Heesterbeek, & Britton, 2012; Keeling & Rohani, 2011; Martcheva, 2015). However, COVID-19 brought a challenge to account for previously unprecedented social distancing measures, taken by most countries. When included, these effects have been, up to now, accounted for by the direct changes in the transmissibility term (Chowell, Sattenspiel, Bansal, & Viboud, 2016; Tian et al., 2020), which, however, corresponds to introducing a phenomenological dependence in otherwise mechanistic models. That is, to be included consistently in the model, social distancing should move individuals from one compartment to the other, just as vaccination and quarantine are usually implemented. On the other hand, it is necessary to construct a minimal mechanistic model in terms of the ability to explain the data with the smallest number of parameters, so that relevant infection progression properties can be inferred without overfitting. With this goal in mind, we used our systems biology background to develop a minimal model that accounts for all the main qualitative features of the SARS-CoV-2 infection spread under epidemic mitigation measures. As outlined above, we opt for a deterministic model due to the robust and computationally less demanding parameter inference (Wilkinson, 2018).

To describe the COVID-19 epidemic, we developed SPEIRD model depicted schematically in Fig. 3. It assumes that healthy persons susceptible to infection (S), can be infected, but in the case of this (and many other) viruses they do not immediately become contagious to other people, but first spend some time in the compartment E (**E**xposed to the virus) and then develop symptoms and pass to the compartment I. Infected persons can either recover at home, moving to the compartment R, or they can be diagnosed with SARS-CoV-2 virus infection (**A**ctive detected cases). A (Active) cases can, further, either become healed (H) or die from the disease (F). To consistently implement the social distancing within this model structure, we

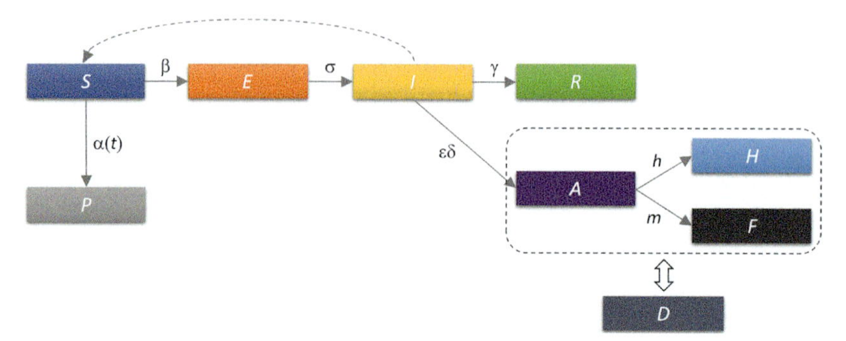

Fig. 3 Schematic representation of the SPEIRD model. Compartments and the transition rates are as indicated in the text, where transitions between different compartments are marked by arrows. The time-dependent transition rate from susceptible to protected category $\alpha(t)$ is indicated by the solid arrow. The infected can transition to the recovered category either without being diagnosed (transition to *R*), or being diagnosed and then transitioning to confirmed healed or fatality cases. The dashed rectangle indicates that *A*, *H*, and *F* categories in the starting model are substituted for the cumulative case counts (*D*), which removes *h* and *m* from the analysis, where *D* is fitted to the observed data.

included a compartment *P* (**P**rotected) in the model, which contains susceptible persons who are protected from exposure to the virus as a result of the epidemic mitigation measures, such as self-imposed isolation, social distancing, and advised changes in individual behavior.

The following differential equations describe how different categories change with time:

$$dS/dt = -\beta \cdot I \cdot S/N - \alpha(t) \cdot S \quad (1)$$

$$dE/dt = \beta \cdot I \cdot S/N - \sigma \cdot E \quad (2)$$

$$dI/dt = \sigma \cdot E - \gamma \cdot I - \varepsilon \cdot \delta \cdot I \quad (3)$$

$$dA/dt = \varepsilon \cdot \delta \cdot I - h \cdot A - m \cdot A \quad (4)$$

$$dH/dt = h \cdot A \quad (5)$$

$$dF/dt = m \cdot A \quad (6)$$

where β is the infection rate in a fully susceptible population; $\alpha(t)$, the time-dependent protection rate, i.e., the rate at which the population moves from susceptible to the protected category, quantifying the impact of the social protection measures; σ, the inverse of the exposed period; γ, the inverse of the infectious period; δ, the inverse of the period of the infection diagnosis; ε, the detection efficiency; h, the healing rate of diagnosed cases; m, the mortality rate.

The probability that an infected person will meet a susceptible person is proportional to S/N, where N is the total number of individuals in the population. The rate at which individuals move from S to E is obtained when the product of I and S/N is multiplied by the infection rate, β, which quantifies the efficiency of transmission of a particular virus in the population with certain demographic characteristics and meteorological conditions, and it does not depend on epidemic suppression measures. Thus, β is a characteristic of the virus, the population, and the external conditions in which the virus is transmitted. Since the compartment S is being emptied, the corresponding rate in the first equation is specified with the minus sign.

S also decays by moving the individuals to P with a protection rate that may vary with time. While mitigation measures are commonly accounted for by models with time-independent terms (Martcheva, 2015), we note that the social distancing term should depend on time, as this measure is introduced at a certain point in epidemics and may also evolve gradually. We denote the time point (more specifically, the date) of the onset of the social distancing measures in the examined population with t_0. The protection rate $\alpha(t)$ is then taken as 0 before t_0 and a constant value α afterwards.

One may notice a direct parallel between the model outlined above, and e.g., modeling gene expression regulation in systems biology with a step function that approximates the activity of a promoter to which repressor proteins are highly cooperatively bound: the promoter is initially silenced and upon receiving a signal which leads to the abrupt removal of repression, promoter activity rises sharply to its maximum value. We notice that the step function is a satisfactory approximation of the dynamics of social distancing, i.e., it may not be necessary to further increase the number of parameters by applying the Hill function (which describes a more gradual activation), since governments quickly introduced these measures, together with their effective implementation. Note however that in (Djordjevic et al., 2021) we introduced a more complex model with Hill function, and provided analytical results for key properties of this model.

Compartment E is filled by infecting the susceptibles and emptied by moving the individuals to I, with the rate σ representing the inverse value of the latent period during which the person is not contagious. While compartment I is filled with individuals from E, it is depleted through two channels. Individuals move to R with the rate γ, which is the inverse of the period of contagiousness, and to A with the rate δ, which is the inverse of the time required for diagnosis, multiplied by ε, reflecting that only a fraction (likely small due to many asymptomatic infections) of the total infected are detected. Note that case detection reduces the number of individuals in I that can infect susceptibles: the model assumes that the detected cases are quarantined and thus isolated from the general population. The numbers in compartments A, H, and F change following the same logic described for the other compartments.

We can further simplify the analysis by looking at the total number of detected cases (D), which is the sum of A, H, and F. By adding the Eqs. (4)–(6), we obtain:

$$dD/dt = \varepsilon \cdot \delta \cdot I, \tag{7}$$

and thus lose two parameters, h and m. The total number of detected cases in time is a measurable quantity from which we can determine the dynamics of other model compartments since this is the data that is available for various different regions and countries. Thereby, we assume that before t_0 social distancing does not take effect, and the measures introduced at t_0 will take effect on $D \sim 10$ days later, as this is about the time that elapses between infection and detection/diagnosis (Feng et al., 2020). Consequently, for the first $t_0 + 10$ days, the D curve reflects disease transmission without epidemic suppression measures.

3.1 Virus transmission in the early stages of epidemics

We will now focus on the dynamics of the infection spread at the very beginning of the epidemic, i.e., on the period before the introduction and practice of any control measures (Salom et al., 2021). Regarding the model, we assume that there is no social distancing (no transition from S to P), there is no quarantine, and almost the entire population consists of people susceptible to infection, so $S/N = 1$. This gives us an even simpler mathematical model which appears to be very useful because it allows analytical derivation of the expressions we need. Our system of Eqs. (1)–(3) and (7) is reduced to two linear differential equations that we can write in matrix form

$$\frac{d}{dt}\begin{pmatrix} E \\ I \end{pmatrix} = \begin{pmatrix} -\sigma & \beta \\ \sigma & -\gamma \end{pmatrix}\begin{pmatrix} E \\ I \end{pmatrix} = A\begin{pmatrix} E \\ I \end{pmatrix}, \tag{8}$$

determine the eigenvalues and eigenvectors of the matrix and, subsequently, the solutions of the system, $E(t)$ and $I(t)$. Specifically, the cumulative number of infected in time, $I(t)$, is obtained according to the following equation:

$$I(t) = C_1 \exp(\lambda_+ t) + C_2 \exp(\lambda_- t), \tag{9}$$

where λs are eigenvalues of the matrix. Since one of the eigenvalues, here denoted by λ_-, is negative, the corresponding term of the Eq. (9) will decrease over time, and I(t) will be effectively described by the first term, already after few days from the epidemic outbreak (Salom et al., 2021). We can further derive this equation for the dependence of the logarithm of the number of detected cases in time:

$$\log(D(t)) = \log(\varepsilon \cdot \delta \cdot I(0)/\lambda_+) + \lambda_+ \cdot t \tag{10}$$

This is the straight line equation whose slope is given by the value of λ_+ (the dominant, positive eigenvalue of the matrix in Eq. (8)).

Once we know λ_+, we can calculate the value of the so-called basic reproduction number, $R_{0,\text{free}}$, by fixing mean values of the latency period and the infectivity period ($\gamma = 0.4$ days^{-1}, $\sigma = 0.2$ days^{-1}), which are known from the literature and characterize the fundamental infection processes (Kucharski et al., 2020; Li et al., 2020):

$$R_{0,free} = \frac{\beta}{\gamma} = 1 + \frac{\lambda_+(\gamma + \sigma) + \lambda_+^2}{\gamma\sigma} \tag{11}$$

$R_{0,\text{free}}$ is an important epidemiological parameter that characterizes the inherent biological transmission of the virus in a completely unprotected population. In particular, it is the mean number of secondarily infected by one infected person introduced in a completely susceptible population. It depends on the biology of the specific virus, as well as the demographic characteristics of the population and the environmental conditions, while it does not depend on the applied infection control measures (Brauer, 2008). In Salom et al. (2021) we utilized a bioinformatics analysis, akin to those often used to understand complex data in systems biology, to pinpoint demographic and meteorological factors that affect $R_{0,\text{free}}$ (i.e., inherent virus transmissibility in population). This furthermore underlines that a rich array of techniques developed and/or widely used within systems biology can be successfully employed within infectious disease modeling.

4. Parameter analysis and inference

$R_{0,\text{free}}$, α, t_0, two initial conditions (I_0 and E_0), and the detection efficiency ε, are unknown and may differ between the provinces. Is it possible to determine these unknown parameters from different properties of the D curve? Early in the infection, almost the entire population is susceptible ($S \approx N$), so Eqs. (2) and (3) become linear, and decoupled from the rest of the system, as discussed in the previous section. This sets the ratio of I_0 to E_0, through the eigenvector components with the dominant (positive) eigenvalue of the Jacobian for this subsystem. This eigenvalue, corresponding to the initial slope of the *log*(*D*) curve, sets the value of λ_+ and subsequently, of $R_{0,\text{free}}$ (see Eq. 11). From Eq. (7) one can see that the product of I_0 and $\varepsilon{\cdot}\delta$ is set by dD/dt at the initial time ($t=0$). Later dynamics of the D curve is determined solely by the combination $t_\alpha = t_0 + 1/\alpha$ (which we denote as protection time), setting the time at which ~½ of the population moves to the protected category. We also numerically checked this, and confirmed that t_0 can be lowered at the expense of increasing $1/\alpha$, without affecting the fit quality. We allowed for t_0 to vary in reasonable proximity of January 23, as the social distancing was generally introduced close to Wuhan closure (e.g., on that date, all major events in Beijing were canceled) (Chen et al., 2020; Du et al., 2020), but we cannot be sure when the measures effectively took place. Our inferred t_0 values are within a week from Wuhan closure, appearing as reasonable. The remaining independent parameter (I_0) is then left to be determined from D curve properties at the late infection stage, such as its saturation time. The number of characteristic dynamics features is thus at least equal to the number of fit parameters, leading to constrained numerical analysis, so that overfitting is not expected. For few provinces, we however observed that I_0 can be decreased compared to the best fit value, without noticeably affecting the fit quality. For these provinces, we chose the lowest I_0 value that still leads to a comparably good fit. This allows obtaining the most conservative (i.e., as high as possible while still consistent with data) IFR estimate, as the reported fatality counts for provinces other than Hubei is surprisingly low.

Parameter inference and uncertainties are estimated separately for each province. However, within a given province, demographic, special, or population activity (network effects) heterogeneities (Britton, Ball, & Trapman, 2020; Diekmann et al., 2012), or seasonality effects (Wong et al., 2020), are not taken into account. These are potentially important, particularly for projections (longer-term predictions of infection dynamics under different scenarios), and can be readily included in our model. Such extensions would however complicate parameter inference, due to an increase in parameter number, as this may either lead to overfitting or require special/additional data that may be available only for a limited number of countries/regions (which would limit the generality of our proposed method). A more complex model structure may also obscure a straightforward relationship between the model parameters and distinct dynamical features of the confirmed case count curve analyzed above. While the inclusion of additional effects is left for future work, we here employ the model structure and parameter inference introduced above on widely available case count data, as proof of the principle for the generality of our proposed approach. Moreover, a major advantage of our approach is that it allows consistent analysis for all provinces with the same model, numerical procedure, and parameter set, allowing an objective comparison of the obtained results.

Our model was numerically solved by the Runge-Kutta method (Dormand & Prince, 1980) for each parameter combination. Parameter values were inferred by exhaustive search over a wide parameter range, to avoid reaching a local minimum of the objective function (R^2). To infer the unknown parameters, we fit (by minimizing R^2) the model to the observed total number of detected D for each province. As an alternative to exhaustive search, some of many optimization techniques used in epidemics modeling, such as the Markov chain Monte Carlo (MCMC) approach, can be used instead (Keeling & Rohani, 2011; Wong et al., 2020)—exhaustive search is however straightforward, guarantees that the global minimum is reached, and is in this case not computationally demanding. Errors were estimated through Monte-Carlo simulations (Press, Flannery, Teukolsky, & Vetterling, 1986), individually for each province with the assumption that count numbers follow the Poisson distribution. Monte-Carlo simulations were found as the most reliable estimate of the fit parameter uncertainties for a non-linear fit (Cunningham, 1993). This also serves as an independent check for overfitting, as in that case, data point perturbations would lead to large parameter uncertainties. We find no indication of this in the results reported below, as the inferred uncertainties (consistently indicated with all results) are reasonably small. In particular, the differences in the inferred parameter values, which are relevant for the reported results/conclusions, are statistically highly significant. P values for extracted parameter differences between provinces are estimated by the t-test.

5. Analysis of COVID-19 transmission in China

We used our SPEIRD model with the parameter inference described above, to analyze all Mainland China provinces, except Tibet, where only one COVID-19 case was reported. Parameters were estimated separately for each of the 30 provinces by the same model and parameter set, which enables an impartial comparison of the results presented below. To allow for a straightforward comparison of the infection progression between different provinces, the starting date (i.e., $t=0$) in our analysis is the same for all the provinces and corresponds to January 23 (when the data for all the provinces became publically available and continuously tracked (Hu et al., 2020)).

In Fig. 4A and B, we show that our model can robustly explain the observed D, in the cases of large outburst (Hubei on Fig. 4A), as well as for all other provinces, where D is in the range from intermediate (e.g., Guangdong) to low (e.g., Inner Mongolia). Provinces in Fig. 2B were selected to cover the entire range of observed D (from lower to higher counts), while comparably good fits were obtained for other provinces, which were all included in the further analysis. Our method is also robust to data perturbations (which might be frequent), e.g., in the case of Hubei (Wuhan), a large number of counts was added on February 12, based on clinical diagnosis (CT scan) (Feng et al., 2020), which is apparent as a discontinuity in observed D in Fig. 4A. The model however interpolates this discontinuity, finding a reasonable description of the overall data.

We backpropagated the dynamics inferred for Hubei, to estimate that January 5 (± 4 days) was the onset of the infection's exponential growth in the population (not to be confused with the appearance of first infections, which likely happened in December (Feng et al., 2020)). This agrees well with (Feng et al., 2020) (cf. Fig. 3A), which tracked cases according to their symptom onset (shifted for $\sim$12 days with respect to detection/diagnosis, cf. Fig. 3B), and coincides with WHO reports on social media that there is a cluster of pneumonia cases—with no deaths—in Wuhan (WHO, 2020). Since our analysis does not directly use any information before January 23, this agreement provides confidence in our I_0 estimate. Note that we infer I_0 separately for each province of interest, through which we also take into account different times of the infection onset in different provinces (so that earlier onset time would generally lead to a larger number of infected on January 23).

Key parameters inferred from our analysis are summarized in Fig. 4C–F, with individual results and errors for all the provinces shown in Table 1. Fig. 4C shows the distribution of $R_{0,\text{free}}$. Note that $R_{0,\text{free}}$ might depend on demographic (population density, etc.) and climate factors (temperature, humidity…), which are not controllable, but are unrelated to the applied social distancing measures (see above). It is known that the R_0 value can strongly depend on the model, e.g., the number of introduced compartments (Keeling & Rohani, 2011); accordingly, a wide range of R_0 values

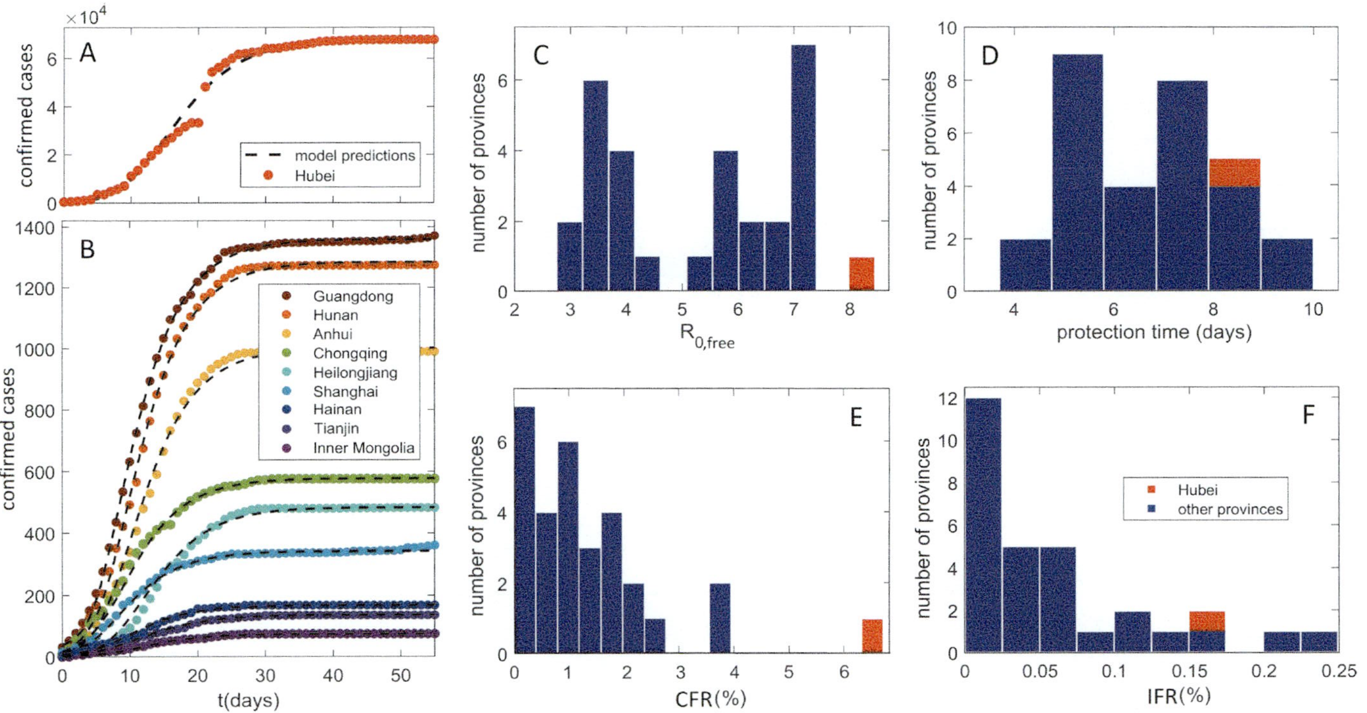

Fig. 4 Model predictions: comparison with data and key parameter estimates. Predictions (compared to data) of detected infection counts for (A) Hubei, (B) other Mainland China provinces. Zeros on the horizontal axis correspond to January 23, which is the initial time in our numerical analysis for all the provinces. The observed counts are shown by dots and our model predictions by dashed curves. Names of the provinces are indicated in the legend, with provinces selected to cover the full range of the observed total detected counts. The distribution with respect to provinces of (C) the basic reproduction number in the absence of social distancing, $R_{0,free}$, (D) the protection time t_{α}. (E) Case Fatality Rate, calculated directly from the reported data. (F) Infected Fatality Rate. The values for Hubei are indicated by the red bars.

Table 1 Inferred COVID-19 infection progression parameters for Mainland China provinces

Province	t_α (days)	R_0	E_0	I_0	IFR (%)	CFR (%)	Detected (%)
Anhui	6.6 ± 0.5	5.5 ± 0.8	920 ± 30	220 ± 20	0.04 ± 0.02	0.6 ± 0.3	6 ± 3
Beijing	7.9 ± 0.5	3.5 ± 0.4	610 ± 20	180 ± 10	0.12 ± 0.05	1.7 ± 0.7	7 ± 3
Chongqing	7.0 ± 0.2	3.5 ± 0.2	1900 ± 40	560 ± 20	0.04 ± 0.03	1.0 ± 0.5	4 ± 2
Fujian	3.7 ± 0.4	7 ± 2	1660 ± 40	360 ± 20	0.007 ± 0.003	0.3 ± 0.4	2 ± 1
Gansu	5 ± 1	6 ± 3	630 ± 20	150 ± 10	0.03 ± 0.04	1 ± 1	2 ± 3
Guangdong	5.0 ± 0.1	7 ± 1	1360 ± 40	290 ± 20	0.04 ± 0.01	0.6 ± 0.2	7 ± 2
Guangxi	7 ± 1	3.8 ± 0.8	1000 ± 30	290 ± 20	0.02 ± 0.02	0.8 ± 0.6	3 ± 3
Guizhou	8.1 ± 0.6	7 ± 1	53 ± 7	11 ± 3	0.06 ± 0.03	1 ± 1	4 ± 2
Hainan	7.6 ± 0.8	3.3 ± 0.7	300 ± 20	90 ± 10	0.21 ± 0.09	4 ± 2	6 ± 3
Hebei	6.0 ± 0.6	7 ± 2	240 ± 20	52 ± 7	0.11 ± 0.03	1.8 ± 0.8	6 ± 2
Heilongjiang	7 ± 1	6 ± 2	260 ± 20	59 ± 7	0.15 ± 0.07	2.9 ± 0.9	5 ± 3
Henan	7.0 ± 0.3	4.5 ± 0.5	1780 ± 40	460 ± 20	0.09 ± 0.04	1.7 ± 0.4	5 ± 2
Hubei	8.3 ± 0.2	8.2 ± 0.4	$31,900 \pm 400$	6600 ± 200	0.15 ± 0.09	6.5 ± 0.1	2 ± 2
Hunan	5.1 ± 0.1	6.8 ± 0.8	1430 ± 40	310 ± 20	0.02 ± 0.01	0.4 ± 0.2	5 ± 2
I. Mongolia	10.0 ± 0.8	2.8 ± 0.4	940 ± 30	300 ± 20	0.01 ± 0.03	1 ± 1	1 ± 3
Jiangsu	5.5 ± 0.5	7 ± 2	500 ± 20	110 ± 10	0 ± 0	0 ± 0	6 ± 2

Jiangxi	7.0 ± 0.2	5.6 ± 0.9	890 ± 30	210 ± 10	0.005 ± 0.002	0.1 ± 0.1	5 ± 2
Jilin	10.0 ± 0.7	4.0 ± 0.8	270 ± 20	76 ± 9	0.02 ± 0.02	1 ± 1	1 ± 2
Liaoning	7 ± 1	2.9 ± 0.7	1240 ± 40	390 ± 20	0.02 ± 0.04	2 ± 2	1 ± 2
Ningxia	5.3 ± 0.9	7 ± 3	72 ± 9	15 ± 4	0 ± 0	0 ± 0	6 ± 23
Qinghai	6.1 ± 0.6	4.0 ± 0.5	2260 ± 50	640 ± 30	0 ± 0	0 ± 0	0 ± 2
Shaanxi	5.2 ± 0.5	6 ± 1	380 ± 20	90 ± 10	0.07 ± 0.03	1.3 ± 0.8	6 ± 2
Shandong	9 ± 1	3.5 ± 0.5	900 ± 30	260 ± 20	0.06 ± 0.01	1.0 ± 0.4	6 ± 1
Shanghai	5.0 ± 0.4	6 ± 1	1570 ± 40	370 ± 20	0.02 ± 0.02	0.8 ± 0.5	2 ± 3
Shanxi	5.2 ± 0.5	6 ± 2	1600 ± 40	370 ± 20	0 ± 0	0 ± 0	1 ± 2
Sichuan	7.7 ± 0.8	3.7 ± 0.5	990 ± 30	280 ± 20	0.03 ± 0.02	0.6 ± 0.3	5 ± 3
Tianjin	7 ± 2	4 ± 2	170 ± 10	46 ± 7	0.14 ± 0.06	2 ± 1	7 ± 3
Xinjiang	7.3 ± 0.9	6 ± 1	42 ± 7	10 ± 3	0.25 ± 0.09	3 ± 2	8 ± 2
Yunnan	4.0 ± 0.2	7 ± 2	360 ± 20	76 ± 9	0.06 ± 0.03	1.2 ± 0.9	5 ± 2
Zhejiang	5.0 ± 0.1	7.2 ± 0.8	1340 ± 40	290 ± 20	0.005 ± 0.002	0.1 ± 0.1	7 ± 3

t_α, protection time; $R_{0,\text{free}}$, basic reproduction number; E_0, initial exposed; I_0, initial infected; *IFR*, Infected Fatality Rate; *CFR*, Case Fatality Rate; *detected %*, fraction of the infected population that has been detected. Error of the quantities correspond to one standard deviation.

were reported for China in the literature (Sanche et al., 2020; Wu, Leung, Bushman, et al., 2020). Consequently, a clear advantage of our study is that parameters for all China provinces were determined from the same model and data set, which allows direct comparisons. Our obtained average $R_{0,\text{free}}$ for provinces outside of Hubei is 5.3 ± 0.3, in a reasonable agreement with a recent estimate (≈ 5.7) (Sanche et al., 2020). Furthermore, we observe that $R_{0,\text{free}}$ for Hubei is a far outlier with a value of 8.2 ± 0.4, which is notably larger than for other provinces with $p \sim 10^{-11}$. This then strongly suggests that demographic and climate factors that determine $R_{0,\text{free}}$, played a decisive role in a large outburst in Hubei vs other provinces, which we further address below.

The distribution of protection time t_α for the provinces is shown in Fig. 4D, with the value for Hubei indicated in red. The mean for the other provinces is 6.6 ± 0.2 days. That is, we observe that the suppression measures were efficiently implemented, with ~½ of the population moving to the protected category within a week from Wuhan closure. The protection time for Hubei of 8.3 ± 0.2 days was longer, which is statistically significant at the $p \sim 10^{-11}$ level. The estimated less efficient protection in the case of Hubei may also be an important contributing factor in the surprising difference in Hubei vs other provinces, which we further investigate below.

CFR distribution, based on the fatality numbers reported for Hubei and other provinces is shown in Fig. 4E. These numbers are not based on the model predictions, i.e., can be straightforwardly obtained by dividing the total number of fatalities by the total number of detected cases. CFR for other provinces with a mean of $1.2 \pm 0.4\%$ is significantly smaller compared to CFR for Hubei, which was 4.6% before the correction on April 17, and 6.5% after the correction (with 1290 fatalities added to Wuhan). This large difference in CFR between Hubei and other provinces further accentuates the differences noted in Fig. 1.

IFR is harder to determine than CFR, as a majority of COVID-19 infections correspond to asymptomatic or mild cases that are by large not diagnosed (Day, 2020). We consequently calculate IFR as the total number of fatalities divided by the total number of infections (cumulative incidence) for the entire outburst, where cumulative incidence is estimated from our model. As the infections precede fatalities, both the total number of fatalities and the cumulative incidence in our estimate correspond to the entire outburst, so that all the infections had a sufficient time to recover or lead to fatalities—this is directly feasible for the provinces in China, where all detected case counts reached saturation. Note that IFR calculated in this way corresponds to an averaged quantity so that it does not capture possible time-dependent change over the outburst interval (in fact, for Wuhan it is known that the fatality rate was larger at the very beginning of the outburst). Nevertheless, the estimated IFR's present a reasonable measure of COVID-19 mortality across China provinces.

IFR distribution, which provides a much less biased measure of the infection mortality, is shown in Fig. 4F. In distinction to CFR, estimated IFR shows a much smaller difference between Hubei ($0.15 \pm 0.09\%$) and other provinces ($0.056 \pm 0.007\%$). Therefore, while Hubei is a clear outlier with respect to CFR, we observe similar IFR values for all Mainland China provinces, where few provinces have even higher IFR than Hubei. The ratio of IFR to CFR equals the fraction of all infected that got detected (*detection coverage*). We estimate that the mean detection coverage for all provinces except Hubei is higher than detection coverage for Hubei ($4.5 \pm 0.9\%$ vs $2 \pm 2\%$). This difference is responsible for a decrease by a factor of two from CFR to IFR for Hubei, compared to the other provinces, and consequently for more uniform mortality estimates at the IFR level. Xinjiang has the highest IFR of $0.25 \pm 0.09\%$ so that Hubei is not an outlier anymore. Estimated IFR's of up to 0.3% in China provinces are in general agreement with the estimates reported elsewhere (see e.g., (Bar-On et al., 2020; Djordjevic et al., 2021; Mizumoto, Kagaya, & Chowell, 2020)).

In Fig. 5A, two key infection progression parameters are plotted against each other: protection time t_α vs basic reproduction number $R_{0,\text{free}}$. Unexpectedly, there is a high negative correlation, with Pearson correlation coefficient $R = -0.70$, which is statistically highly significant $p \sim 10^{-5}$, where these two are a priori unrelated (see above). Actually, stronger social distancing measures—which by definition are not included in $R_{0,\text{free}}$—would lead to a decrease in *effective* transmissibility. This would then lead to a tendency of transmissibility to positively correlate with t_α, oppositely from the strong negative correlation observed in Fig. 5A. Therefore, higher basic reproduction number is genuinely related to a shorter protection time (larger effect of the suppression measures). Intuitively, this could be understood as a negative feedback loop, commonly observed in systems biology (Alon, 2019; Phillips, Kondev, Theriot, & Garcia, 2012), where larger $R_{0,\text{free}}$ leads to steeper initial growth in the infected numbers, which may elicit stronger measures and better observing of these measures by the population faced with a more serious outbreak. Interestingly, similar negative feedback was also obtained in the context of epidemics research other than COVID-19 (Wang, Andrews, Wu, Wang, & Bauch, 2015).

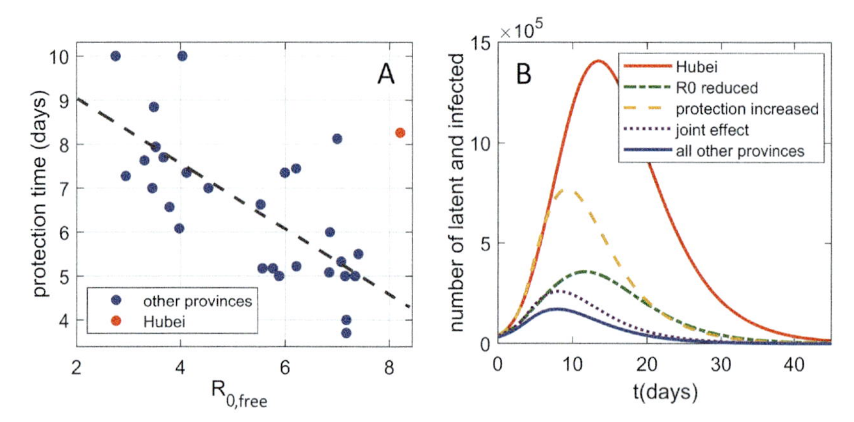

Fig. 5 The interplay of transmissibility and effective social distancing. (A) The correlation plot of t_α vs $R_{0,\text{free}}$ for all provinces, where the point corresponding to Hubei is marked in red. (B) The effect (on the Hubei dynamics of infected and latent cases) of reducing $R_{0,\text{free}}$ and t_α to the mean values of other Mainland China provinces. Both the unperturbed Hubei dynamics and the sum of infected and latent cases for all other provinces are included as references.

The two main properties of the Hubei outburst are therefore higher $R_{0,\text{free}}$ and t_α compared to other provinces. In Fig. 5B, we investigate how these two properties separately affect the Wuhan outburst for latent and infected cases, where unperturbed Hubei dynamics is shown by the red full curve. We first reduce only $R_{0,\text{free}}$ from the Hubei value, to the mean value for all other provinces (the dash-dotted green curve). We see that this reduction substantially lowers the peak of the curve, though it still remains wide. Next, instead of decreasing $R_{0,\text{free}}$, we decrease the protection time t_α to the mean value for all other provinces (dashed orange curve). While reducing t_α also significantly lowers the peak of the curve, its main effect is in narrowing the curve, i.e., reducing the outburst time. Finally, when $R_{0,\text{free}}$ and t_α are jointly reduced, we obtain the (dotted purple) curve that is both significantly lower and narrower than the original Hubei progression. This curve comes quite close to the curve that presents the sum of all other provinces (full blue curve)—the dotted curve remains somewhat above this sum, mainly because the initial number of latent and infected cases is somewhat higher for Hubei compared to the sum of all other provinces. This synergy between the transmissibility and the control measures will be further discussed below.

6. Conclusions

In this study, we applied a systems biology approach to develop a novel method of COVID-19 transmission dynamics. The model includes (time-dependent) social distancing measures in a simple manner, consistent with the compartmental mechanistic nature of the underlying process. The model has a major advantage that it is independent of the specific transmission process considered, and requires only commonly available count data as an input. The model allows extracting key infection parameters from the data that are readily available and publicly accessible (both for China and other countries), so that, in a nutshell, our approach is of wide applicability. To our best knowledge, such parameters (necessary to assess any future COVID-19 risks), were not extracted by other computational approaches.

The developed method is subsequently applied to the problem that appears highly non-trivial, i.e., to understand the puzzle created by the drastic differences in the infection and fatality counts between Hubei and the rest of Mainland China. The goal was to determine if it is possible to consistently explain such drastic differences by the same model, and what are the resulting numerical estimates and conclusions. We found that Hubei was a suitable ground for infection transmission, being an outlier with respect to two key infection progression parameters: having significantly larger $R_{0,\text{free}}$, and a longer time needed to move a sizable fraction of the population from susceptible to a protected category. While stricter measures were formally introduced in Hubei, the initial phase of the outburst put a large strain on the system, arguably leading to less effective measures compared to other provinces.

The fact that the initial epidemic in Hubei was not followed by similar outbursts in the rest of Mainland China may be understood as a serendipitous interplay of the two factors noted above. While both smaller $R_{0,\text{free}}$ and lower half-protection time (more efficient measures) significantly suppress the infection curve, their effect is also qualitatively different. While lowering $R_{0,\text{free}}$ more significantly suppresses the peak, decreasing the half-protection time significantly reduces the outburst duration. Consequently, the synergy of these two effects appears to lead to drastically suppressed infection dynamics in other Mainland China provinces compared to Hubei. The number of detected (diagnosed) cases in the entire Mainland China is, therefore, though unintuitive, well consistent with the model, and is explainable by a seemingly reasonable combination of circumstances. Our obtained negative feedback between transmissibility and effects of social distancing may be understood in terms of larger transmissibility triggering more stringent social distancing measures, where a similar conclusion was also obtained through entirely different means (a combination of real-time human mobility data and regression analysis) (Kraemer et al., 2020).

In summary, we showed that unintuitive dissimilarity in the infection progression for Hubei vs other Mainland China provinces is consistent with our model, and can be attributed to the interplay of transmissibility and effective protection, demonstrating that regional differences may drastically shape the infection outbursts. This also shows that comparisons in terms of the confirmed cases, or fatality counts (even when normalized for population size), between COVID-19 and other infectious diseases, or between different regions for COVID-19, are not feasible, and that parameter inference from quantitative models (individually for different affected regions) is necessary. Consequently, this paper illustrates that utilization of uncommon strategies, such as systems biology application to mathematical epidemiology, may significantly advance our understanding of COVID-19 and other infectious diseases.

Acknowledgment

This work was supported by the Ministry of Education, Science and Technological Development of the Republic of Serbia.

References

Alon, U. (2019). *An introduction to systems biology: Design principles of biological circuits.* CRC press.

Alon, U., Mino, R., & Yashiv, E. (2020). *10–4: How to reopen the economy by exploiting the coronavirus's weak spot.* The New York Times.

Bar-On, Y. M., Flamholz, A., Phillips, R., & Milo, R. (2020). Science forum: SARS-CoV-2 (COVID-19) by the numbers. *eLife, 9,* e57309.

Brauer, F. (2008). Compartmental models in epidemiology. In *Mathematical epidemiology* (pp. 19–79). Springer.

Britton, T., Ball, F., & Trapman, P. (2020). A mathematical model reveals the influence of population heterogeneity on herd immunity to SARS-CoV-2. *Science, 369*(6505), 846–849.

Chen, S., Yang, J., Yang, W., Wang, C., & Bärnighausen, T. (2020). COVID-19 control in China during mass population movements at New Year. *The Lancet, 395*(10226), 764–766.

Chowell, G., Sattenspiel, L., Bansal, S., & Viboud, C. (2016). Mathematical models to characterize early epidemic growth: A review. *Physics of Life Reviews, 18,* 66–97.

Cunningham, R. W. (1993). Comparison of three methods for determining fit parameter uncertainties for the Marquardt Compromise. *Computers in Physics,* 7(5), 570–576.

Day, M. (2020). Covid-19: Identifying and isolating asymptomatic people helped eliminate virus in Italian village. *British Medical Journal, 368*(m1165)(Online).

Diekmann, O., Heesterbeek, H., & Britton, T. (2012). Mathematical tools for understanding infectious disease dynamics. In (Vol. 7). Princeton University Press.

Djordjevic, M., Djordjevic, M., Ilic, B., Stojku, S., & Salom, I. (2021). Understanding infection progression under strong control measures through universal COVID-19 growth strategies. *Global Challenges*, *5*, 2000101.

Djordjevic, M., Rodic, A., & Graovac, S. (2019). From biophysics to 'omics and systems biology. *European Biophysics Journal*, *48*(5), 413–424.

Dormand, J. R., & Prince, P. J. (1980). A family of embedded Runge-Kutta formulae. *Journal of Computational and Applied Mathematics*, *6*(1), 19–26.

Du, Z., Xu, X., Wang, L., Fox, S. J., Cowling, B. J., Galvani, A. P., et al. (2020). Effects of proactive social distancing on COVID-19 outbreaks in 58 cities, China. *Emerging Infectious Diseases*, *26*(9), 2267.

Eilersen, A., & Sneppen, K. (2020). Cost–benefit of limited isolation and testing in COVID-19 mitigation. *Scientific Reports*, *10*(1), 1–7.

Feng, Z., et al. (2020). The epidemiological characteristics of an outbreak of 2019 novel coronavirus diseases (COVID-19)—China, 2020. *China CDC Weekly*, *2*(8), 113–122.

Hu, T., Weihe, W. G., Zhu, X., Shao, Y., Liu, L., Du, J., et al. (2020). Building an open resources repository for COVID-19 research. *Data and Information Management*, *4*(3), 130–147.

Ingalls, B. P. (2013). *Mathematical modeling in systems biology: An introduction*. MIT Press.

Karin, O., Bar-On, Y. M., Milo, T., Katzir, I., Mayo, A., Korem, Y., et al. (2020). Cyclic exit strategies to suppress COVID-19 and allow economic activity. *medRxiv* 2020.04.04.20053579.

Keeling, M. J., & Rohani, P. (2011). *Modeling infectious diseases in humans and animals*. Princeton University Press.

Kermack, W. O., & McKendrick, A. G. (1927). A contribution to the mathematical theory of epidemics. *Proceedings of the Royal Society of London Series A, Containing Papers of a Mathematical and Physical Character*, *115*(772), 700–721.

Kraemer, M. U., Yang, C.-H., Gutierrez, B., Wu, C.-H., Klein, B., Pigott, D. M., et al. (2020). The effect of human mobility and control measures on the COVID-19 epidemic in China. *Science*, *368*(6490), 493–497.

Kucharski, A. J., Russell, T. W., Diamond, C., Liu, Y., Edmunds, J., Funk, S., et al. (2020). Early dynamics of transmission and control of COVID-19: A mathematical modelling study. *The Lancet Infectious Diseases*, *20*(5), 553–558.

Li, Q., Guan, X., Wu, P., Wang, X., Zhou, L., Tong, Y., et al. (2020). Early transmission dynamics in Wuhan, China, of novel coronavirus–infected pneumonia. *New England Journal of Medicine*, *382*(13), 1199–1207.

Martcheva, M. (2015). An introduction to mathematical epidemiology. In (Vol. 61)Springer.

Mizumoto, K., Kagaya, K., & Chowell, G. (2020). Early epidemiological assessment of the transmission potential and virulence of coronavirus disease 2019 (COVID-19) in Wuhan City, China, January–February, 2020. *BMC Medicine*, *18*(1), 1–9.

Phillips, R., Kondev, J., Theriot, J., & Garcia, H. (2012). *Physical biology of the cell*. Garland Science.

Press, W. H., Flannery, B. P., Teukolsky, S. A., & Vetterling, W. T. (1986). *Numerical recipes: The art of scientific computing*. Cambridge: Cambridge University Press.

Saad-Roy, C. M., Grenfell, B. T., Levin, S. A., Pellis, L., Stage, H. B., van den Driessche, P., et al. (2021). Superinfection and the evolution of an initial asymptomatic stage. *Royal Society Open Science*, *8*(1), 202212.

Salom, I., Rodic, A., Milicevic, O., Zigic, D., Djordjevic, M., & Djordjevic, M. (2021). Effects of demographic and weather parameters on COVID-19 basic reproduction number. *Frontiers in Ecology and Evolution*, *8*, 617841.

Sanche, S., Lin, Y. T., Xu, C., Romero-Severson, E., Hengartner, N., & Ke, R. (2020). High contagiousness and rapid spread of severe acute respiratory syndrome coronavirus 2. *Emerging Infectious Diseases*, *26*(7), 1470–1477.

Tian, H., Liu, Y., Li, Y., Wu, C.-H., Chen, B., Kraemer, M. U., et al. (2020). An investigation of transmission control measures during the first 50 days of the COVID-19 epidemic in China. *Science, 368*(6491), 638–642.

Vilar, J. M., & Saiz, L. (2020). The evolving worldwide dynamic state of the COVID-19 outbreak. *medRxiv*.

Voit, E., Martens, H., & Omholt, S. (2015). 150 years of the mass action law. *PLoS Computational Biology, 11*(1), e1004012.

Wang, Z., Andrews, M. A., Wu, Z.-X., Wang, L., & Bauch, C. T. (2015). Coupled disease–behavior dynamics on complex networks: A review. *Physics of Life Reviews, 15*, 1–29.

WHO. (2020). *WHO timeline—COVID-19*. https://www.who.int/news-room/detail/27-04-2020-who-timeline—covid-19: World Health Organization.

Wilkinson, D. J. (2018). *Stochastic modelling for systems biology* (3rd ed.). London: Chapman and Hall/CRC.

Wong, G. N., Weiner, Z. J., Tkachenko, A. V., Elbanna, A., Maslov, S., & Goldenfeld, N. (2020). Modeling COVID-19 dynamics in Illinois under nonpharmaceutical interventions. *Physical Review X, 10*(4), 041033.

Wu, J. T., Leung, K., Bushman, M., Kishore, N., Niehus, R., de Salazar, P. M., et al. (2020). Estimating clinical severity of COVID-19 from the transmission dynamics in Wuhan, China. *Nature Medicine, 26*(4), 506–510.

Wu, J. T., Leung, K., & Leung, G. M. (2020). Nowcasting and forecasting the potential domestic and international spread of the 2019-nCoV outbreak originating in Wuhan, China: A modelling study. *The Lancet, 395*(10225), 689–697.

CHAPTER TEN

An integrative analysis to distinguish between emphysema (EML) and alpha-1 antitrypsin deficiency-related emphysema (ADL)—A systems biology approach

S. Udhaya Kumar[a], N. Madhana Priya[b], D. Thirumal Kumar[c], V. Anu Preethi[d], Vibhaa Kumar[a], Dhanushya Nagarajan[a], R. Magesh[b], Salma Younes[e], Hatem Zayed[e], and C. George Priya Doss[a,*]

[a]School of BioSciences and Technology, Vellore Institute of Technology, Vellore, Tamil Nadu, India
[b]Department of Biotechnology, Sri Ramachandra Institute of Higher Education and Research (DU), Porur, Chennai, Tamil Nadu, India
[c]Meenakshi Academy of Higher Education and Research, Chennai, Tamil Nadu, India
[d]School of Computer Science and Engineering, Vellore Institute of Technology, Vellore, Tamil Nadu, India
[e]Department of Biomedical Sciences, College of Health and Sciences, QU Health, Qatar University, Doha, Qatar
*Corresponding author: e-mail address: georgepriyadoss@vit.ac.in

Contents

Advances in Protein Chemistry and Structural Biology, Volume 127
ISSN 1876-1623
https://doi.org/10.1016/bs.apcsb.2021.02.004

Abstract

Lung Emphysema is an abnormal enlargement of the air sacs followed by the destruction of alveolar walls without any prominent fibrosis. This study primarily identifies the differentially expressed genes (DEGs), interactions between them, and their significant involvement in the activated signaling cascades. The dataset with ID GSE1122 (five normal lung tissue samples, five of usual emphysema, and five of alpha-1 antitrypsin deficiency-related emphysema) from the gene expression omnibus (GEO) was analyzed using the GEO2R tool. The physical association between the DEGs were mapped using the STRING tool and was visualized in the Cytoscape software. The enriched functional processes were identified with the ClueGO plugin's help from Cytoscape.

Further integrative functional annotation was performed by implying the GeneGo Metacore™ to distinguish the enriched pathway maps, process networks, and GO processes. The results from this analysis revealed the critical signaling cascades that have been either activated or inhibited due to identified DEGs. We found the activated pathways such as immune response IL-1 signaling pathway, positive regulation of smooth muscle migration, BMP signaling pathway, positive regulation of leukocyte migration, NIK/NF-kappB signaling, and cytochrome-c oxidase activity. Finally, we mapped four crucial genes (*CCL5*, *ALK*, *TAC1*, *CD74*, and *HLA-DOA*) by comparing the functional annotations that could be significantly influential in emphysema molecular pathogenesis. Our study provides insights into the pathogenesis of emphysema and helps in developing potential drug targets against emphysema.

1. Introduction

Pulmonary emphysema is a type of chronic obstructive pulmonary disease (COPD), a serious lung disease that portrays the lung air spaces (alveoli) as an irregular and prolonged expansion, accompanied by the destruction of their walls. COPD is a silent killer in low- and middle-income countries. It is reported that about 328 million people worldwide are affected by COPD (Eisner et al., 2010). As per the World Health Organization (WHO), COPD is the primary cause of death worldwide at present, responsible for approximately 6% of total deaths (WHO, 2020). In 15 years, it is predicted to become the leading cause of death (WHO, 2010).

Long-term and severe exposure to toxic substances is the trigger of emphysema, among which smoking cigarettes is the primary cause. It is estimated that 80% to 90% of COPD patients are cigarette smokers

(Pahal, Avula, & Sharma, 2020). There are also less common causes of emphysema attributed to genetics, primarily related to alpha-1 antitrypsin deficiency, an autosomal recessive disease estimated to account for 1–2% of COPD cases. The causes of long-term damage to airways distal to bronchiole are the clinical symptoms of emphysema, known as the acinus, including alveolar sacs, respiratory bronchiole, alveoli, and alveolar ducts. Due to the proteinases' action, there is persistent aberrant dilatation of the air spaces and degradation of their walls. Thus, it leads to a reduction in the surface area of the alveolar and capillary, decreasing the gas exchange, typically manifesting as shortness of breath (Goldklang & Stockley, 2016).

In emphysema, the response of the body's immune system leads to deterioration in the lungs of structural components, particularly elastin, eventually, the developing regions in the lungs that could not normally function. Elastin is a vital constituent of the extracellular matrix expected to manage lung parenchyma integrity and narrow airways (Hizawa et al., 2008; Pahal et al., 2020). The elastase/anti-elastase imbalance makes it susceptible to the pulmonary disruption that leads to airspace expansion. Elastase and Proteinase (i.e., cathepsins and neutrophil-derived proteases) eliminate the connective tissue of lung parenchyma by acting against elastin. The destruction of epithelial cells is due to the release of TNF-α and perforins by Cytotoxic T cells. Cigarette smoking often blocks alveolar macrophage and anti-proteolytic enzymes (Yuan, Chang, Lu, & Deng, 2017). Smoking frequently causes pulmonary oxidative stress, which results in the synthesis of reactive nitrogen, reactive oxygen species, and antioxidant reduction, including vitamin E and A, glutathione, catalase, and superoxide dismutase, are decreased (Domej, Oettl, & Renner, 2014). This oxidant-antioxidant imbalance reduces *NRF2* activity. Oxidative stress, particularly inflammation, aging, and DNA damage, has many direct/indirect effects in downstream signaling (Domej et al., 2014). It also induces epigenetic modifications, including the inactivation of histone deacetylase, and the inactivation of HDAC2 results in continued pro-inflammatory gene expression and emphysema development (Taraseviciene-Stewart & Voelkel, 2008).

A useful technique to examine intricate pathogenic disease mechanisms is to evaluate a microarray-based gene expression. Several studies focused on identifying emphysema pathogenesis; only a few microarray studies allow a significant number of smokers with extreme emphysema to determine the expression profiles of genes from emphysematous lung tissue (Quackenbush, 2001). The genes that are used to distinguish severe emphysema from mild or no emphysema can be done by using computational algorithms, and various studies have utilized systematic analyses to identify biomarkers and hub genes in various diseases (Kumar, Kumar, Siva, Doss, & Zayed, 2019; Udhaya Kumar, Rajan, et al., 2020; Udhaya Kumar, Thirumal Kumar, Bithia, et al., 2020; Wan et al., 2019; Zhao et al., 2020).

To better describe the emphysema molecular basis, we suggested a method to classify emphysema-responsible DEGs and pathways. This bioinformatics approach entailed retrieving DEGs data from the GEO database, analysis using GEO2R, and visualization using volcano plots. Gene interaction networks were mapped and visualized using STRING and Cytoscape tools. Furthermore, protein interactions and pathways were explored to gain a better understanding of the complex mechanisms involved in emphysema. Our analysis demonstrates the pathophysiology associated with DEGs and the mechanism related to the pathogenesis of emphysema, which could provide a unique perspective into the establishment of a novel strategy for clinical therapy.

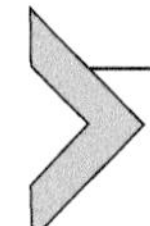

2. Methods

2.1 Screening of DEG's

The emphysema dataset (GSE1122) was obtained from the GEO database. "Lung Emphysema" and "*Homo sapiens*" were used as keywords to identify the dataset. The dataset contains normal lung tissue (NML), usual emphysema (EML), and alpha-1 antitrypsin deficiency-related emphysema (ADL), comprising 15 samples in total (Golpon et al., 2004). Further, the GEO2R tool was used to determine the DEGs from the emphysema dataset based on the logFC and *P*-values (Davis & Meltzer, 2007; Smyth, 2005). Along with the GEO2R tool, the Benjamin and Hochberg and *t*-test methods were also used to identify the *P*-values and logFC. The values $P < 0.05$ and $\text{logFC} \leq -2$ were considered cutoff criteria to identify the DEGs (Aubert, Bar-Hen, Daudin, & Robin, 2004). Identified DEGs were further used to construct a volcano plot using the RStudio software. The results were collected and used for further analysis.

2.2 Protein-physical interaction network

Protein-protein interaction (PPI) networks were drawn with the help of Search Tool for the Retrieval of Interacting Genes/Proteins (STRING) (Szklarczyk et al., 2015), a web-tool with detailed documentation and potential interactions, to establish the interacting connections between the DEG-encoded macromolecules. To avoid inaccurate PPI networks, a cutoff of ≥ 0.4 was set to get the significant PPIs. Results are visualized in Cytoscape software (v 3.7.2) (Shannon et al., 2003).

2.3 Backbone network analysis

The Network Analyzer add-on app in Cytoscape was used to identify the DEGs' backbone networks (Saito et al., 2012). Using powerful network methods, this module evaluates a detailed description of static and dynamic computational complexity parameters. A full set of topological conditions for undirected and directed connections is computed by the NetworkAnalyzer, which includes edges, nodes numbers, characteristic path length, closeness and betweenness, shortest path lengths, and topological coefficients (Assenov, Ramírez, Schelhorn, Lengauer, & Albrecht, 2008; Brandes, 2001; Newman, 2005; Stelzl et al., 2005).

2.4 ClueGO enrichment analysis

ClueGO is a Cytoscape app that facilitates the biological interpretation of large lists of genes and proteins by selecting representative Gene Ontology (GO) terms and pathways from multiple ontologies and visualizes them into functionally organized networks. ClueGO is strengthened by CluePedia, which enables a thorough analysis of the signaling cascades. Results are illustrated as a functionally grouped network, as plots and tables. Results are automatically mapped on the network in different visual styles (Bindea et al., 2009). A significant *P*-value <0.05 with a kappa score of 0.4 was considered for the present study (Fu, Zhang, Yang, Huang, & Xin, 2020; Li, Zhao, & You, 2019; Mishra et al., 2021; Udhaya Kumar, Thirumal Kumar, Siva, et al., 2020; Yan et al., 2019).

2.5 GeneGo analysis

The Metacore and Cortellis Solution software was used to categorize statistically significant DEGs to delineate the GO function and pathway enrichment analyses identified from ClueGO. In terms of hypergeometric distribution, we utilized the pathway maps method to define the activated signaling cascades involving DEGs with FDR $P < 0.005$. Graphical representations of the signaling cascades between the DEGs were created with a *P*-value <0.05. Based on ClueGO and GeneGo, we made the PPI network for the crucial genes with a higher connectivity degree. Based on the GO, AN from GeneGo, and KEGG pathway analyses, the hub genes were selected. The GeneMANIA web-server was used to make an interrelation analysis for the identified hub genes (Warde-Farley et al., 2010).

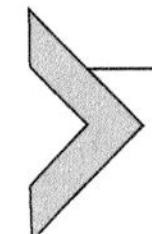

3. Results

3.1 Data acquisition

The dataset GSE1122 was retrieved from the GEO database, which includes five normal lung tissue samples (NML), five of 'usual' emphysema (EML), and five of alpha-1 antitrypsin deficiency-related emphysema (ADL). The dataset was generated with the help of the GPL80 platform (Affymetrix Human Full Length HuGeneFL Array) (Golpon et al., 2004). The Gene expression profiling of lung tissue comprised a total of 15 samples from three subsets was provided in Table 1. Further analysis was conducted in two sets, the NML vs. ADL and the NML vs. EML. By implying the GEO2R tool, the top DEGs were found by comparing the ADL and EML samples with the NML control group.

3.2 Identification of DEGs

We used the GEO2R to analyze the dataset GSE1122 for the top DEGs for both the ADL and EML phenotypes individually. A total of 6434 DEGs were identified for both ADL and EML subsets. Out of the total, DEG's top 250 were selected based on the cutoff criteria ($P < 0.05$ and $\log FC \leq -1$) and are shown in Tables 1 and 2 in Supplementary material in the online version at https://doi.org/10.1016/bs.apcsb.2021.02.004 for ADL and

Table 1 Information of GSE1122 from GEO dataset.

Group	Accession	Disease state	Organism
NML	GSM18403 GSM18404 GSM18405 GSM18406 GSM18407	Undiseased	*Homo sapiens*
ADL	GSM18408 GSM18409 GSM18410 GSM18411 GSM18412	Alpha-1 Antitrypsin Deficiency-related	*Homo sapiens*
EML	GSM18413 GSM18414 GSM18415 GSM18416 GSM18417	Usual emphysema	*Homo sapiens*

Table 2 Upregulated and downregulated DEGs of ADL & EML from GSE1122 dataset.

Probes	Gene name	*P*-value	logFC
ADL			
L49169_at	*FOSB*	0.00186378	3.192
U37529_at	*TAC1*	0.0003588	2.998
L19778_at	*HIST1H2AH/HIST1H2AG*	0.00400876	2.756
U25997_at	*STC1*	0.00285447	2.698
M29335_s_at	*HLA-DOA*	0.00143758	2.605
X12517_at	*SNRPC*	0.0000011	−2.539
L06147_at	*GOLGA2*	0.00000946	−2.303
EML			
M58600_rna1_at	*SERPIND1*	0.0060497	3.226
M29335_s_at	*HLA-DOA*	0.0003241	3.197
M87860_at	*LGALS2*	0.0009677	2.948
X93510_at	*PDLIM4*	0.0048253	2.73
D89050_at	*OLR1*	0.0029742	2.727
D83657_at	*S100A12*	0.0099451	−3.375
Z49205_at	*P2RY1*	0.0001037	−2.665
M58460_at	*EXOSC9*	0.0002225	−2.428
U31116_at	*SGCB*	0.007054	−2.115
X74819_at	*TNNT2*	0.0056765	−2.031

EML, respectively. Out of the top 250 DEGs, the most upregulated and downregulated genes were identified (Table 2). Using RStudio, we constructed a Volcano plot for the DEGs identified from this study with the logFC value as X-axis and *P*-values as Y-axis by comparing the ADL, EML, and NML (Fig. 1). The volcano plot in Figs. 1A (NML vs. ADL) and B (NML vs. EML) describes all the DEGs with a logFC against—log10 (*P*-value) between the three groups.

3.3 PPI network construction and analysis of backbone

We utilized the STRING tool to evaluate the PPI network of DEGs from all three groups (ADL, EML, and NML) to assess the physical associations

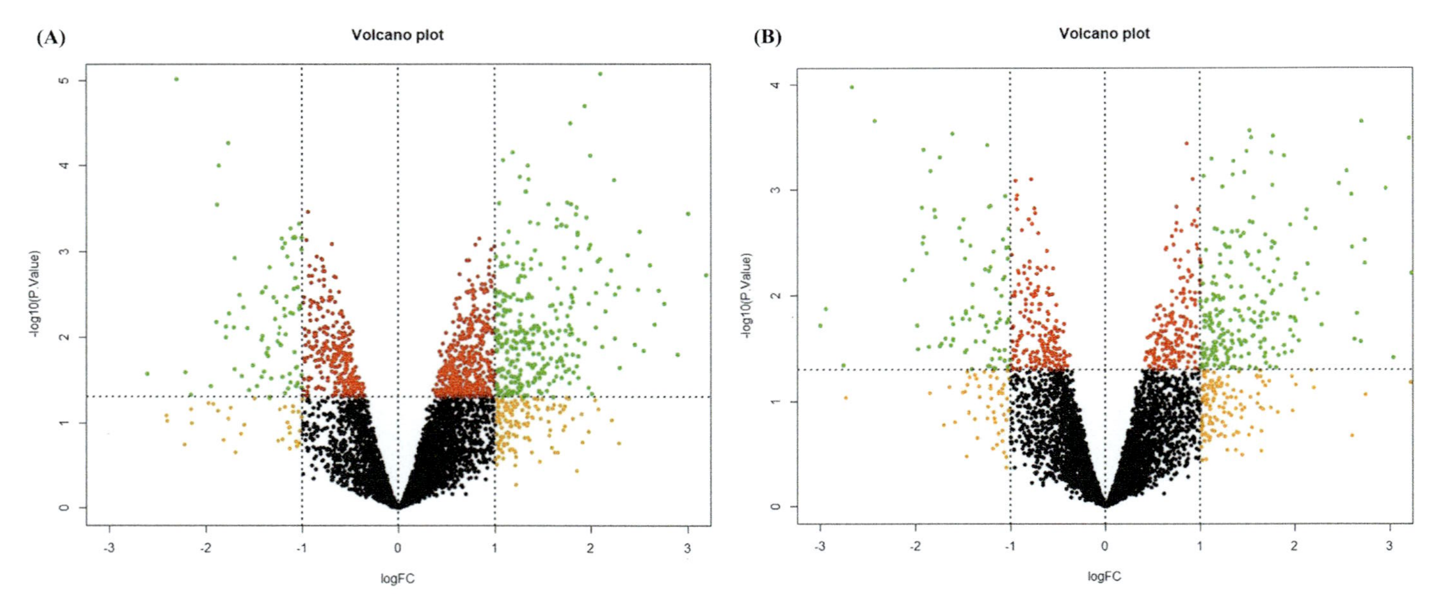

Fig. 1 Volcano Plot of the DEG's in lung emphysema from the GSE1122 dataset (A: NML vs. ADL; B: NML vs. EDL) visualized using Rstudio. The plot shows the logFC value on the x-axis and the *P*-value on the y-axis. The red dots exhibit the *P*-values <0.05; the yellow dots show logFC values ≤ -1 and ≥ 1. The green dots and black dots indicate the significant and non-significant genes, respectively, present in the dataset. The genes are significant when both logFC and *P*-value are satisfied. Each array dataset has both upregulated and downregulated genes; based on its expression, the right panel represents upregulated genes, whereas left panel represents downregulated genes.

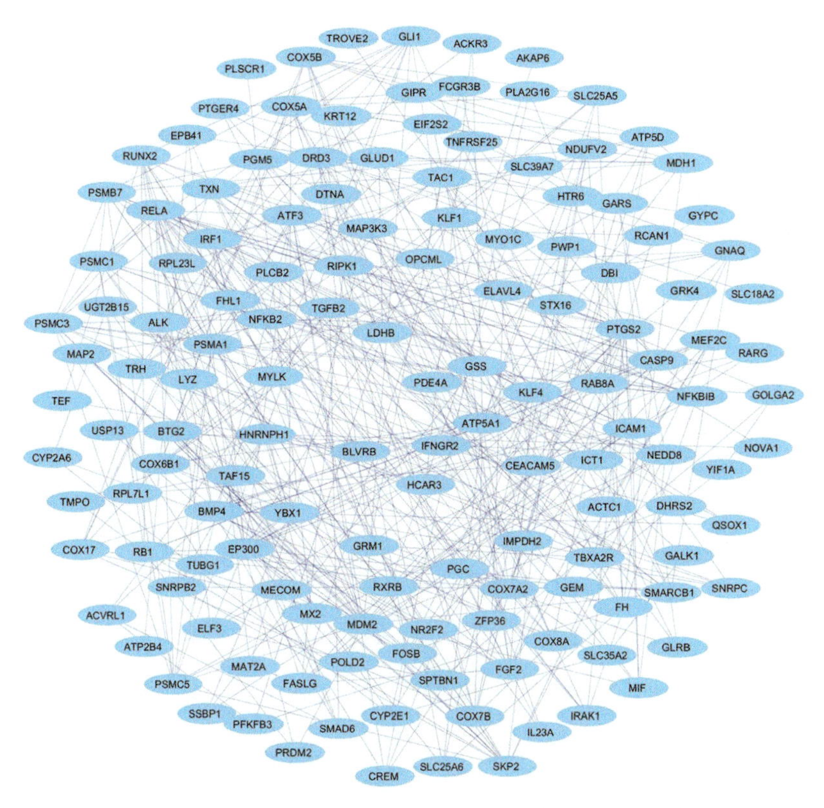

Fig. 2 The retrieved genes were plotted as a network using the STRING and Cytoscape software. The PPI network shows the interaction of NML vs. ADL group. The nodes are represented in spheres and the edges as lines. Nodes are colored in blue and edges in gray.

and their functions. The criteria of ≥0.4 score were considered to be an essential interaction for the nodes. The PPI networks of ADL and EML are represented graphically in Figs. 2 and 3, respectively. The PPI backbone of ADL and EML consists of 142 & 156 nodes, respectively, with calculated clustering coefficients of 0.156 and 0.158. Each node represents the protein, and interactions are shown as edges, which counts 420 and 523 for ADL and EML, respectively. The Network Analyzer plugin of Cytoscape was applied to analyze the backbone of the ADL and EML PPI, and the topological network parameters are shown in Table 3. The topological parameters of the ADL and EML PPI, along with the topological coefficient, betweenness centrality, topological components, shortest path length distribution, distribution of node degree, and closeness centrality, is given in Fig. 4A and B.

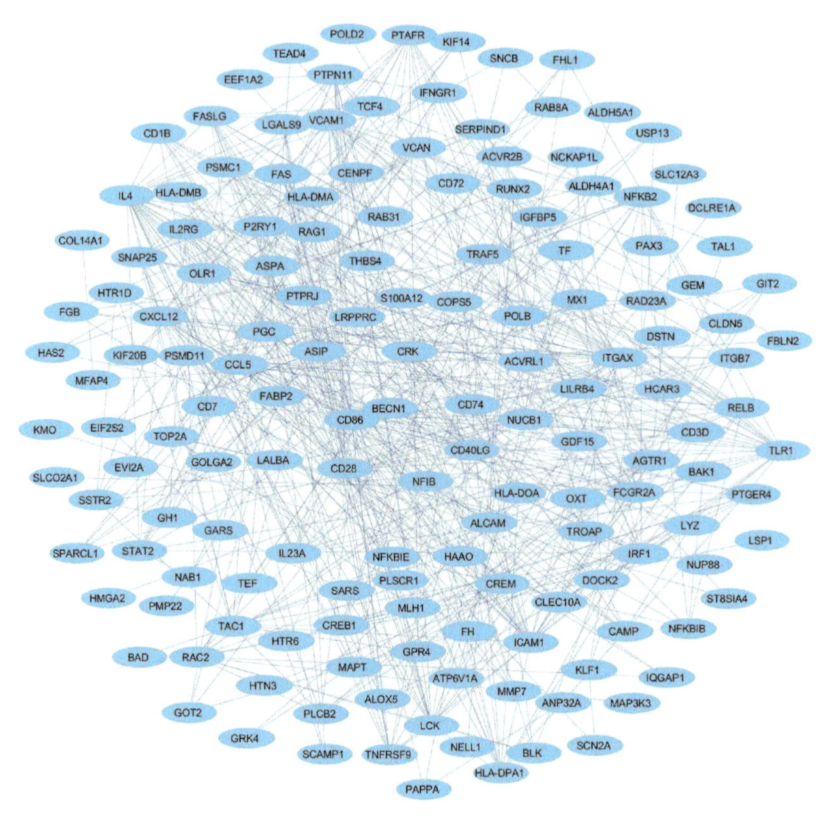

Fig. 3 The PPI network of NML vs. EML group was constructed with genes from the STRING using Cytoscape. Each node is represented in spheres and the edges as lines. The network demonstrates the PPI between the DEGs identified from GSE1122 using Cytoscape. The nodes are in blue ellipse and edges as gray lines.

Table 3 Topological parameters for ADL & EML PPI network.

		Comprehend values	
S. No	**Topological parameters**	**NML vs. ADL**	**NML vs. EML**
1.	Number of nodes	142	156
2.	Number of Eedges	420	523
3.	Clustering co-efficient	0.156	0.158
4.	Network density	0.042	0.043
5.	Network centralization	0.144	0.198
6.	Network heterogeneity	0.808	1.002
7.	Average path length	2.497	2.231
8.	Average number of neighbors	5.915	6.705

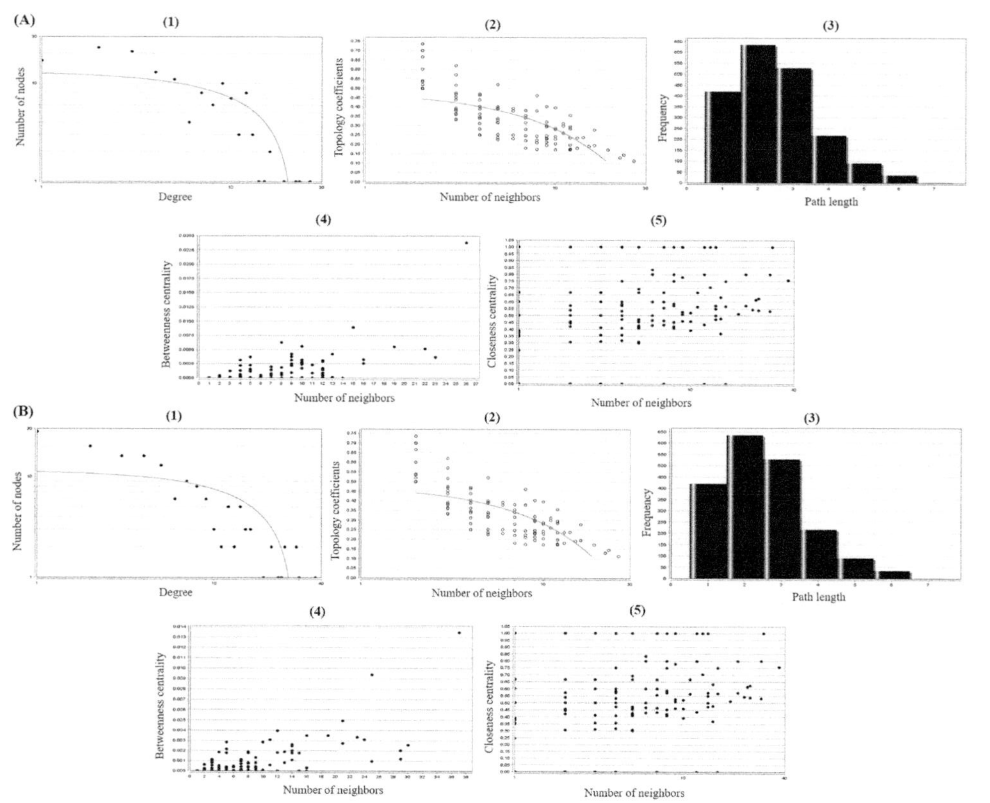

Fig. 4 The topology parameters for DEGs backbone network. (A) The network of NML vs. ADL DEGs. (B) The network of NML vs. EML DEGs. (1) Distribution of the node degree (2) Topological coefficients (3) Shortest path length distribution (4) Betweenness centrality (5) Closeness centrality.

3.4 ClueGO enrichment analysis

The ClueGO/CluePedia plugin of Cytoscape was employed to assess and examine the functional annotation of the identified DEGs. The statistical options for functional enrichment analysis were set with $P \leq 0.05$, Benjamini-Hochberg correction, and kappa score ≥ 0.4 on a two-sided hypergeometric test as required criteria. The DEGs of ADL were significantly found to be enriched in regulation of blood vessel endothelial cell migration (GO:0043535), NIK/NF-kappaB signaling (GO:0038061), cytochrome-c oxidase activity (GO:0004129), regulation of leukocyte adhesion to vascular endothelial cell (GO:1904994), positive regulation of transcription regulatory region DNA binding (GO:2000679), interleukin-12 production (GO:0032735), filopodium assembly (GO:0051491), cartilage development (GO:0061036), smooth muscle cell proliferation (GO:0048661), negative regulation of cellular response to the drug (GO:2001039), response to isoquinoline alkaloid (GO:0014072), regulation of ubiquitin-protein ligase activity (GO:1904666), cytochrome-c oxidase activity (GO:0004129), regulation of chondrocyte differentiation (GO:0032330), and metallo carboxypeptidase activity (GO:0004181) (Fig. 5). The DEGs from EML are enriched in metaphase plate congression (GO:0051310), positive regulation of tumor necrosis factor superfamily cytokine production (GO:1903557), cell junction assembly (GO:1901890), smooth cell migration (GO:0014911), negative regulation of integrin activation (GO:0033624), T cell differentiation involved in immune response (GO:0002292), lymphocyte differentiation (GO:0030098), BMP signaling pathway (GO:0030509), quinolinate biosynthetic process (GO:0019805), dicarboxylic acid metabolic process (GO:0043648), positive regulation of isotype switching to IgG isotypes (GO:0048304), leukocyte differentiation (GO:0002521), regulation of epithelial cell apoptotic process (GO:1904035), positive regulation of peptidyl-tyrosine phosphorylation (GO:0050731), leukocyte cell-cell adhesion (GO:1903039), activation of Janus kinase activity (GO:0010536), MHC class II protein complex assembly (GO:0002399) and regulation of corticosteroid hormone secretion (GO:2000846) (Fig. 6). Taken together, the identified results from the ClueGO analysis revealed that both the emphysema types are not commonly enriched between the pathways. This clearly states that ADL and EML undergo activation of non-identical pathways and results in the progression of lung emphysema.

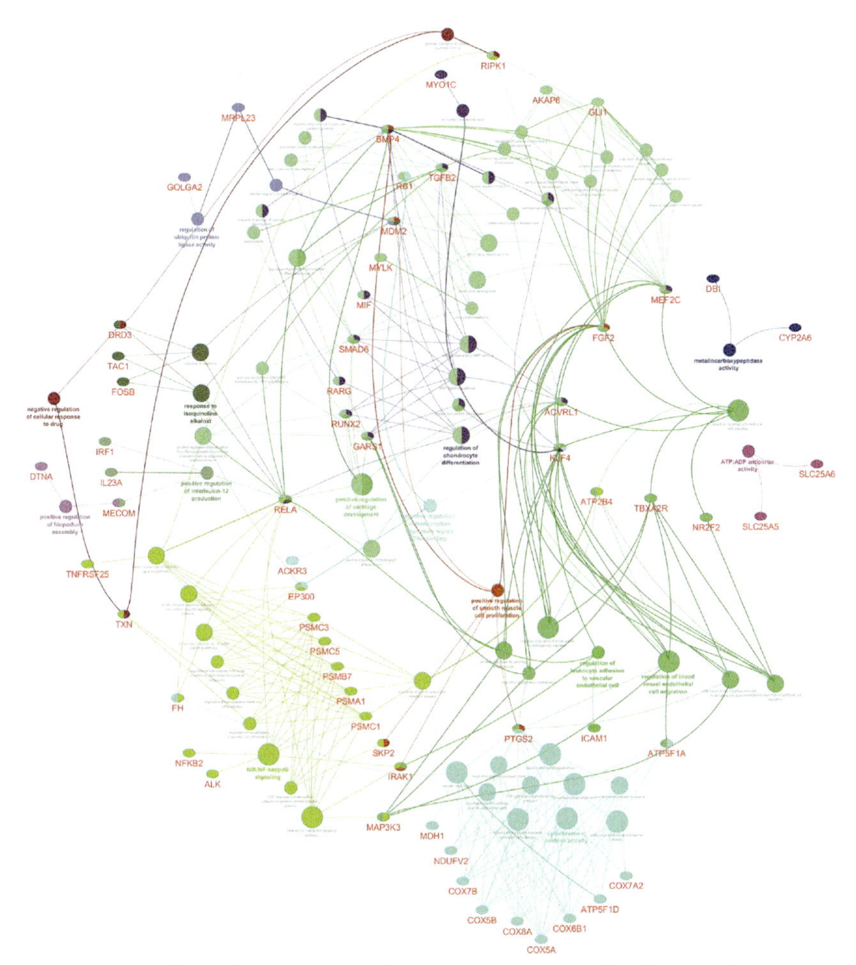

Fig. 5 Functional enrichment analysis was analyzed using the ClueGO. The essential molecular functions and biological processes involved between the DEGs of 3 groups were seen. The interconnection between the DEGs of ADL is described based on the kappa score = 0.4.

3.5 Enrichment analysis by GeneGo

To understand the DEGs set between ADL and EML, the MetacoreTM GeneGo software was used to annotate the enriched pathways, MF, and BP. The strongly enriched pathways with hypergeometric fisher FDR < 0.05 and P-value ≤ 0.05 were taken. A low P-value indicates more significance to the DEGs, signifying a better score than the entity, and IDs

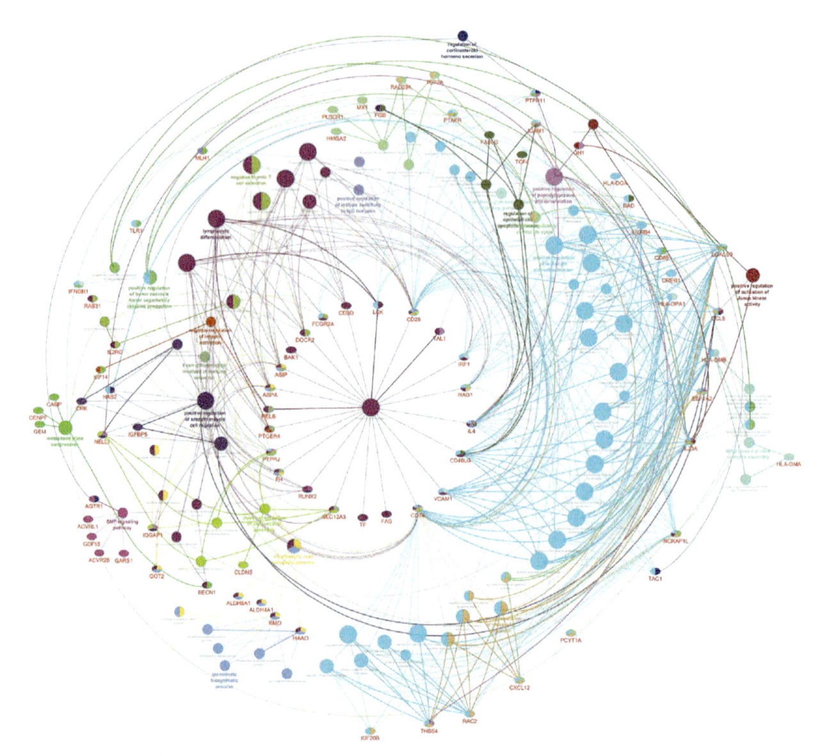

Fig. 6 Functional enrichment analysis was analyzed using the ClueGO. The essential molecular function and biological processes involved between the DEGs of 3 groups were seen. The interconnection between the DEGs of EML is described based on the kappa score = 0.4.

have been identified for specific genes associated with target signaling cascades. The statistically significant results depended on a low *P*-value for each category. With the help of low *P*-value, the top-scored enriched pathway profiles are immune response IL-1 signaling pathway, the role of IFN-beta in the improvement of blood-brain barrier integrity in multiple sclerosis, TNF-alpha-induced inflammatory signaling in normal and asthmatic airway epithelium, immune response MIF-mediated glucocorticoid regulation, apoptosis and survival role of PKR in stress-induced apoptosis (Fig. 7A). The top-scored GO process with significant *P*-values are a response to oxygen-containing compound, response to an organic substance, cell activation, cellular response to chemical stimulus, response to chemical (Fig. 7B). Similarly, the top-scored process networks based on low *P*-value are inflammation protein C signaling, inflammation amphoterin signaling, inflammation histamine signaling, immune response phagocytosis, and inflammation neutrophil activation (Fig. 7C). The top-scored distribution

Fig. 7 (A) Top 10 pathway profiles (B) Top 10 GO processes (C) Top 10 process networks. The top 10 were annotated with the help of GeneGo functional annotation. The genes from ADL and EML were indicated in orange and blue color. The top10 pathways indicate the significant pathways log (*P*-values). Each category represents the main pathways involved individually and in common as a horizontal histogram.

of the pathway map among the DEGs are shown in Fig. 8A, followed by the second low *P*-value pathway map is depicted in Fig. 8B; similarly, the third and fourth-least *P*-value pathway maps are depicted in Fig. 8C and D, respectively. In Fig. 8A–D, the individual symbols represent the proteins. Their interactions are shown as arrows; thermometer-like structures signify the expression level of the gene, where the red ones indicate the upregulated genes. The enriched genes from the pathway maps were used to generate biological networks using Analyze Network (AN) algorithm. Depending on the number of signaling cascades' fragments, the networks are prioritized on the process networks (Table 4). The crucial process networks common between NML vs. ADL were identified, including ATM, Rb protein, Plexin A3,

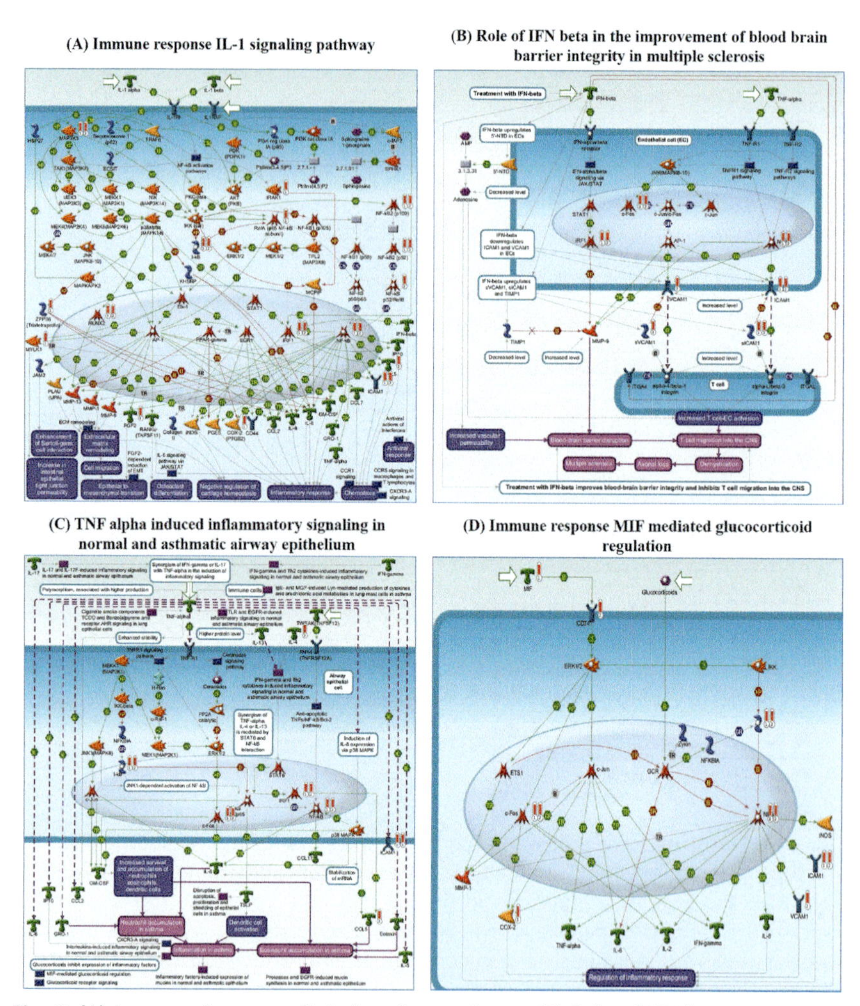

Fig. 8 (A) Immune Response IL-1 signaling pathway; (B) Role of IFN beta to improve blood-brain barrier integrity in multiple sclerosis; (C) TNF alpha-induced signaling in normal and asthmatic airway epithelium; (D) Immune response MIF-mediated glucocorticoid regulation. These top-scored pathways were activated in the emphysema dataset. In the pathways, the upward thermometers are marked red color and indicate upregulated signals, and the downward thermometer in blue shows downregulated expression levels of the genes. The arrows indicate the link between the proteins and their effects (inhibitory and stimulatory). The hub genes essential for each pathway is given in scarlet circles.

Table 4 The process networks from top-sored pathway maps with the help of Analyze Network algorithm.

S·No	Network name	Processes	Size	Target	Pathways	*P*-value	zScore	gScore
NML vs. ADL								
1.	ATM, Rb protein, Plexin A3, Plakophilin 4, SLC25A26	Signal transduction involved in DNA damage checkpoint (20.4%), signal transduction involved in DNA integrity checkpoint (20.4%), signal transduction involved in cell cycle checkpoint (20.4%), mitotic cell cycle checkpoint (24.5%), DNA damage response, signal transduction by p53 class mediator (20.4%)	50	18	2	3.47e-42	82.41	84.91
2.	mGluR1, ALK, SLC25A5, PMCA4, NDUFV2	Negative regulation of organelle organization (34.8%), nucleosome positioning (15.2%), negative regulation of chromatin silencing (15.2%), regulation of organelle organization (50.0%), positive regulation of biological process (87.0%)	50	17	0	2.16e-39	77.82	77.82
3.	FBXO46, PBXIP1, RPL17, COX17, FLJ23867	Cyclic nucleotide catabolic process (17.8%), cGMP catabolic process (13.3%), purine ribonucleotide catabolic process (17.8%), ribonucleotide catabolic process (17.8%), purine nucleotide catabolic process (17.8%)	50	16	0	8.31e-37	73.97	73.97
NML vs. EML								
4	TAL1, Fibrinogen beta, Claudin-5, ITGAD, MHC class II	Glial cell activation (25.0%), neuroinflammatory response (25.0%), leukocyte activation involved in inflammatory response (22.9%), microglial cell activation (22.9%), positive regulation of defense response (43.8%)	50	14	2	4.31e-31	63.20	65.70
5	TAL1, ATP6V0A1, LRP130, KIF14, HLA-DMA	Antigen processing and presentation of peptide or polysaccharide antigen via MHC class II (26.5%), positive regulation of T cell activation (30.6%), positive regulation of leukocyte cell-cell adhesion (30.6%), regulation of leukocyte cell-cell adhesion (32.7%), regulation of T cell activation (32.7%)	50	14	0	2.22e-31	64.51	64.51
6	TAL1, CD74, CAMP, HLA-DM, Beta-tryptase 1	Antigen processing and presentation of exogenous peptide antigen via MHC class II (65.3%), antigen processing and presentation of peptide antigen via MHC class II (65.3%), antigen processing and presentation of peptide or polysaccharide antigen via MHC class II (65.3%), antigen processing and presentation of exogenous peptide antigen (65.3%), antigen processing and presentation of exogenous antigen (65.3%)	50	12	0	2.82e-26	55.85	55.85

Plakophilin 4, SLC25A26, mGluR1, ALK, SLC25A5, PMCA4, NDUFV2, FBXO46, PBXIP1, RPL17, COX17, and FLJ23867. Similarly, TAL1, Fibrinogen beta, Claudin-5, ITGAD, MHC class II, ATP6V0A1, LRP130, KIF14, HLA-DMA, CD74, CAMP, and Beta-tryptase 1 process networks are identified and typical between NML vs. EML (Table 4). Taken together, our systematic identified genes such as *FOSB*, *TAC1*, *ALK*, *SLC25A5*, *GOlGA2*, *OLR1*, *CCL5*, *CD74*, *VCAM-1*, *IL-4*, and *HLA-DOA*, are considered as the crucial genes involved in the disease mechanism. These notable pathways and genes possibly dysregulated in the emphysema patients due to change in their expression level. Nevertheless, functional validations are required to validate our bioinformatics observations.

4. Discussion

Microarray analysis has become an important method to identify differentially expressed genes and the PPI for many diseases. This study focuses on the DEGs of ADL and EML in comparison with the NML from the GSE1122 dataset. A total of 6435 and 3630 genes were found for ADL and EML groups, respectively. The top 250 DEGs were selected for further analysis. A gene falls into differentially expressed if a difference is observed or a change in reading counts or expression levels between two experimental conditions is statistically significant. By implying the 250 identified DEGs, PPI was constructed to identify the major proteins involved in the essential interactions within them; the proteins are referred to as nodes and interactions as edges (Figs. 2 and 3). From the identified PPIs, we explored the essential interactions, which gives an overview of the DEGs' role in various canonical pathways. The individual networks were analyzed based on integrated local components with the topological coefficient, distribution node of degree, shortest path length, and betweenness centrality followed by closeness centrality as analysis parameters (Fig. 4 and Table 3) (Assenov et al., 2008).

To find the relevant pathways that the DEGs were involved, we utilized the ClueGO/CluePedia inbuilt Cytoscape app. The GO analysis was mainly performed to govern the MF, BP, and the pathways in which the DEGs are involved based on the *P*-value. The DEGs of ADL were significantly enriched in various signaling cascades, including NIK/NF-kappaB signaling, positive regulation of interleukin-12 production, and cytochrome-c oxidase activity (Fig. 5). Interestingly, the increased levels of NF-kappaB were previously associated with asthma and COPD patients from inflammatory cells and bronchial biopsies (Caramori et al., 2003; Di Stefano et al., 2002; Schuliga, 2015). Moreover, Cero et al. reported that IL-12 and IL-18 were expressed in high levels and exhibited a synergistic activity on the lungs by inducing cytokines, MMPs, and IL-6 (Cero et al., 2012).

Similarly, the increased expression levels of cytochrome-c oxidase were observed with COPD patients from respiratory and skeletal muscles (Nam, Izumchenko, Dasgupta, & Hoque, 2017; Taivassalo & Hussain, 2016). The DEGs of EML were significantly enriched in the BMP signaling pathway (Fig. 6), quinolinate biosynthetic process, dicarboxylic acid metabolic process, and positive regulation of activation Janus kinase activity. Further investigations are required to decipher the molecular mechanisms of these signaling cascades in the pathogenesis of lung emphysema. Thus, our systematic analyses provided the involvement of identified DEGs in signaling cascades, and induction of these signaling cascades may promote lung emphysema in patients.

We used GeneGo Metacore to demarcate further the molecular pathways and interactions from the ClueGO analysis. GeneGo has a substantial array of regulatory and metabolic process details and includes signaling networks accurately curated. To acquire a detailed image of the DEGs participating in the emphysema, the GeneGo Metacore™ was utilized to explore the DEGs and significant signaling cascades. The top four enriched pathways were selected from the pathway profiles, including immune response IL-1 signaling pathway, the role of IFN-beta in the improvement of blood-brain barrier integrity in multiple sclerosis, TNF-alpha-induced inflammatory signaling in normal and asthmatic airway epithelium, and immune response MIF-mediated glucocorticoid regulation (Fig. 7A). The top four GO processes were explored, including response to oxygen-containing compound, response to an organic substance, cell activation, and cellular response to chemical stimulus (Fig. 7B). The top four process networks were identified, including inflammation protein C signaling, inflammation amphoterin signaling, inflammation histamine signaling, and immune response phagocytosis (Fig. 7C). Also, there were four regulated signaling cascades involved in the ADL and EML groups that were enriched in immune response IL-1 signaling pathway, the role of IFN-beta in the improvement of blood-brain barrier integrity in multiple sclerosis, TNF-alpha-induced inflammatory signaling in normal and asthmatic airway epithelium, and immune response MIF-mediated glucocorticoid regulation (Fig. 8). Furthermore, by using the AN algorithm in GeneGo, we analyzed the pathway networks of the DEGs for ADL and EML. The results revealed that various DEGs in the pathway networks significantly affected in the ADL and EML groups (Table 4). Several crucial genes were incorporated in these networks' elements, including *ALK*, *SLC25A5*, *CD74*, *IL-4*, and *HLA-DOA*. By comparing the upregulating and downregulating genes from ADL and EML (Table 2) along with the results from AN algorithm (Table 4), we speculated the crucial genes including *FOSB*, *TAC1*, *ALK*, *SLC25A5*, *GOlGA2*, *OLR1*, *CCL5*, *CD74*, *VCAM-1*, *IL-4*, and *HLA-DOA* to be significantly involved in the signal transduction by p53 class mediator (20.4%), nucleosome positioning (15.2%), negative regulation of chromatin silencing (15.2%), ribonucleotide catabolic process (17.8%), glial cell activation (25%), regulation of T-cell activation (32.7%), and antigen processing and presentation of exogenous peptide antigen via MHC class II (65.3%), and regulation of leukocyte cell-cell adhesion (32.7%) (Table 4).

Also, we performed an interrelation analysis for the selected genes to explicate the possible direct gene-gene interactions and indirect gene relationships. As a result, we found five genes, *CCL5*, *ALK*, *TAC1*, *CD74*, and *HLA-DOA* that were involved in the positive regulation of mononuclear cell proliferation, leukocyte proliferation, B-cell proliferation, and regulation of lymphocyte proliferation (Fig. 9).

CCL5 is a critical chemokine in recruiting CD8 T cells to the lung and has been implicated in classical IFN-γ dominant Th1 responses. *CCL5* has been previously identified as part of a common pathway in the pathogenesis of late-onset asthma and COPD with mild emphysema (Hizawa et al., 2008). In contrast, *CCL5* is linked to Th2 cells-driven eosinophilic disease. In association with specific cytokines released by T cells, such as IL-2 and IFN-γ, the stimulation and proliferation of specific natural-killer cells to produce CC chemokine-activated killer cells is also triggered by CCL5. (Ignatov, Robert, Gregory-Evans, & Schaller, 2006; Maghazachi, Al-Aoukaty, & Schall, 1996; Proudfoot et al., 2001). CCL5's activity is regulated by its binding to CCR1, CCR3, and CCR5 (Pakianathan, Kuta, Artis, Skelton, & Hébert, 1997). In the present study, a reduced expression of CCL5 was observed that might contribute to emphysema pathogenesis (Table 2 in Supplementary material in the online version at https://doi.org/10.1016/bs.apcsb.2021.02.004).

The *ALK* gene provides instruction to produce a protein called anaplastic lymphoma kinase, which is part of a family of receptor tyrosine kinases (RTKs). The activation of *ALK* continues through a series of proteins in its signaling cascades (Palmer, Vernersson, Grabbe, & Hallberg, 2009). From the ADL group, the ALK was significantly upregulated

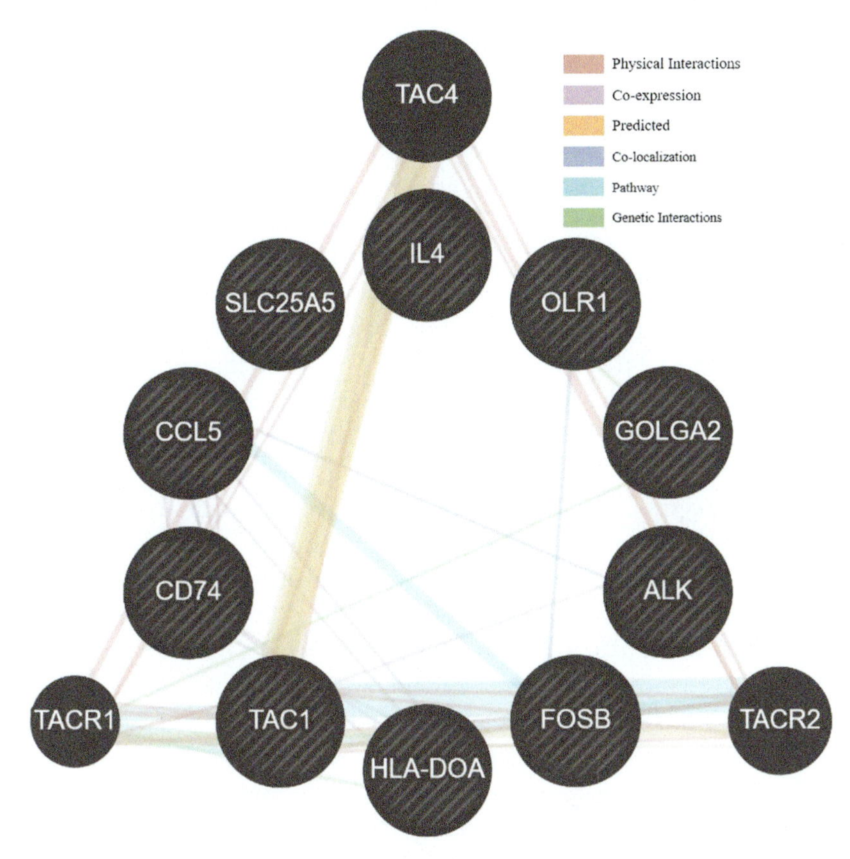

Fig. 9 Interrelation analysis of the identified crucial genes and their physical interactions, co-expression, pathway, predicted, genetic interactions, and co-localization are shown as edges. In contrast, genes are depicted in a black circle.

compared to NML (Table 1 in Supplementary material in the online version at https://doi.org/10.1016/bs.apcsb.2021.02.004). The ALK might undergo rearrangements-harboring mutations within them, and testing ALK rearrangements are recommended for select patients for targeted drug therapy. ALK rearrangement and EGFR and KRAS mutations frequently occurred in non-small-cell lung cancer patients (Gridelli et al., 2014; Leighl et al., 2014; Lim et al., 2015).

Preprotachykinin A is encoded by the *TAC1* gene, which produces two different precursors, namely neurokinin A (NKA) and substance P (SubP) (Atanasova & Reznikov, 2018). NKA and SubP are found below and within the airway epithelium, across the blood vessels, and to a lesser extent inside the airway smooth muscle, in the sensory nerves (Barnes, Drazen, Rennard, & Thomson, 2008; Chuaychoo, Hunter, Myers, Kollarik, & Undem, 2005). It is noteworthy to mention that *TAC1* was previously found to be associated with sputum eosinophilia and confined to patients with extreme asthma, recurrent exacerbations and severe airflow obstruction (Pavlidis et al., 2017). Significantly, our analyses observed the higher expression level of *TAC1* in the ADL-group (Table 2). Genes specified in the TAC profiles were predominantly expressed in bronchial biopsies of COPD patients. The presence of an eosinophilic phenotype consistent with the T2 pathway is supported by high enrichment of the TAC1 (high eosinophilia/Th2 high) in COPD (Faiz et al., 2020).

CD74 was discovered recently by the MIF receptor and observed to be the invariant chain of the HLA class II. A multi-functional cytokine that is over-expressed in lung cancer is the macrophage migration inhibitory factor (MIF) (Sauler et al., 2014). Upon oncogenic progression, *CD74* expression was reported in the lung in macrophages and pneumocytes of type II (Marsh et al., 2009; Mathew et al., 2013). A recent study by Kim et al. (2020) found that *CD74* is overexpressed in COPD patients. Using MIF20, a novel molecule and MIF agonist, prevented smoking-induced lung emphysema (Kim et al., 2020). Our systematic analysis revealed the increased expression of *CD74* in the EML-group; it may undergo dysregulation of CD74-dependent signaling cascades (Table 2 in Supplementary material in the online version at https://doi.org/10.1016/bs.apcsb.2021.02.004).

The human leukocyte antigen (HLA) class I and class II loci are the most phenotypic genes in the genome, with firmly structured and quilt sequence patterns. The heterodimer, HLA-DO, is located in B cells in lysosomes and controls the loading of HLA-DM-mediated peptides on MHC class II molecules (Nagarajan et al., 2002). This gene demonstrates limited sequence variation compared to classical HLA class II proteins, particularly at the protein level (Fagerberg et al., 2014). Yucesoy et al. (2014) reported that genetic polymorphisms occurring in *HLA-DO* and related genes contributed to the possibility of diisocyanate-induced asthma (Yucesoy et al., 2014). Additionally, the recent study by Kubysheva et al. (2018) found increased levels of HLA class II in COPD patients (Kubysheva et al., 2018).

Similarly, our study found a significant increase in the *HLA-DOA* in both the groups (ADL and EML), suggesting that these could contribute to the pathogenesis of emphysema in patients. Taken together, identified DEGs that are present in the ADL and EML in comparison to NML activated several important signaling cascades in the patients. Our study limits to the small patient groups, and studying larger patient populations with biochemical analysis would help in confirming our identification.

5. Conclusion

Emphysema is a lung condition with shortness of breath and damaging the air sacs, either as an EML condition or as ADL. Therefore treatment for this disease is urgently needed; mapping targets for this disease will help develop therapeutics for this disease. In this study, we mapped five DEGs (*CCL5*, *ALK*, *TAC1*, *CD74*, and *HLA-DOA*) that either upregulated or downregulated in expression and aid in developing emphysema. The study of expression and the association of several genes would certainly help understand such genes' roles in the development of emphysema. Furthermore, the identified crucial DEGs might be potential targets for developing emphysema-based treatments and drug discovery. However, our hypothesis can be tested with large emphysema samples, and functional studies must validate these findings.

Acknowledgments

Mr. Udhaya Kumar. S, one of the authors, gratefully acknowledges the Indian Council of Medical Research (ICMR), India, for providing him a Senior Research Fellowship [ISRM/11(93)/2019]. The authors would like to thank the Vellore Institute of Technology, India, and Qatar University, Qatar, for providing the necessary research facilities and encouragement to carry out this work.

Conflicts of interest

The authors have declared that no conflicts of interest exist.

Author contributions

UKS, HZ, and GPDC, contributed to designing the study and data acquisition, analysis, and interpretation. UKS, TKD, MPN, VAP, VK, DN, SY, and MR are involved in the acquisition, analysis, and interpreting of the results. UKS, MPN, TKD, MR, VAP, VK, DN, SY, and CGPD contributed to data interpretation, conducted, and drafting the manuscript. CGPD and HZ supervised the entire study and studied, acquiring, analyzing, understanding the data, and drafting the manuscript. The manuscript was reviewed and approved by all the authors.

Role of funding source

No funding agency is involved in the present study.

References

Leighl, N. B., Rekhtman, N., Biermann, W. A., Huang, J., Mino-Kenudson, M., Ramalingam, S. S., et al. (2014). Molecular testing for selection of patients with lung cancer for epidermal growth factor receptor and anaplastic lymphoma kinase tyrosine kinase inhibitors: American Society of Clinical Oncology endorsement of the College of American Pathologists/International Association for the study of lung cancer/association for molecular pathology guideline. *Journal of Clinical Oncology: Official Journal of the American Society of Clinical Oncology*, *32*(32), 3673–3679. https://doi.org/10.1200/JCO.2014.57.3055.

Mishra, S., Shah, M. I., Udhaya Kumar, S., Thirumal Kumar, D., Gopalakrishnan, C., Al-Subaie, A. M., et al. (2021). Chapter eleven—Network analysis of transcriptomics data for the prediction and prioritization of membrane-associated biomarkers for idiopathic pulmonary fibrosis (IPF) by bioinformatics approach. In R. Donev (Ed.), *Vol. 123*. *Advances in protein chemistry and structural biology* (pp. 241–273). Academic Press. https://doi.org/10.1016/bs.apcsb.2020.10.003.

Assenov, Y., Ramírez, F., Schelhorn, S.-E., Lengauer, T., & Albrecht, M. (2008). Computing topological parameters of biological networks. *Bioinformatics (Oxford, England)*, *24*(2), 282–284. https://doi.org/10.1093/bioinformatics/btm554.

Atanasova, K. R., & Reznikov, L. R. (2018). Neuropeptides in asthma, chronic obstructive pulmonary disease, and cystic fibrosis. *Respiratory Research*, *19*(1), 149. https://doi.org/10.1186/s12931-018-0846-4.

Aubert, J., Bar-Hen, A., Daudin, J. J., & Robin, S. (2004). Determination of the differentially expressed genes in microarray experiments using local FDR. *BMC Bioinformatics*, *5*, 125. https://doi.org/10.1186/1471-2105-5-125.

Barnes, P., Drazen, J., Rennard, S., & Thomson, N. (2008). *Asthma and COPD: Basic mechanisms and clinical management* (2nd ed.). Elsevier. https://www.elsevier.com/books/asthma-and-copd/barnes/978-0-12-374001-4.

Bindea, G., Mlecnik, B., Hackl, H., Charoentong, P., Tosolini, M., Kirilovsky, A., et al. (2009). ClueGO: A Cytoscape plugin to decipher functionally grouped gene ontology and pathway annotation networks. *Bioinformatics*, *25*(8), 1091–1093. https://doi.org/10.1093/bioinformatics/btp101.

Brandes, U. (2001). A faster algorithm for betweenness centrality. *The Journal of Mathematical Sociology*, *25*(2), 163–177. https://doi.org/10.1080/0022250X.2001.9990249.

Caramori, G., Romagnoli, M., Casolari, P., Bellettato, C., Casoni, G., Boschetto, P., et al. (2003). Nuclear localisation of p65 in sputum macrophages but not in sputum neutrophils during COPD exacerbations. *Thorax*, *58*(4), 348–351. https://doi.org/10.1136/thorax.58.4.348.

Cero, F. T., Hillestad, V., Løberg, E. M., Christensen, G., Larsen, K.-O., & Skjønsberg, O. H. (2012). IL-18 and IL-12 synergy induces matrix degrading enzymes in the lung. *Experimental Lung Research*, *38*(8), 406–419. https://doi.org/10.3109/01902148.2012.716903.

Chuaychoo, B., Hunter, D. D., Myers, A. C., Kollarik, M., & Undem, B. J. (2005). Allergen-induced substance P synthesis in large-diameter sensory neurons innervating the lungs. *The Journal of Allergy and Clinical Immunology*, *116*(2), 325–331. https://doi.org/10.1016/j.jaci.2005.04.005.

Davis, S., & Meltzer, P. S. (2007). GEOquery: A bridge between the gene expression omnibus (GEO) and BioConductor. *Bioinformatics*, *23*(14), 1846–1847. https://doi.org/10.1093/bioinformatics/btm254.

Di Stefano, A., Caramori, G., Oates, T., Capelli, A., Lusuardi, M., Gnemmi, I., et al. (2002). Increased expression of nuclear factor-kappaB in bronchial biopsies from smokers and patients with COPD. *The European Respiratory Journal*, *20*(3), 556–563. https://doi.org/10.1183/09031936.02.00272002.

Domej, W., Oettl, K., & Renner, W. (2014). Oxidative stress and free radicals in COPD—Implications and relevance for treatment. *International Journal of Chronic Obstructive Pulmonary Disease*, *9*, 1207–1224. https://doi.org/10.2147/COPD.S51226.

Eisner, M. D., Anthonisen, N., Coultas, D., Kuenzli, N., Perez-Padilla, R., Postma, D., et al. (2010). An official American Thoracic Society public policy statement: Novel risk factors and the global burden of chronic obstructive pulmonary disease. *American Journal of Respiratory and Critical Care Medicine*, *182*(5), 693–718. https://doi.org/10.1164/rccm.200811-1757ST.

Fagerberg, L., Hallström, B. M., Oksvold, P., Kampf, C., Djureinovic, D., Odeberg, J., et al. (2014). Analysis of the human tissue-specific expression by genome-wide integration of transcriptomics and antibody-based proteomics. *Molecular & Cellular Proteomics*, *13*(2), 397–406. https://doi.org/10.1074/mcp.M113.035600.

Faiz, A., Pavlidis, S., Kuo, C. S., Rowe, A., Hiemstra, P. S., Timens, W., et al. (2020). Novel type 2-high gene clusters associated with corticosteroid sensitivity in COPD. *BioRxiv*. 2020.01.22.912022. https://doi.org/10.1101/2020.01.22.912022.

Fu, D., Zhang, B., Yang, L., Huang, S., & Xin, W. (2020). Development of an immune-related risk signature for predicting prognosis in lung squamous cell carcinoma. *Frontiers in Genetics*, *11*, 1–19. Article 978 https://doi.org/10.3389/fgene.2020.00978.

Goldklang, M., & Stockley, R. (2016). Pathophysiology of emphysema and implications. *Chronic Obstructive Pulmonary Diseases*, *3*(1), 454–458. https://doi.org/10.15326/jcopdf.3.1.2015.0175.

Golpon, H. A., Coldren, C. D., Zamora, M. R., Cosgrove, G. P., Moore, M. D., Tuder, R. M., et al. (2004). Emphysema lung tissue gene expression profiling. *American Journal of Respiratory Cell and Molecular Biology*, *31*(6), 595–600. https://doi.org/10.1165/rcmb.2004-0008OC.

Gridelli, C., Peters, S., Sgambato, A., Casaluce, F., Adjei, A. A., & Ciardiello, F. (2014). ALK inhibitors in the treatment of advanced NSCLC. *Cancer Treatment Reviews*, *40*(2), 300–306. https://doi.org/10.1016/j.ctrv.2013.07.002.

Hizawa, N., Makita, H., Nasuhara, Y., Hasegawa, M., Nagai, K., Ito, Y., et al. (2008). Functional single nucleotide polymorphisms of the CCL5 gene and nonemphysematous phenotype in COPD patients. *The European Respiratory Journal*, *32*(2), 372–378. https://doi.org/10.1183/09031936.00115307.

Ignatov, A., Robert, J., Gregory-Evans, C., & Schaller, H. C. (2006). RANTES stimulates Ca2+ mobilization and inositol trisphosphate (IP3) formation in cells transfected with G protein-coupled receptor 75. *British Journal of Pharmacology*, *149*(5), 490–497. https://doi.org/10.1038/sj.bjp.0706909.

Kim, S.-J., Wan, F., Zhang, X., Zhang, Y., Ifedigbo, E., Leng, L., et al. (2020). MIF-CD74 Signaling protects against endothelial senescence in chronic obstructive pulmonary disease. *The FASEB Journal*, *34*(S1), 1. https://doi.org/10.1096/fasebj.2020.34.s1.04905.

Kubysheva, N., Soodaeva, S., Novikov, V., Eliseeva, T., Li, T., Klimanov, I., et al. (2018). Soluble HLA-I and HLA-II molecules are potential prognostic markers of progression of systemic and local inflammation in patients with COPD. *Disease Markers*, *2018*. Hindawi, Article ID 3614341, 1–7 November 27, [Research Article] https://doi.org/10.1155/2018/3614341.

Kumar, S. U., Kumar, D. T., Siva, R., Doss, C. G. P., & Zayed, H. (2019). Integrative bioinformatics approaches to map potential novel genes and pathways involved in ovarian cancer. *Frontiers in Bioengineering and Biotechnology*, 7, 1–15. Article 391 https://doi.org/10.3389/fbioe.2019.00391.

Li, N., Zhao, X., & You, S. (2019). Identification of key regulators of pancreatic ductal adenocarcinoma using bioinformatics analysis of microarray data. *Medicine*, *98*(2), e14074. https://doi.org/10.1097/MD.0000000000014074.

Lim, J. U., Yeo, C. D., Rhee, C. K., Kim, Y. H., Park, C. K., Kim, J. S., et al. (2015). Chronic obstructive pulmonary disease-related non-small-cell lung cancer exhibits a low prevalence of EGFR and ALK driver mutations. *PLoS One*, *10*(11), e0142306. https://doi.org/10.1371/journal.pone.0142306.

Maghazachi, A. A., Al-Aoukaty, A., & Schall, T. J. (1996). CC chemokines induce the generation of killer cells from CD56 + cells. *European Journal of Immunology*, *26*(2), 315–319. https://doi.org/10.1002/eji.1830260207.

Marsh, L. M., Cakarova, L., Kwapiszewska, G., von Wulffen, W., Herold, S., Seeger, W., et al. (2009). Surface expression of CD74 by type II alveolar epithelial cells: A potential mechanism for macrophage migration inhibitory factor-induced epithelial repair. *American Journal of Physiology. Lung Cellular and Molecular Physiology*, *296*(3), L442–L452. https://doi.org/10.1152/ajplung.00525.2007.

Mathew, B., Jacobson, J. R., Siegler, J. H., Moitra, J., Blasco, M., Xie, L., et al. (2013). Role of migratory inhibition factor in age-related susceptibility to radiation lung injury via NF-E2-related factor-2 and antioxidant regulation. *American Journal of Respiratory Cell and Molecular Biology*, *49*(2), 269–278. https://doi.org/10.1165/rcmb.2012-0291OC.

Nagarajan, U. M., Lochamy, J., Chen, X., Beresford, G. W., Nilsen, R., Jensen, P. E., et al. (2002). Class II transactivator is required for maximal expression of HLA-DOB in B cells. *Journal of Immunology (Baltimore, Md.: 1950)*, *168*(4), 1780–1786. https://doi.org/10.4049/jimmunol.168.4.1780.

Nam, H.-S., Izumchenko, E., Dasgupta, S., & Hoque, M. O. (2017). Mitochondria in chronic obstructive pulmonary disease and lung cancer: Where are we now? *Biomarkers in Medicine*, *11*(6), 475–489. https://doi.org/10.2217/bmm-2016-0373.

Newman, M. E. J. (2005). A measure of betweenness centrality based on random walks. *Social Networks*, *27*(1), 39–54. https://doi.org/10.1016/j.socnet.2004.11.009.

Pahal, P., Avula, A., & Sharma, S. (2020). Emphysema. In B. Abai, A. Abu-Ghosh, A. B. Acharya, U. Acharya, S. G Adhia, & T. C Aeby (Eds.), *StatPearls* (pp. 1–13). StatPearls Publishing LLC. http://www.ncbi.nlm.nih.gov/books/NBK482217/.

Pakianathan, D. R., Kuta, E. G., Artis, D. R., Skelton, N. J., & Hébert, C. A. (1997). Distinct but overlapping epitopes for the interaction of a CC-chemokine with CCR1, CCR3 and CCR5. *Biochemistry*, *36*(32), 9642–9648. https://doi.org/10.1021/bi970593z.

Palmer, R. H., Vernersson, E., Grabbe, C., & Hallberg, B. (2009). Anaplastic lymphoma kinase: Signalling in development and disease. *The Biochemical Journal*, *420*(3), 345–361. https://doi.org/10.1042/BJ20090387.

Pavlidis, S., Faiz, A., Postma, D., Timens, W., Berg, M. V. D., Kuo, S. C., et al. (2017). Gene signatures from U-BIOPRED transcriptomic-associated clusters exist in COPD. *European Respiratory Journal*, *50*(Suppl. 61). https://doi.org/10.1183/1393003.congress-2017.OA1492.

Proudfoot, A. E., Fritchley, S., Borlat, F., Shaw, J. P., Vilbois, F., Zwahlen, C., et al. (2001). The BBXB motif of RANTES is the principal site for heparin binding and controls receptor selectivity. *The Journal of Biological Chemistry*, *276*(14), 10620–10626. https://doi.org/10.1074/jbc.M010867200.

Quackenbush, J. (2001). Computational analysis of microarray data. *Nature Reviews. Genetics*, *2*(6), 418–427. https://doi.org/10.1038/35076576.

Saito, R., Smoot, M. E., Ono, K., Ruscheinski, J., Wang, P.-L., Lotia, S., et al. (2012). A travel guide to cytoscape plugins. *Nature Methods*, *9*(11), 1069–1076. https://doi.org/10.1038/nmeth.2212.

Sauler, M., Leng, L., Trentalange, M., Haslip, M., Shan, P., Piecychna, M., et al. (2014). Macrophage migration inhibitory factor deficiency in chronic obstructive pulmonary disease. *American Journal of Physiology. Lung Cellular and Molecular Physiology*, *306*(6), L487–L496. https://doi.org/10.1152/ajplung.00284.2013.

Schuliga, M. (2015). NF-kappaB signaling in chronic inflammatory airway disease. *Biomolecules*, *5*(3), 1266–1283. https://doi.org/10.3390/biom5031266.

Shannon, P., Markiel, A., Ozier, O., Baliga, N. S., Wang, J. T., Ramage, D., et al. (2003). Cytoscape: A software environment for integrated models of biomolecular interaction networks. *Genome Research*, *13*(11), 2498–2504. https://doi.org/10.1101/gr.1239303.

Smyth, G. K. (2005). Limma: Linear models for microarray data. In R. Gentleman, V. J. Carey, W. Huber, R. A. Irizarry, & S. Dudoit (Eds.), *Bioinformatics and computational biology solutions using r and bioconductor* (pp. 397–420). Springer. https://doi.org/10.1007/0-387-29362-0_23.

Stelzl, U., Worm, U., Lalowski, M., Haenig, C., Brembeck, F. H., Goehler, H., et al. (2005). *******A human protein-protein interaction network: A resource for annotating the proteome. *Cell*, *122*(6), 957–968. https://doi.org/10.1016/j.cell.2005.08.029.

Szklarczyk, D., Franceschini, A., Wyder, S., Forslund, K., Heller, D., Huerta-Cepas, J., et al. (2015). STRING v10: Protein-protein interaction networks, integrated over the tree of life. *Nucleic Acids Research*, *43*(Database issue), D447–D452. https://doi.org/10.1093/nar/gku1003.

Taivassalo, T., & Hussain, S. N. A. (2016). Contribution of the mitochondria to Locomotor muscle dysfunction in patients with COPD. *Chest*, *149*(5), 1302–1312. https://doi.org/10.1016/j.chest.2015.11.021.

Taraseviciene-Stewart, L., & Voelkel, N. F. (2008). Molecular pathogenesis of emphysema. *The Journal of Clinical Investigation*, *118*(2), 394–402. https://doi.org/10.1172/JCI31811.

Udhaya Kumar, S., Rajan, B., Thirumal Kumar, D., Preethi, V. A., Abunada, T., Younes, S., et al. (2020). Involvement of essential signaling cascades and analysis of gene networks in Diabesity. *Genes*, *11*(11), 1256. https://doi.org/10.3390/genes11111256.

Udhaya Kumar, S., Thirumal Kumar, D., Bithia, R., Sankar, S., Magesh, R., Sidenna, M., et al. (2020). Analysis of differentially expressed genes and molecular pathways in familial hypercholesterolemia involved in atherosclerosis: A systematic and bioinformatics approach. *Frontiers in Genetics*, *11*, 1–16. Article 734 https://doi.org/10.3389/fgene.2020.00734.

Udhaya Kumar, S., Thirumal Kumar, D., Siva, R., George Priya Doss, C., Younes, S., Younes, N., et al. (2020). Dysregulation of signaling pathways due to differentially expressed genes from the B-cell transcriptomes of systemic lupus erythematosus patients—A bioinformatics approach. *Frontiers in Bioengineering and Biotechnology*, *8*, 1–17. Article 276 https://doi.org/10.3389/fbioe.2020.00276.

Wan, J., Jiang, S., Jiang, Y., Ma, W., Wang, X., He, Z., et al. (2019). Data mining and expression analysis of differential lncRNA ADAMTS9-AS1 in prostate cancer. *Frontiers in Genetics*, *10*, 1–11. Article 1377 https://doi.org/10.3389/fgene.2019.01377.

Warde-Farley, D., Donaldson, S. L., Comes, O., Zuberi, K., Badrawi, R., Chao, P., et al. (2010). The GeneMANIA prediction server: Biological network integration for gene prioritization and predicting gene function. *Nucleic Acids Research*, *38*(suppl_2), W214–W220. https://doi.org/10.1093/nar/gkq537.

WHO. (2010). *WHO | Global status report on NCDs*. WHO; World Health Organization. http://www.who.int/chp/ncd_global_status_report/en/.

WHO. (2020). *The top 10 causes of death*. December 9 https://www.who.int/news-room/fact-sheets/detail/the-top-10-causes-of-death.

Yan, H., Zheng, G., Qu, J., Liu, Y., Huang, X., Zhang, E., et al. (2019). Identification of key candidate genes and pathways in multiple myeloma by integrated bioinformatics analysis. *Journal of Cellular Physiology*, *234*(12), 23785–23797. https://doi.org/10.1002/jcp.28947.

Yuan, C., Chang, D., Lu, G., & Deng, X. (2017). Genetic polymorphism and chronic obstructive pulmonary disease. *International Journal of Chronic Obstructive Pulmonary Disease*, *12*, 1385–1393. https://doi.org/10.2147/COPD.S134161.

Yucesoy, B., Johnson, V., Lummus, Z., Kashon, M., Rao, M., Bannerman-Thompson, H., et al. (2014). Genetic variants in the major histocompatibility complex class I and class II genes are associated with diisocyanate-induced asthma. *Journal of Occupational and Environmental Medicine*, *56*, 382–387. https://doi.org/10.1097/JOM.0000000000000138.
Zhao, Z., Li, J., Li, H., Yuan Wu, N.-Y., Ou-Yang, P., Liu, S., et al. (2020). Integrative bioinformatics approaches to screen potential prognostic immune-related genes and drugs in the cervical cancer microenvironment. *Frontiers in Genetics*, *11*, 1–13. Article 727 https://doi.org/10.3389/fgene.2020.00727.

A systemic approach to explore the mechanisms of drug resistance and altered signaling cascades in extensively drug-resistant tuberculosis

S. Udhaya Kumar[a], Aisha Saleem[a], D. Thirumal Kumar[b], V. Anu Preethi[c], Salma Younes[d], Hatem Zayed[e], Iftikhar Aslam Tayubi[f], and C. George Priya Doss[a,*]

[a]School of Biosciences and Technology, Vellore Institute of Technology, Vellore, Tamil Nadu, India
[b]Department of Bioinformatics, Saveetha School of Engineering, Saveetha Institute of Medical and Technical Sciences, Chennai, Tamil Nadu, India
[c]School of Computer Science and Engineering, Vellore Institute of Technology, Vellore, Tamil Nadu, India
[d]Translational Research Institute, Women's Wellness and Research Center, Hamad Medical Corporation, Doha, Qatar
[e]Department of Biomedical Sciences, College of Health and Sciences, QU Health, Qatar University, Doha, Qatar
[f]Faculty of Computing and Information Technology, King Abdul-Aziz University, Rabigh, Saudi Arabia
*Corresponding author: e-mail address: georgepriyadoss@vit.ac.in

Contents

Advances in Protein Chemistry and Structural Biology, Volume 127
ISSN 1876-1623
https://doi.org/10.1016/bs.apcsb.2021.02.002

Abstract

Background and aim: The persistence of extensively drug-resistant (XDR) strains of *Mycobacterium tuberculosis* (MTB) continue to pose a significant challenge to the treatment and control of tuberculosis infections worldwide. XDR-MTB strains exhibit resistance against first-line anti-TB drugs, fluoroquinolones, and second-line injectable drugs. The mechanisms of drug resistance of MTB remains poorly understood. Our study aims at identifying the differentially expressed genes (DEGs), associated gene networks, and signaling cascades involved in rendering this pathogen resistant to multiple drugs, namely, isoniazid, rifampicin, and capreomycin. *Methods*: We used the microarray dataset GSE53843. The GEO2R tool was used to prioritize the most significant DEGs (top 250) of each drug exposure sample between XDR strains and non-resistant strains. The validation of the 250 DEGs was performed using volcano plots. Protein-protein interaction networks of the DEGs were created using STRING and Cytoscape tools, which helped decipher the relationship between these genes. The significant DEGs were functionally annotated using DAVID and ClueGO.

The concomitant biological processes (BP) and molecular functions (MF) were represented as dot plots. *Results and conclusion*: We identified relevant molecular pathways and biological processes, such as cell wall biogenesis, lipid metabolic process, ion transport, phosphopantetheine binding, and triglyceride lipase activity. These processes indicated the involvement of multiple interconnected mechanisms in drug resistance. Our study highlighted the impact of cell wall permeability, with the dysregulation of the mur family of proteins, as essential factors in the inference of resistance. Additionally, upregulation of genes responsible for ion transport such as *ctpF*, *arsC*, and *nark3*, emphasizes the importance of transport channels and efflux pumps in potentially driving out stress-inducing compounds. This study investigated the upregulation of the Lip family of proteins, which play a crucial role in triglyceride lipase activity. Thereby illuminating the potential role of drug-induced dormancy and subsequent resistance in the mycobacterial strains. Multiple mechanisms such as carboxylic acid metabolic process, NAD biosynthetic process, triglyceride lipase activity, phosphopantetheine binding, organic acid biosynthetic process, and growth of symbiont in host cell were observed to partake in resistance of XDR-MTB. This study ultimately provides a platform for important mapping targets for potential therapeutics against XDR-MTB.

1. Introduction

Mycobacterium tuberculosis (MTB) is the etiological agent of Tuberculosis (TB) (Arnold, 2007; Lillebaek et al., 2016). Around one-third of the global population has been infected by TB, causing over 60 million deaths since 2000, rendering it a cause of global concern (WHO, 2020). TB can exist in the lungs in the form of Pulmonary TB or extrapulmonary TB, which infect the lymph nodes, bones, abdomen, and meninges (Lee, 2015; Singh et al., 2020). Immunocompromised individuals are often more susceptible to the disease (WHO, 2020). The prevalence of this disease has increased tremendously with the emergence of antibiotic resistance within the bacterium strains. Together with the alarming rate of occurrence of multi-drug resistance, the extended persistence of MTB infection enables the implementation of novel and successful anti-TB therapeutics, an international health concern (Karlas et al., 2010; Kumar et al., 2010). Individuals infected with non-resistant strains of the bacterium can be treated with standard first-line drugs such as isoniazid, rifampicin, ethambutol, and pyrazinamide (Luo et al., 2020; Singh et al., 2020). However, strains of TB have gained resistance to most of these drugs. Monoresistance is often exhibited against isoniazid and is the most common form of drug-resistant TB in the world (Chien et al., 2015; Furin, Cox, & Pai, 2019). Multidrug-resistant strains of TB (MDR–TB) harbor resistance against rifampicin and isoniazid. The existence of additional fluoroquinolone resistance and at least one injectable second-line drug (capreomycin, kanamycin, and amikacin) renders these MDR–TB strains drug-resistant (XDR-MTB) (Coll et al., 2018; Seung, Keshavjee, & Rich, 2015) extensively.

Various mechanisms have been attributed to the development of resistance. Growing evidence reveals that the sequential inclusion of genetic mutations in the genes that code for drug processing enzymes or drug targets are the main contributors to drug resistance acquisition (Coll et al., 2018). In resistant clinical TB isolates, a large number of mutations are being reported, most of which were located in the exon region, which implies that structural changes to the target protein of drugs play a crucial role in the resistance development (Sandgren et al., 2009; Zhang et al., 2013). Nevertheless, recognizing some resistance-related intergenic mutations also indicates a correlation between resistance and the regulation of gene expression (Zhang et al., 2013). It has been previously reported that resistance against isoniazid occurs due to mutations in the *katG* gene, responsible for converting the prodrug to active (INH). Pro-drug INH is activated upon forming an adduct with NAD. This adduct target the inhA protein, which plays an active role in cell wall synthesis. Mutations in inhA protein impart resistance to the bacterium as well (Hassan et al., 2020). Other prime contributors to drug resistance are cell envelope impermeability, cell wall lipid changes, which infer Rifampicin resistance, and overexpression of efflux pumps (Luo et al., 2020; Singh et al., 2020).

Mycobacterial virulence has been attributed to a wide range of virulence determinants and host responses at different pathogenesis stages (Madacki, Mas Fiol, & Brosch, 2019). To date, over 400 genes have been identified as being essential for in vivo growth and survival of the bacterium (Zhang et al., 2013). Genes determined as potential virulent factors show immense diversity in their roles; for example, genes involved in primary cell metabolism, complex lipids of the cell envelope and components of ESX secretion systems are known to play equally significant roles in pathogenesis (Madacki et al., 2019). The convoluted pathways involved in the pathogenesis of MTB and mechanisms which infer drug resistance can be mapped via the identification of genetic determinants of the bacterium (Coll et al., 2018). It is the need of the hour to decipher the conundrum of drug resistance, which has rendered MTB a global threat. High throughput screening technologies such as microarray analysis can be employed to study diseases with complex pathogenesis (Alam et al., 2019; Cui & He, 2014; Kumar, Kumar, Siva, Doss, & Zayed, 2019; Mishra et al., 2020; Penn et al., 2018; Udhaya Kumar, Rajan, et al., 2020; Udhaya Kumar, Thirumal Kumar, Bithia et al., 2020; Udhaya Kumar, Thirumal Kumar, Siva, et al., 2020). In an attempt to ascertain the genetic determinants inferring drug resistance, we investigated the differentially expressed genes (DEGs) between non-resistant and extremely drug-resistant strains of MTB when treated with isoniazid, rifampicin, and capreomycin (GSE53843) (Guohua & Yu, 2014). This bioinformatics approach entailed retrieving DEGs data from the GEO database and its analysis using the web-tool GEO2R and visualization through the creation of volcano plots. Gene interaction networks were visualized using STRING and Cytoscape tools. Furthermore, protein interactions and pathways that are unique to drug resistance were explored to gain a better understanding of the complex mechanisms involved in extreme drug resistance.

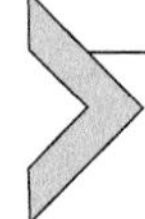

2. Materials and methods

2.1 Data procurement

For this study, the dataset GSE53843 was acquired from NCBI's GEO database. Gene expression profiles of diverse experiments were stored in the GEO database and accessed by users for further analysis (Barrett et al., 2013). The keywords "*Mycobacterium tuberculosis*," "Drug-resistant," and "microarray" were utilized to extract related gene expression profile array datasets from the database. Samples of GSE53843 expression profiles comprise a reference strain H37Rv and two extensively drug-resistant (XDR) strains of MTB (Guohua & Yu, 2014). The three strains were treated with three anti-TB drugs and water as control, resulting in a sample size of twelve. The platform GPL18119 Agilent-027618 *Mycobacterium tuberculosis* H37Rv 8X15K microarray was employed to obtain the above sample set.

2.2 Processing of data and screening for DEGs

The statistical tool GEO2R was used to explore the DEGs upon different drug exposures between the XDR strains and the non-resistant reference strains. The GEO2R tool is incorporated with R/Bioconductor and Limma package (Ritchie et al., 2015). The *t*-test and Benjamini and Hochberg (False discovery rate), which are inbuilt statistical tools of GEO2R, aid in calculating *P*-value and FDR and play an essential role in determining the DEGs (Aubert, Bar-Hen, Daudin, & Robin, 2004). A series of four tests were carried out using GEO2R to extract DEGs on treatment with water (control), capreomycin, isoniazid, and rifampicin, resulting in four distinct datasets. We considered principal standard as $P \leq 0.05$ and $|\log(\text{fold change})| > 2$ to extract significant DEGs from the dataset. Upregulated genes were demarcated by a log fold change value of ≥ 2 and downregulated genes by log $FC \leq -2$. The raw data and the top 250 DEGs were obtained from the dataset and processed via the removal of unidentified gene entries and hypothetical proteins. The selected DEGs were subjected to further investigation by constructing a volcano plot using RStudio and library Calibrate package, followed by identifying PPI networks.

2.3 Identification of PPI networks

Protein-protein interactions networks (PPIs) allow for the elucidation of relationships between proteins, which may be correlated through specific functional interactions (Childs & Larremore, 2021). In this study, we used STRING v.11 to identify the networks that existed among the DEGs. STRING is a web-based tool, which integrates information from multiple protein-protein association databases and renders computational predictions of these interactions in the form of networks (Szklarczyk et al., 2019). We eliminated interactions, which were inconsistent with the dataset by setting the cut off to an interaction score of >0.4. We further visualized the networks obtained from STRING on Cytoscape software. The plug-in GeneMANIA was employed to obtain intersected clusters from the PPI network for the identified DEGs based on integrated scored values (Franz et al., 2018).

2.4 Functional annotation and enrichment analysis

To evaluate the functional determinants of the gene sets that have been significantly upregulated or downregulated during drug exposure, we performed an enrichment analysis using the DAVID v6.8. The Database for Annotation, Visualization, and Integrated Discovery (DAVID), incorporates information from a multitude of public bioinformatics databases and provides biological meaning to lists of genes (Huang et al., 2007). Using this tool, we determined the functional classification of GO terms and performed an enrichment analysis.

For comprehensive visualization and functional annotation of the Genes of GO terms and KEGG pathways, we utilized ClueGo software along with DAVID. ClueGo is a Cytoscape Plugin, which incorporates KEGG or BioCarta pathways and Gene Ontology (GO) terms to create a network of functionally grouped genes or pathways, representing the analysis results (Bindea et al., 2009). ClueGo v2.5.5 was used on the top 250 DEGs to obtain the GO terms and pathways involved in drug resistance against each of the three drugs.

To further elucidate the role of DEGs in the enriched biological pathways, dotplots were generated for each drug exposure dataset using GoPlot and ggplot2 R packages (Walter, Sánchez-Cabo, & Ricote, 2015; Wickham, 2016). DAVID database was utilized to functionally annotate the DEGs, whose regulatory levels in concomitant enriched pathways were visualized as dotplots.

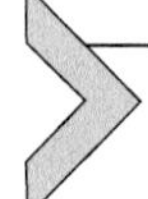

3. Results

3.1 Data procurement and screening for DEGs

The dataset GSE53843 comprises gene expression profiles of two XDR strains of MTB and a non-resistant MTB strain as reference. Twelve samples were created due to the treatment of three strains with three anti-TB drugs and water as a control (Table 1). The GPL18119 platform (Agilent-027618 *Mycobacterium tuberculosis* H37Rv 8X15K microarray) was used in this study. The XDR strains treated with a specific anti-TB drug were compared against the reference strain treated with the same drug-using the NCBI-inbuilt GEO2R tool. The cut-off values of $|\text{logFC}| \geq 2.0$ and P-values ≤ 0.05 for the DEGs were computed using the integrated limma package v3.26.8 and R/Bioconductor. The significant 250 DEGs between the reference strain and XDR strains were subsequently identified. The identified

DEGs from the XDR-MTB strains treated with water, capreomycin, isoniazid, rifampicin is provided in Supplementary material in the online version at https://doi.org/10.1016/bs.apcsb.2021.02.002 in Supplementary Tables 1–4, respectively. Volcano plots created through R studio and calibrate (library) package was utilized to visualize these DEGs. Fig. 1 comprises the volcano plots of each of the three Anti TB drug-treated sets and water treated set as control. The significant genes satisfying all the statistical values ($P<0.05$, logFC ≥ 2, and ≤ -2) are demarcated as green dots.

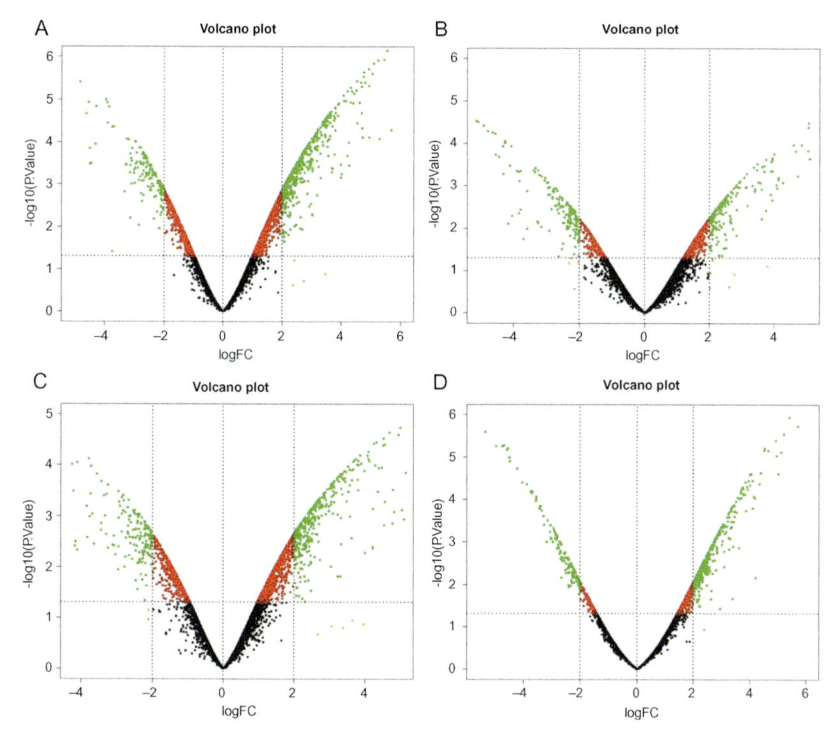

Fig. 1 Volcano plots comprising of DEGs visualized using R studio. Four plots were created, comparing the DEGs between control H37Rv strain and XDR strains treated with Water, capreomycin, isoniazid, and rifampicin. The legend is as follows: x-axis LogFC; y-axis –log10 of *P*-values. Upregulated genes are present in the right panel of the plot, and downregulated genes on the left. The red dots indicate *P*-values ≤ 0.05 and yellow dots indicate logFC ≥ 2 and logFC ≤ 2. The green dots signify DEGs that satisfy both logFC and *P*-value. The genes in the array that did not show any notable changes in expression are displayed as black dots.

Table 1 The MTB strains information from GSE53843 are provided.

Group	Accession	Title	Source name	Strain	Drug
1.	GSM1302015	H37Rv_CAP	1 × MIC CAP, 6 h	H37Rv	Capreomycin (CAP)
2.	GSM1302016	H37Rv_INH	10 × MIC INH, 6 h	H37Rv	Isoniazid (INH)
3.	GSM1302017	H37Rv_Ctrl	Water, 6 h	H37Rv	Water (control)
4.	GSM1302018	H37Rv_RIF	1 × MIC RIF, 6 h	H37Rv	Rifampicin (RIF)
5.	GSM1302019	XDR1219_CAP	1 × MIC CAP, 6 h	Extensively drug-resistant strain XDR1219	Capreomycin (CAP)
6.	GSM1302020	XDR1219_INH	10 × MIC INH, 6 h	Extensively drug-resistant strain XDR1219	Isoniazid (INH)
7.	GSM1302021	XDR1219_Ctrl	Water, 6 h	Extensively drug-resistant strain XDR1219	Water (control)
8.	GSM1302022	XDR1219_RIF	1 × MIC RIF, 6 h	Extensively drug-resistant strain XDR1219	Rifampicin (RIF)
9.	GSM1302023	XDR1221_CAP	1 × MIC CAP, 6 h	Extensively drug-resistant strain XDR1221	Capreomycin (CAP)
10.	GSM1302024	XDR1221_INH	10 × MIC INH, 6 h	Extensively drug-resistant strain XDR1221	Isoniazid (INH)
11.	GSM1302025	XDR1221_Ctrl	Water, 6 h	Extensively drug-resistant strain XDR1221	Water (control)
12.	GSM1302026	XDR1221_RIF	1 × MIC RIF, 6 h	Extensively drug-resistant strain XDR1221	Rifampicin (RIF)

3.2 Identification of PPI networks

A map depicting the functional and physical linkages among the DEGs' proteins was created using STRING and Cytoscape tools. Initial data for PPIs was procured in the tabular text through STRING and then further visualized using Cytoscape v3.7.1. Fig. 2 comprises the PPI networks of DEGs obtained through water-treated strains and capreomycin indicated by blue and pink nodes, respectively. As depicted in the figure, we found 49 nodes and 63 edges for water-treated strains, whereas, for capreomycin-treated strains, we found 34 nodes and 49 edges. Fig. 3 depicts the PPI network of isoniazid and rifampicin exposures in green and purple nodes, respectively. As a result, we found 33 nodes and 30 edges for Isoniazid-treated

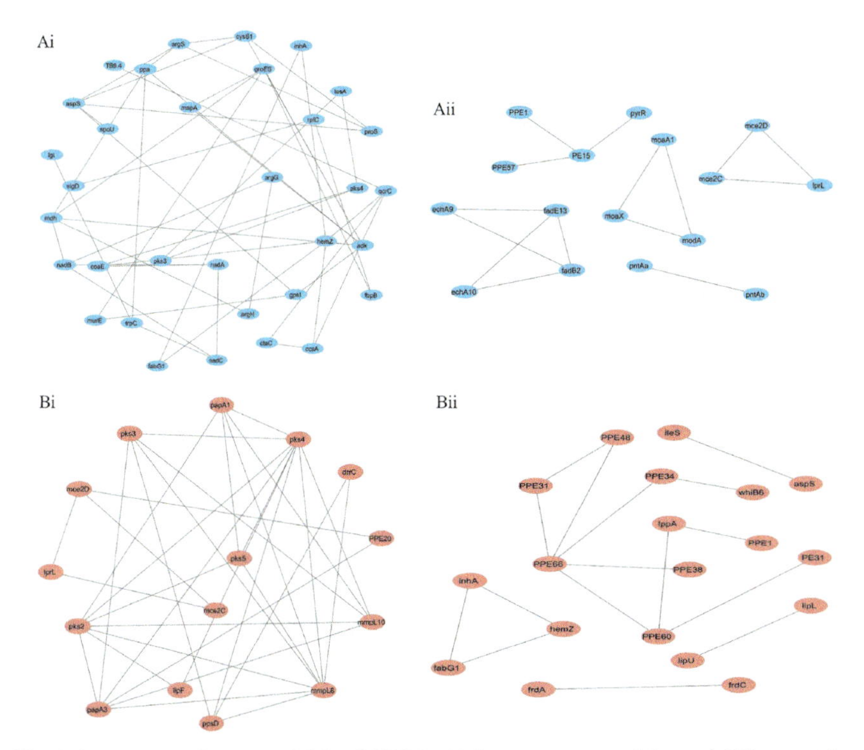

Fig. 2 Interactions between DEGs of GSE53843 dataset portrayed through PPI networks. STRING tool was used to retrieve the nodes and edges, which were then, plotted using Cytoscape software. Blue nodes (Ai and Aii) represent the PPI networks of DEGs of water-treated strains, pink nodes represent DEGs of strains treated with capreomycin (Bi and Bii). Edges are displayed in gray.

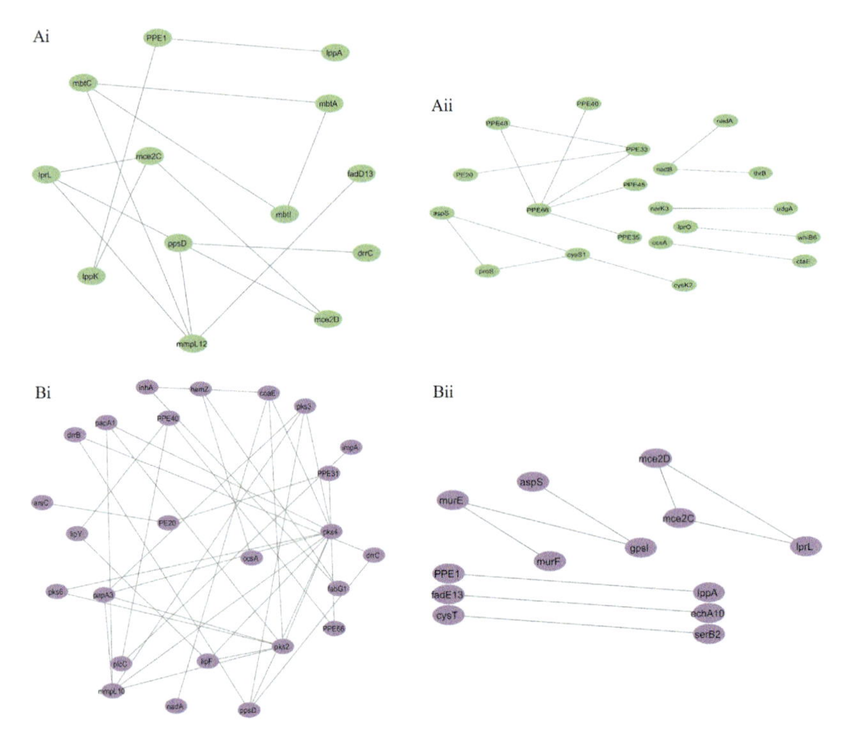

Fig. 3 PPI networks mapping the interactions between DEGs of the GSE53843 dataset. Cytoscape software was used to plot the nodes and edges retrieved from the STRING tool. Green nodes (Ai and Aii) represent the PPI networks of DEGs of isoniazid-treated strains. Purple nodes represent the DEGs of strains treated with rifampicin (Bi and Bii). Edges are displayed in gray.

strains, whereas, for rifampicin-treated strains, we found 38 nodes and 46 edges. The number of proteins are denoted as nodes, and edges represent their interactions.

3.3 Functional annotation and enrichment analysis

The functional enrichment analysis for the DEGs from the dataset was carried out using the Cytoscape plug-in ClueGO. The MF and BP terms of the GO annotation of each drug treatment was set, i.e., control, capreomycin, isoniazid, and rifampicin, which are depicted in Figs. 4–7, respectively. A hypergeometric test, which is two-sided with Benjamini-Hochberg correction, kappa score > 0.4, and $P < 0.05$ as a significant condition, determined the statistical options for the ClueGO enrichment annotation. The DEG's PPI network in the absence of drug exposure (Fig. 4) was enriched

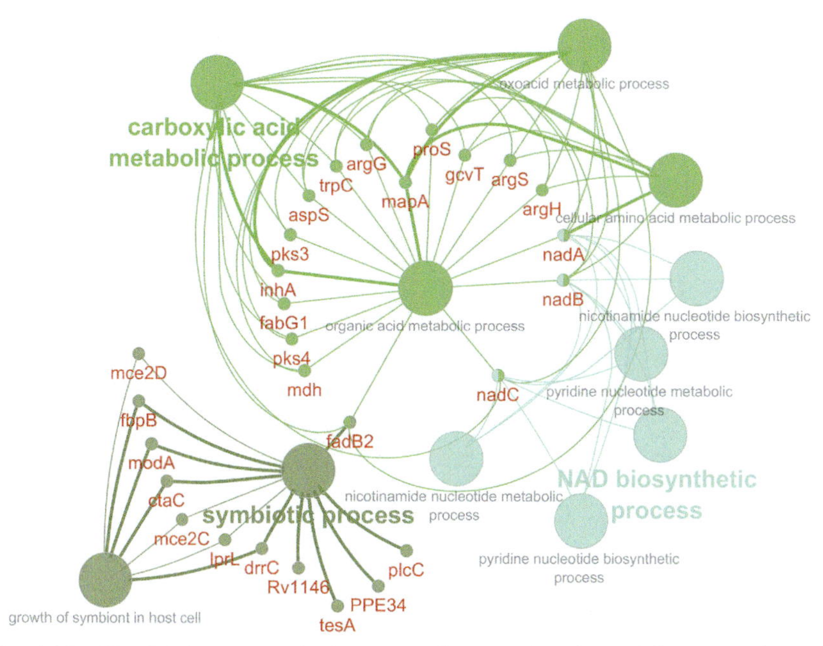

Fig. 4 The ClueGO Cytoscape plugin was used to visualize the enrichment analysis of strains treated with water by Gene Ontology (GO) terms. The visualization tool incorporates KEGG and Reactome pathways, which allow it to display the significant Molecular Functions (MF) and Biological Processes (BP) exhibited by the top 250 DEGs. The node size is dictated by *P*-value $\leq$0.05 and node colors display the distinct yet interconnected MF and BP groups enriched upon treatment. The most important functional GO terms which denominate the MF and BP groups are indicated in bold font. Red font is used to display the names of the DEGs in each group.

in BP such as symbiotic processes (GO:0044403), NAD biosynthetic process (GO:0009435), and carboxylic acid metabolic process (GO:0019752). MF and BP of strains exposed to capreomycin (Fig. 5) were significantly enriched in phosphopantetheine binding (GO:0031177), sulfur compound biosynthetic process (GO:0044272), and triglyceride lipase activity (GO:0004806). For the enrichment analysis of isoniazid exposure (Fig. 6), MF and BP were predominantly enriched in Iron-sulfur cluster binding (GO:0051536), ion transport (GO:0006811), and organic acid biosynthetic process (GO:0016053). Rifampicin treatment (Fig. 7) exhibited enrichment of DEGs involved in transport (GO:0006810), growth of symbiont in a host cell (GO:0044119), and phosphopantetheine binding (GO:0031177). Additionally, GO terms enriched during INH treatment illustrated as dotplots (Fig. 8A) comprised of oxoacid metabolic process (GO:0043436), organonitrogen compound

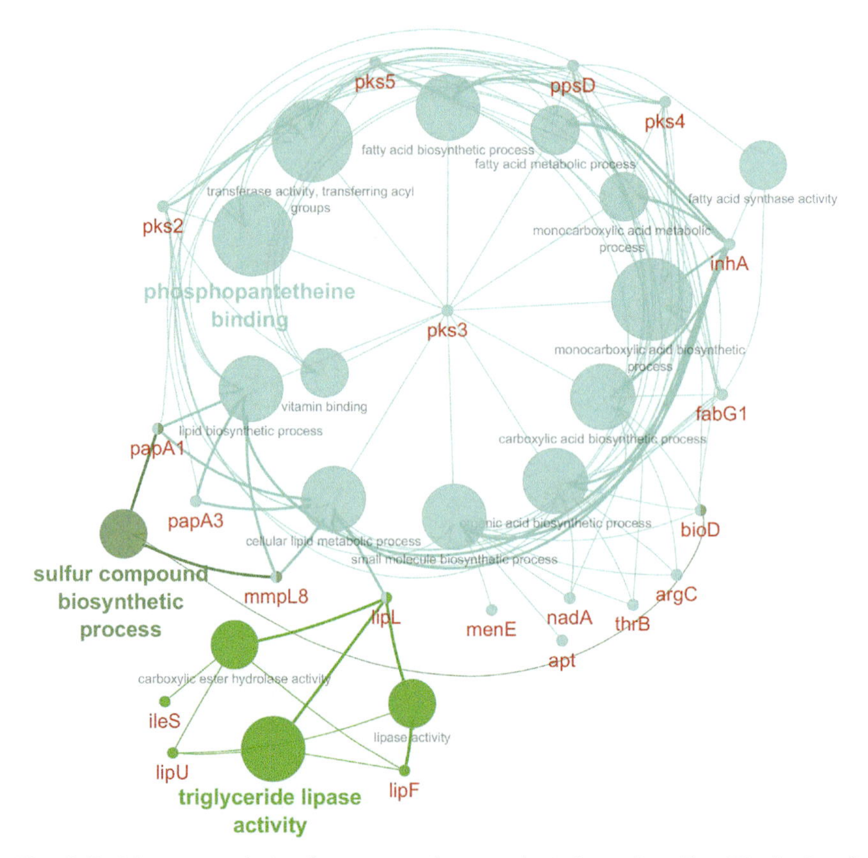

Fig. 5 Enrichment analysis of capreomycin-treated strains using ClueGO plugin of Cytoscape software. A kappa score of 0.4 determines the connectivity of the network between distinct pathways. A *P*-value $\leq$ of 0.5 is required for the pathways to be deemed significant in the enrichment analysis. Node size depends upon the *P*-value and node color specifies the functional class they belong to. The salient GO/pathway terms defining the MF and BP groups are displayed in bold font. Red font signifies the DEGs involved in each group. Triglyceride lipase activity, phosphopantetheine, and sulfur compound biosynthetic processes are significantly enriched among strains treated with capreomycin.

biosynthetic process (GO:1901566), organic acid metabolic process (GO:0006082), organic acid biosynthetic process (GO:0016053), monocarboxylic acid biosynthetic process (GO:0072330), carboxylic acid biosynthetic process (GO:0046394), and carboxylic acid metabolic process (GO:0019752). For RIF treatment, the enriched terms were involved in lipid metabolic process (GO:0006629), lipid biosynthetic process (GO:0008610), peptidoglycan-based cell wall biogenesis (GO:0009273), cell wall biogenesis (GO:0042546),

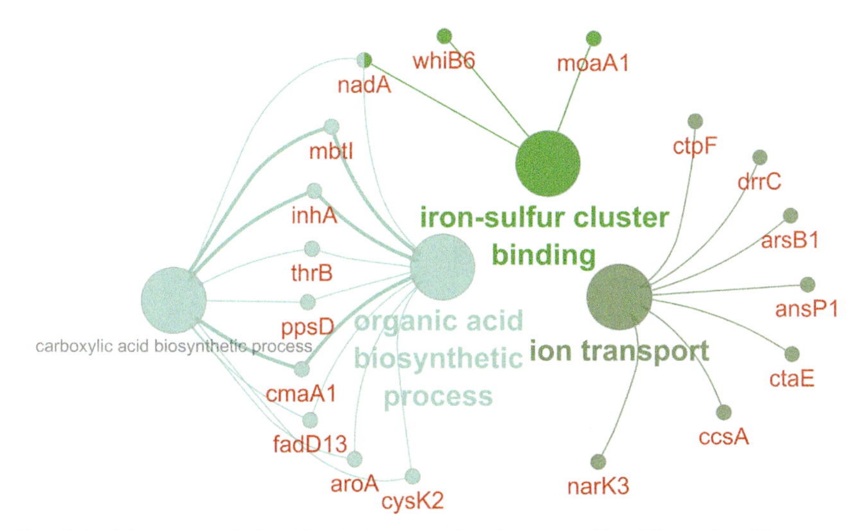

Fig. 6 Enrichment analysis of isoniazid-treated strains using ClueGO plugin of Cytoscape software. The plugin acquires data from KEGG and REACTOME pathways as well and therefore provides a detailed enrichment analysis of the DEGs. Iron-sulfur cluster binding, ion transport and organic acid biosynthetic processes are indicated to play a critical role in the inference of isoniazid resistance in XDR-MTB strains. Red font indicates DEGs involved in the metabolic processes. Significant physiological processes are indicated in bold font. Node size and node color differ with *P*-value and associated functional class, respectively.

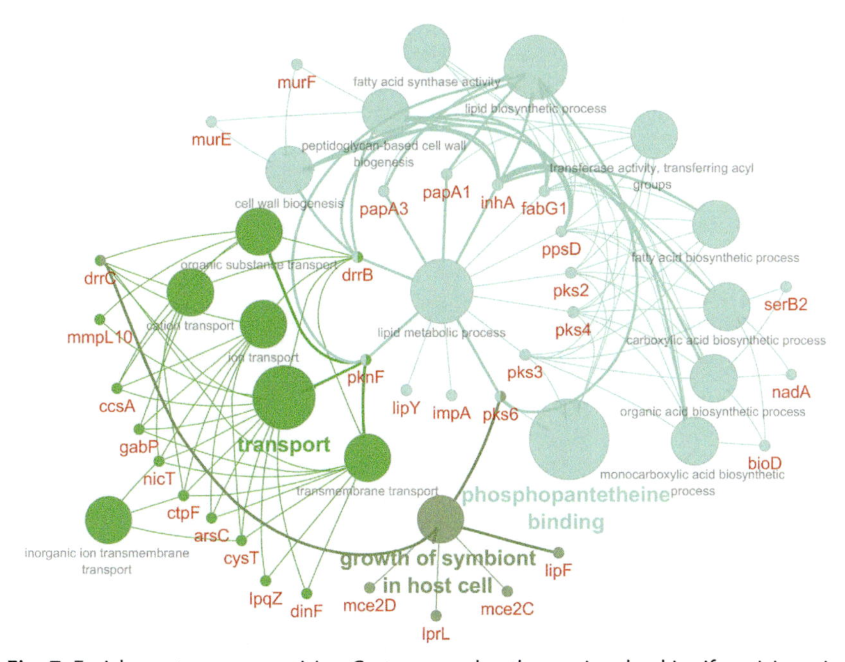

Fig. 7 Enrichment map comprising Go terms and pathways involved in rifampicin resistance visualized using ClueGO plugin of Cytoscape software. Transport mechanisms, growth of symbiont in host cells and phosphopantetheine binding mechanisms are the significantly enriched physiological processes among strains treated with rifampicin. Node size is determined by a *P*-value $\leq$ of 0.05 various MF, and BP are attributed to node colors. The names of the significant GO terms, which define the enriched MF and BP groups, are displayed in bold font. DEGs comprising these groups are denominated in red font.

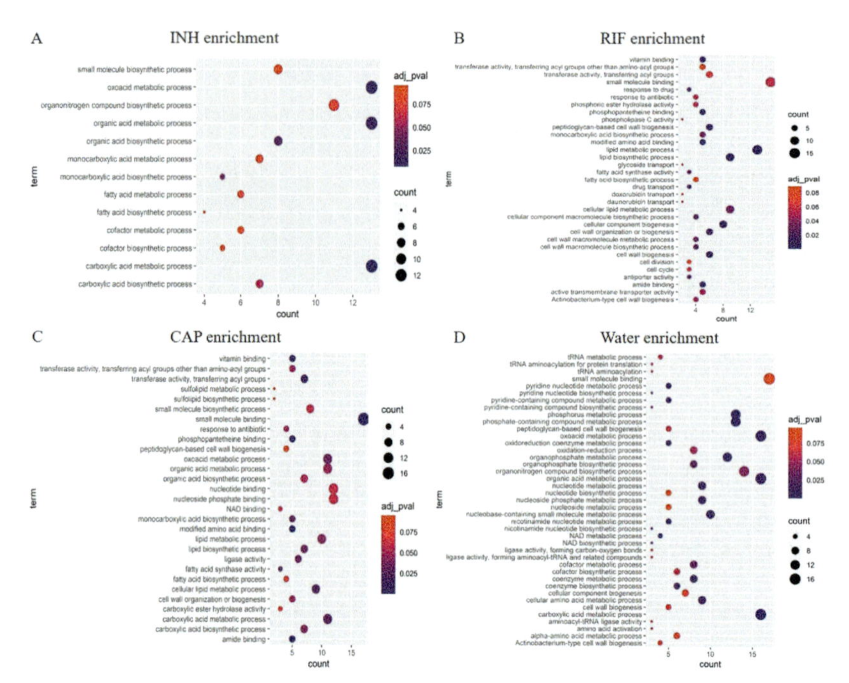

Fig. 8 Graphical representation of enriched GO terms in the form of a DotPlot. The GO terms of BP and MF are plotted against the count of genes. The dots size represents the number of genes among the significant DEGs associated with the GO term. The color of the dots represent their significance in accordance with the adjusted *P*-value. (A) enriched Go terms on exposure to isoniazid, (B) enriched Go terms on exposure to rifampicin, (C) enriched Go terms on exposure to capreomycin, (D) enriched Go terms in control-water.

phosphopantetheine binding (GO:0031177), fatty acid synthase activity (GO:0004312) and small-molecule binding (GO:0036094) (Fig. 8B). In CAP treatment, some of the standard enriched terms were carboxylic acid biosynthetic process (GO:0046394) and organic acid biosynthetic process (GO:0016053). Cellular lipid metabolic process (GO:0044255), carboxylic acid metabolic process (GO:0019752), lipid biosynthetic process (GO:0008610), oxoacid metabolic process (GO:0043436), lipid metabolic process (GO:0006629), and organic acid metabolic process (GO:0006082) were the additional enriched BP terms and phosphopantetheine binding (GO:0031177), modified amino acid binding (GO:0072341), amide binding (GO:0033218), and small molecule binding (GO:0036094) were the enriched molecular functions (Fig. 8C). The standard terms enriched in XDR-MTB during treatment with water were carboxylic acid metabolic process

(GO:0019752), pyridine nucleotide metabolic process (GO:0019362), NAD biosynthetic process (GO:0009435), nicotinamide nucleotide biosynthetic process (GO:0019359), and cellular amino acid metabolic process (GO:0006520). Organophosphate metabolic process (GO:0090407), phosphorus metabolic process (GO:0006793), oxoacid metabolic process (GO:0043436), organic acid metabolic process (GO:0006082), and organonitrogen compound biosynthetic process (GO:1901566) were identified as significantly enriched GO terms as well (Fig. 8D). Through the above results, it can be proclaimed that the XDR strains have adapted diverse mechanisms to overcome lethal dosages of anti–MTB drugs, and thereby, have gained resistance against them.

4. Discussion

Drug resistance in TB is often attributed to genetic mutations, which affect either the bacterial enzymes that activate prodrugs or the drug target itself. However, there exist multiple alternate mechanisms, which facilitate drug resistance in this pathogenic bacterium (Peñuelas-Urquides, Castorena-Torres, SilvaRamírez, & de León, 2017). Transcriptome analysis elucidated through comparison of DEGs, evaluation of pathways associated with these DEGs, and interactions between these processes may give us an insight into the overall mechanism of drug resistance in extremely drug-resistant strains of MTB (Li et al., 2020; Peñuelas-Urquides et al., 2017). In the present study, we investigated gene expression profiles of dataset GSE53843 to identify the interconnected processes and pathways involved in drug resistance. The dataset comprised of microarray results obtained through the exposure of XDR strains of MTB and MTB reference strain to water and inhibitory concentrations of the anti-TB drugs; isoniazid, rifampicin, and capreomycin, which were then scrutinized through the identification of DEGs, evaluation of protein interaction networks, functional annotation, and enrichment of MF, BP and pathways involved.

Rifampicin's known mechanism of bacterial growth inhibition is through the suppression of RNA synthesis by binding of the drug to β subunit of RNA polymerase, RpoB. Resistance to a lethal concentration of rifampicin has been attributed to mutations in the *rpoB* gene (Peñuelas-Urquides, et al., 2017). However, in a recent study carried out by Zhu et al. (2018), bacterial populations exhibiting resistance upon exposure to an inhibitory concentration of rifampicin were upregulating the expression of gene *rpoB* (Zhu et al., 2018). This conforms with the results of our study as well, in which *rpoB* is upregulated in XDR-MTB strains that are exposed to rifampicin. Another factor known to contribute towards rifampicin resistance is the cell wall, a crucial drug target (Luo et al., 2020). ClueGo enrichment analysis identified GO:0009273 peptidoglycan-based cell wall biogenesis, GO:0042546 cell wall biogenesis, and cell wall-related pathways such as GO:0006629. Lipid metabolic process GO:0008610 and lipid biosynthetic process as significantly enriched processes among the strains exposed to rifampicin. Genes *murF* and *murE*, which are considered essential drug targets due to their catalytic activity in peptidoglycan synthesis, were significantly downregulated upon rifampicin exposure; however, upregulation of *murA* gene, which plays a role in cell wall synthesis (Luo et al., 2020), was observed (Fig. 7). It is important to note that the development of low cell permeability is an intrinsic mechanism of drug resistance (Louw et al., 2009); the dysregulation of cell wall-related genes observed in our study may corroborate this mechanistic property. The bactericidal effect of rifampicin has recently been associated with sequential hydroxyl radical formation (Piccaro, Pietraforte, Giannoni, Mustazzolu, & Fattorini, 2014), resulting in the bacterium's lyses. Upregulation of universal stress protein (uspB) was observed in this study with INH treated MTB (with logFC value 3) (see Table 3 in Supplementary material in the online version at https://doi.org/10.1016/bs.apcsb.2021.02.002), which complements the view that bactericidal drugs increase oxidative stress in MTB present within macrophages, and therefore may further validate bactericidal effects of the drug (Piccaro et al., 2014). Additionally, UspB is an integral part of the ATP binding ABC transporter protein, which is essential for the growth of the bacterium in vitro (Fullam, Prokes, Fütterer, & Besra, 2016).

The activation of pro-drug isoniazid occurs through the catalase-peroxidase activity of an enzyme encoded by katG, which target genes involved in mycolic acid synthesis pathway, *inhA* (enoyl ACP reductase) and *kasA* (beta-ketoacyl ACP synthase) (Vilchèze & Jacobs, 2014). A total of 5053 genes were differentially expressed in XDR-MTB strains exposed to a lethal concentration of INH. The key enriched terms were ion transport GO:0006811, iron-sulfur cluster binding GO:0051536, organic acid biosynthetic process GO:0016053 and carboxylic acid biosynthetic process GO:0046394 (Fig. 6). Significant upregulation of gene *inhA* was observed in CAP, INH, and RIF treated MTB (with logFC value >3); however, there was no notable change in the expression of *katG*. A study carried out by Ando et al. (2011) suggested that the downregulation of *katG* is a contributing factor for isoniazid resistance (Ando et al., 2011); similarly, *inhA* upregulation has also been attributed toward increased isoniazid resistance (Vilchèze & Jacobs, 2014). Isoniazid is a hydrazide derivative of a carboxylic acid, and *Mycobacterium tuberculosis* is known to be highly sensitive to weak acids (Maresca, Vullo, Scozzafava, Manole, & Supuran, 2013). Some of the enriched terms in isoniazid treatment were carboxylic acid metabolic processes such as GO:0019752, oxoacid metabolic process GO:0043436, organic acid metabolic process GO:0006082, organic acid biosynthetic process GO:0016053, monocarboxylic acid biosynthetic process GO:0072330, and carboxylic acid biosynthetic process GO:004 6394, which are indicative of the mechanisms involved in overriding this selective pressure.

Transporters and Efflux pumps play a vital role in developing multidrug resistance via partaking in the extrusion of specific compounds and even seemingly unrelated compounds (Louw et al., 2009). DEGs involved in ion transport GO:0006811 were observed in both rifampicin exposure as well as isoniazid exposure. Genes responsible for calcium pumping (*ctpF*) during oxidative stress conditions (Maya-Hoyos, Rosales, Novoa-Aponte, Castillo, & Soto, 2019), nitrite extrusion (*nark3*), and arsenic transport (*arsC*) were observed to be significantly upregulated in response to isoniazid and rifampicin exposures, throwing light upon the involvement of drug metabolism and transport in the development of multidrug resistance.

Upon treatment with capreomycin, a total of 5053 genes were differentially expressed, which were primarily involved in phosphopantetheine binding GO:0031177, sulfur compound biosynthetic process GO:0044272 and triglyceride lipase activity GO:0004806 (Fig. 5). Although previous studies have identified ribosomes to be the target of CAP, the proteins that interact with this drug are not yet known (Lin et al., 2014). However, studies have investigated the processes of lipase activity and fatty acid synthesis in contributing to virulence factors and determinants of resistance in MTB. Stress factors induce the production of lipid bodies in the organism, known to bring about a state of dormancy in-vivo and consequently infer resistance (Deb et al., 2009). To enter this state of dormancy, MTB stores energy in the form of triacylglycerols (Deb et al., 2009). Significant involvement of gene *lipU* was observed in our study (Fig. 5), which is known to hydrolyze short-chain carbon substrates (Li et al., 2017). Similarly, *lipY* carries out hydrolysis of long-chain stored triglycerides, leading to stored energy utilization (Deb et al., 2009). In our study, the genes *lipF*, *lipL*, and *lipU* were observed to play essential roles in triglyceride lipase activity, indicating the possibility of acquiring dormancy upon capreomycin exposure (Fig. 5).

The enrichment of Phophopantetheine GO:0031177 binding was observed in both rifampicin exposure and capreomycin treatment (Figs. 5 and 7). Phosphopantetheine adenylyltransferase (PPAT) plays a crucial role in the CoA biosynthesis pathway (Wubben & Mesecar, 2011). The DEGs involved in the enriched Phosphopantetheine binding process belonged to the polyketide synthase encoding genes (*pks*). *Pks* genes decorate the cell envelope with unique lipid polyketides, which are critical in virulence and add to the cell envelope's complexity (Parvez et al., 2018). We observed a significant downregulation of these genes upon rifampicin and capreomycin exposure, which indicates a crucial role of this process and associated cell envelope proteins in developing resistance. MTB is known to possess multiple virulence factors and, consequently, multiple mechanisms to evade anti-TB drugs (Madacki et al., 2019). Our study highlighted some of the mechanisms that played vital roles in the XDR strain's survival upon drug exposure. The key factors that were identified as contributors to drug resistance were modifications in the cell wall and its related components, upregulation of universal stress proteins, upregulation of transport channels, metabolism of stress-inducing components such as carboxylic acid derivatives, and transition to the stage of dormancy via the production of lipid storage molecules and subsequent energy utilization through hydrolysis of these molecules.

5. Conclusion

Our study investigated drug resistance mechanisms in XDR-MTB strains, specifically against Capreomycin, isoniazid, and rifampicin, using the microarray dataset GSE53843. Upon analyses of the DEGs present between the drug-exposed control strains, we identified multiple factors as contributors to drug resistance in XDR-MTB strains, some of which were in accordance with previously identified determinants involvement of cell wall-associated proteins, transport channels, and dysregulation of known drug targets. The peculiar findings of this study, which require further speculation, are processes involved in the stress-induced response, such as in the presence of carboxylic acids, the importance of phosphopantetheine binding in mechanistic drug resistance, and the acquirement of dormancy upon capreomycin exposure. These observations although peculiar and exciting, require functional studies to identify their appropriate role in the drug-induced response and possible role in the acquirement of resistance. Ultimately, our findings provide a platform for mapping the numerous interconnected mechanisms involved in drug resistance and distinguishing novel therapeutic targets for the treatment of XDR-MTB.

Acknowledgments

Mr. Udhaya Kumar. S, one of the authors, gratefully acknowledges the Indian Council of Medical Research (ICMR), India, for providing him a Senior Research Fellowship [ISRM/11(93)/2019]. The authors would like to take this opportunity to thank the management of VIT for providing the necessary facilities and encouragement to carry out this work.

Conflict of interest

The authors have declared that no conflicts of interest exist.

Author contributions

SUK and CGPD were involved in the design of the study. SUK was involved in the data collection and experiment. SUK, AS, DTK, VAP, SY, IAT, and HZ involved in the acquisition, analysis, and interpreting of the results. SUK and AS drafted the manuscript. CGPD supervised the entire study and was involved in study design, the acquisition, analysis, understanding of the data, and critically reviewed the manuscript. All authors edited and approved the submitted version of the article.

Role of funding source

No funding agency was involved in the present study.

References

Alam, A., Imam, N., Ahmed, M. M., Tazyeen, S., Tamkeen, N., Farooqui, A., et al. (2019). Identification and classification of differentially expressed genes and network meta-analysis reveals potential molecular signatures associated with tuberculosis. *Frontiers in Genetics*, *10*, 932. https://doi.org/10.3389/fgene.2019.00932.

Ando, H., Kitao, T., Miyoshi-Akiyama, T., Kato, S., Mori, T., & Kirikae, T. (2011). Downregulation of katG expression is associated with isoniazid resistance in Mycobacterium tuberculosis. *Molecular Microbiology*, *79*(6), 1615–1628. https://doi.org/10.1111/j.1365-2958.2011.07547.x.

Arnold, C. (2007). Molecular evolution of Mycobacterium tuberculosis. *Clinical Microbiology and Infection*, *13*(2), 120–128. https://doi.org/10.1111/j.1469-0691.2006.01637.x.

Aubert, J., Bar-Hen, A., Daudin, J. J., & Robin, S. (2004). Determination of the differentially expressed genes in microarray experiments using local FDR. *BMC Bioinformatics*, *5*, 125. https://doi.org/10.1186/1471-2105-5-125.

Barrett, T., Wilhite, S. E., Ledoux, P., Evangelista, C., Kim, I. F., Tomashevsky, M., et al. (2013). NCBI GEO: Archive for functional genomics data sets—Update. *Nucleic Acids Research*, *41*(Database issue), D991–D995. https://doi.org/10.1093/nar/gks1193.

Bindea, G., Mlecnik, B., Hackl, H., Charoentong, P., Tosolini, M., Kirilovsky, A., et al. (2009). ClueGO: A cytoscape plug-in to decipher functionally grouped gene ontology and pathway annotation networks. *Bioinformatics*, *25*(8), 1091–1093. https://doi.org/10.1093/bioinformatics/btp101.

Chien, J.-Y., Chen, Y.-T., Wu, S.-G., Lee, J.-J., Wang, J.-Y., & Yu, C.-J. (2015). Treatment outcome of patients with isoniazid mono-resistant tuberculosis. *Clinical Microbiology and Infection*, *21*(1), 59–68. https://doi.org/10.1016/j.cmi.2014.08.008.

Childs, L. M., & Larremore, D. B. (2021). Network models for malaria: Antigens, dynamics, and evolution over space and time. In O. Wolkenhauer (Ed.), *Systems medicine* (pp. 277–294). Academic Press. https://doi.org/10.1016/B978-0-12-801238-3.11512-0.

Coll, F., Phelan, J., Hill-Cawthorne, G. A., Nair, M. B., Mallard, K., Ali, S., et al. (2018). Genome-wide analysis of multi- and extensively drug-resistant Mycobacterium tuberculosis. *Nature Genetics*, *50*(2), 307–316. https://doi.org/10.1038/s41588-017-0029-0.

Cui, T., & He, Z.-G. (2014). Improved understanding of pathogenesis from protein interactions in Mycobacterium tuberculosis. *Expert Review of Proteomics*, *11*(6), 745–755. https://doi.org/10.1586/14789450.2014.971762.

Deb, C., Lee, C.-M., Dubey, V. S., Daniel, J., Abomoelak, B., Sirakova, T. D., et al. (2009). A novel in vitro multiple-stress dormancy model for Mycobacterium tuberculosis generates a lipid-loaded, drug-tolerant, dormant pathogen. *PLoS One*, *4*(6), e6077. https://doi.org/10.1371/journal.pone.0006077.

Franz, M., Rodriguez, H., Lopes, C., Zuberi, K., Montojo, J., Bader, G. D., et al. (2018). GeneMANIA update 2018. *Nucleic Acids Research*, *46*(W1), W60–W64. https://doi.org/10.1093/nar/gky311.

Fullam, E., Prokes, I., Fütterer, K., & Besra, G. S. (2016). Structural and functional analysis of the solute-binding protein UspC from Mycobacterium tuberculosis that is specific for amino sugars. *Open Biology*, *6*(6), 160105. https://doi.org/10.1098/rsob.160105.

Furin, J., Cox, H., & Pai, M. (2019). Tuberculosis. *The Lancet*, *393*(10181), 1642–1656. https://doi.org/10.1016/S0140-6736(19)30308-3.

Guohua, Y., & Yu, L. (2014). *Expression profiling of extensively drug-resistant tuberculosis isolates and H37Rv exposed to anti-TB compounds*. GEO Accession Viewer. https://www.ncbi.nlm.nih.gov/geo/query/acc.cgi?acc=GSE53843.

Hassan, S., Sudhakar, V., Mary, M. B. N., Babu, R., Doble, M., Dadar, M., et al. (2020). Computational approach identifies protein off-targets for isoniazid-NAD adduct: Hypothesizing a possible drug resistance mechanism in Mycobacterium tuberculosis. *Journal of Biomolecular Structure and Dynamics*, *38*(6), 1697–1710. https://doi.org/10.1080/07391102.2019.1615987.

Huang, D. W., Sherman, B. T., Tan, Q., Kir, J., Liu, D., Bryant, D., et al. (2007). DAVID bioinformatics resources: Expanded annotation database and novel algorithms to better extract biology from large gene lists. *Nucleic Acids Research*, *35*(Web Server issue), W169–W175. https://doi.org/10.1093/nar/gkm415.

Karlas, A., Machuy, N., Shin, Y., Pleissner, K.-P., Artarini, A., Heuer, D., et al. (2010). Genome-wide RNAi screen identifies human host factors crucial for influenza virus replication. *Nature*, *463*(7282), 818–822. https://doi.org/10.1038/nature08760.

Kumar, S. U., Kumar, D. T., Siva, R., Doss, C. G. P., & Zayed, H. (2019). Integrative bioinformatics approaches to map potential novel genes and pathways involved in ovarian cancer. *Frontiers in Bioengineering and Biotechnology*, 7, 391. https://doi.org/10.3389/fbioe.2019.00391.

Kumar, D., Nath, L., Kamal, M. A., Varshney, A., Jain, A., Singh, S., et al. (2010). Genome-wide analysis of the host intracellular network that regulates survival of Mycobacterium tuberculosis. *Cell*, *140*(5), 731–743. https://doi.org/10.1016/j.cell.2010.02.012.

Lee, J. Y. (2015). Diagnosis and treatment of extrapulmonary tuberculosis. *Tuberculosis Respiratory Disease*, *78*(2), 47–55. https://doi.org/10.4046/trd.2015.78.2.47.

Li, H., Chen, T., Yu, L., Guo, H., Chen, L., Chen, Y., et al. (2020). Genome-wide DNA methylation and transcriptome and proteome changes in Mycobacterium tuberculosis with Para-aminosalicylic acid resistance. *Chemical Biology & Drug Design*, *95*(1), 104–112. https://doi.org/10.1111/cbdd.13625.

Li, C., Li, Q., Zhang, Y., Gong, Z., Ren, S., Li, P., et al. (2017). Characterization and function of Mycobacterium tuberculosis H37Rv lipase Rv1076 (LipU). *Microbiological Research*, *196*, 7–16. https://doi.org/10.1016/j.micres.2016.12.005.

Lillebaek, T., Norman, A., Rasmussen, E. M., Marvig, R. L., Folkvardsen, D. B., Andersen, Å. B., et al. (2016). Substantial molecular evolution and mutation rates in prolonged latent Mycobacterium tuberculosis infection in humans. *International Journal of Medical Microbiology*, *306*(7), 580–585. https://doi.org/10.1016/j.ijmm.2016.05.017.

Lin, Y., Li, Y., Zhu, N., Han, Y., Jiang, W., Wang, Y., et al. (2014). The antituberculosis antibiotic capreomycin inhibits protein synthesis by disrupting interaction between ribosomal proteins L12 and L10. *Antimicrobial Agents and Chemotherapy*, *58*(4), 2038–2044. https://doi.org/10.1128/AAC.02394-13.

Louw, G. E., Warren, R. M., van Pittius, N. C. G., McEvoy, C. R. E., Helden, P. D. V., & Victor, T. C. (2009). A balancing act: efflux/influx in mycobacterial drug resistance. *Antimicrobial Agents and Chemotherapy*, *53*(8), 3181–3189. https://doi.org/10.1128/AAC.01577-08.

Luo, X., Pan, J., Meng, Q., Huang, J., Wang, W., Zhang, N., et al. (2020). High-throughput screen for cell wall synthesis network module in *Mycobacterium tuberculosis* based on integrated bioinformatics strategy. *Frontiers in Bioengineering and Biotechnology*, *8*, 607. https://doi.org/10.3389/fbioe.2020.00607.

Madacki, J., Mas Fiol, G., & Brosch, R. (2019). Update on the virulence factors of the obligate pathogen Mycobacterium tuberculosis and related tuberculosis-causing mycobacteria. *Infection, Genetics and Evolution*, *72*, 67–77. https://doi.org/10.1016/j.meegid.2018.12.013.

Maresca, A., Vullo, D., Scozzafava, A., Manole, G., & Supuran, C. T. (2013). Inhibition of the β-class carbonic anhydrases from Mycobacterium tuberculosis with carboxylic acids. *Journal of Enzyme Inhibition and Medicinal Chemistry*, *28*(2), 392–396. https://doi.org/10.3109/14756366.2011.650168.

Maya-Hoyos, M., Rosales, C., Novoa-Aponte, L., Castillo, E., & Soto, C. Y. (2019). The P-type ATPase CtpF is a plasma membrane transporter mediating calcium efflux in Mycobacterium tuberculosis cells. *Heliyon*, *5*(11), e02852. https://doi.org/10.1016/j.heliyon.2019.e02852.

Mishra, S., Shah, M. I., Udhaya Kumar, S., Thirumal Kumar, D., Gopalakrishnan, C., Al-Subaie, A. M., et al. (2020). Network analysis of transcriptomics data for the prediction and prioritization of membrane-associated biomarkers for idiopathic pulmonary fibrosis (IPF) by bioinformatics approach. *Advances in Protein Chemistry and Structural Biology*, *120*, 349–377. Academic Press https://doi.org/10.1016/bs.apcsb.2020.10.003.

Parvez, A., Giri, S., Giri, G. R., Kumari, M., Bisht, R., & Saxena, P. (2018). Novel type III polyketide synthases biosynthesize methylated polyketides in Mycobacterium marinum. *Scientific Reports*, *8*(1), 6529. https://doi.org/10.1038/s41598-018-24980-1.

Penn, B. H., Netter, Z., Johnson, J. R., Von Dollen, J., Jang, G. M., Johnson, T., et al. (2018). An Mtb-human protein-protein interaction map identifies a switch between host antiviral and antibacterial responses. *Molecular Cell*, *71*(4), 637–648.e5. https://doi.org/10.1016/j.molcel.2018.07.010.

Peñuelas-Urquides, K., Castorena-Torres, F., SilvaRamírez, B., & de León, M. B. (2017). Drug resistance in Mycobacterium tuberculosis. *Mycobacterium—Research and Development*, 117–129. https://doi.org/10.5772/intechopen.69656.

Piccaro, G., Pietraforte, D., Giannoni, F., Mustazzolu, A., & Fattorini, L. (2014). Rifampin induces hydroxyl radical formation in Mycobacterium tuberculosis. *Antimicrobial Agents and Chemotherapy*, *58*(12), 7527–7533. https://doi.org/10.1128/AAC.03169-14.

Ritchie, M. E., Phipson, B., Wu, D., Hu, Y., Law, C. W., Shi, W., et al. (2015). Limma powers differential expression analyses for RNA-sequencing and microarray studies. *Nucleic Acids Research*, *43*(7), e47. https://doi.org/10.1093/nar/gkv007.

Sandgren, A., Strong, M., Muthukrishnan, P., Weiner, B. K., Church, G. M., & Murray, M. B. (2009). Tuberculosis drug resistance mutation database. *PLoS Medicine*, *6*(2), e1000002. https://doi.org/10.1371/journal.pmed.1000002.

Seung, K. J., Keshavjee, S., & Rich, M. L. (2015). Multidrug-resistant tuberculosis and extensively drug-resistant tuberculosis. *Cold Spring Harbor Perspectives in Medicine*, *5*(9), a017863. https://doi.org/10.1101/cshperspect.a017863.

Singh, R., Dwivedi, S. P., Gaharwar, U. S., Meena, R., Rajamani, P., & Prasad, T. (2020). Recent updates on drug resistance in Mycobacterium tuberculosis. *Journal of Applied Microbiology*, *128*(6), 1547–1567. https://doi.org/10.1111/jam.14478.

Szklarczyk, D., Gable, A. L., Lyon, D., Junge, A., Wyder, S., Huerta-Cepas, J., et al. (2019). STRING v11: Protein–protein association networks with increased coverage, supporting functional discovery in genome-wide experimental datasets. *Nucleic Acids Research*, *47*(D1), D607–D613. https://doi.org/10.1093/nar/gky1131.

Udhaya Kumar, S., Rajan, B., Thirumal Kumar, D., Preethi, V. A., Abunada, T., Younes, S., et al. (2020). Involvement of essential signaling cascades and analysis of gene networks in diabesity. *Genes*, *11*(11), 1256. https://doi.org/10.3390/genes11111256.

Udhaya Kumar, S., Thirumal Kumar, D., Bithia, R., Sankar, S., Magesh, R., Sidenna, M., et al. (2020). Analysis of differentially expressed genes and molecular pathways in familial hypercholesterolemia involved in atherosclerosis: A systematic and bioinformatics approach. *Frontiers in Genetics*, *11*, 734. https://doi.org/10.3389/fgene.2020.00734.

Udhaya Kumar, S., Thirumal Kumar, D., Siva, R., George Priya Doss, C., Younes, S., Younes, N., et al. (2020). Dysregulation of signaling pathways due to differentially expressed genes from the B-cell transcriptomes of systemic lupus erythematosus patients—A bioinformatics approach. *Frontiers in Bioengineering and Biotechnology*, *8*, 276. https://doi.org/10.3389/fbioe.2020.00276.

Vilchèze, C., & Jacobs, W. R., Jr. (2014). Resistance to isoniazid and ethionamide in Mycobacterium tuberculosis: Genes, mutations, and causalities. *Microbiology Spectrum*, *2*(4), 1–21. https://doi.org/10.1128/microbiolspec.MGM2-0014-2013.

Walter, W., Sánchez-Cabo, F., & Ricote, M. (2015). GOplot: An R package for visually combining expression data with functional analysis. *Bioinformatics*, *31*(17), 2912–2914. https://doi.org/10.1093/bioinformatics/btv300.

WHO. (2020). *Global tuberculosis report*. World Health Organization. https://www.who.int/teams/global-tuberculosis-programme/data.

Wickham, H. (2016). *ggplot2: Elegant graphics for data analysis* (2nd ed.). Springer International Publishing. https://doi.org/10.1007/978-3-319-24277-4.

Wubben, T., & Mesecar, A. D. (2011). Structure of Mycobacterium tuberculosis phosphopantetheine adenylyltransferase in complex with the feedback inhibitor CoA reveals only one active-site conformation. *Acta Crystallographica Section F: Structural Biology and Crystallization Communications*, *67*(Pt 5), 541–545. https://doi.org/10.1107/S1744309111010761.

Zhang, Y. J., Reddy, M. C., Ioerger, T. R., Rothchild, A. C., Dartois, V., Schuster, B. M., et al. (2013). Tryptophan biosynthesis protects mycobacteria from CD4 T-cell-mediated killing. *Cell*, *155*(6), 1296–1308. https://doi.org/10.1016/j.cell.2013.10.045.

Zhu, J.-H., Wang, B.-W., Pan, M., Zeng, Y.-N., Rego, H., & Javid, B. (2018). Rifampicin can induce antibiotic tolerance in mycobacteria via paradoxical changes in rpoB transcription. *Nature Communications*, *9*(1), 4218. https://doi.org/10.1038/s41467-018-06667-3.